Methods in Molecular Biology™

Series Editor
John M. Walker
School of Life Sciences
University of Hertfordshire
Hatfield, Hertfordshire, AL10 9AB, UK

For further volumes:
http://www.springer.com/series/7651

Progenitor Cells

Methods and Protocols

Edited by

Kimberly A. Mace

The Healing Foundation Centre, Faculty of Life Sciences, University of Manchester, Manchester, UK

Kristin M. Braun

Centre for Cutaneous Research, Blizard Institute, Barts and the London School of Medicine and Dentistry, Queen Mary, University of London, London, UK

Editors
Kimberly A. Mace
The Healing Foundation Centre
Faculty of Life Sciences
University of Manchester
Manchester, UK

Kristin M. Braun
Centre for Cutaneous Research
Blizard Institute, Barts and the London
School of Medicine and Dentistry
Queen Mary, University of London
London, UK

ISSN 1064-3745 ISSN 1940-6029 (electronic)
ISBN 978-1-61779-979-2 ISBN 978-1-61779-980-8 (eBook)
DOI 10.1007/978-1-61779-980-8
Springer New York Heidelberg Dordrecht London

Library of Congress Control Number: 2012943274

Printed on acid-free paper

Humana Press is a brand of Springer
Springer is part of Springer Science+Business Media (www.springer.com)

Preface

Progenitor cells have emerged as important players in regenerative medicine therapies as a result of their potential to differentiate into many cell types. This capability, and understanding how to regulate it, will undoubtedly provide the basis for future cell therapies aimed at correcting tissue and organ dysfunction as a result of disease or injury. Progenitor cells encompass a wider range of cell types than those strictly regarded as stem cells and in general are more numerous and more easily obtained for manipulation. Thus progenitor cells are well suited to studies focused on understanding cell plasticity and mechanisms underlying cell fate decisions. A growing body of evidence suggests that most tissues and organs in the body have resident stem and progenitor cells and that these cells not only contribute to tissue homeostasis but also play a significant role in tissue repair and regeneration in response to injury. In addition, many parallels exist between normal progenitor cells and "cancer stem cells," a subpopulation of cells that retain the ability to self-renew and fuel tumor growth. Therefore, learning about progenitor cell biology may have important clinical implications not only for regenerative medicine but also for suggesting novel cancer therapies.

In this volume we have taken a multisystems approach to organizing a variety of protocols that detail methods of investigating progenitor cells. Our intention is to highlight the fundamental questions pertaining to, and principles governing, progenitor cell biology. A cross-species view encompassing planarians, *Drosophila*, *Caenorhabditis elegans*, zebrafish, *Xenopus*, chicken, mice, and humans provides insights into common themes in progenitor cell biology throughout the animal kingdom. Each species provides different experimental advantages and we have included examples of both in vitro (mostly in Parts I and IV) and in vivo studies (Parts II–IV), demonstrating the flexibility and strengths of each individual approach. The topics covered range broadly from how manipulation of physical forces shapes progenitor cell behavior to using fluorescence-activated cell sorting to isolate progenitor cells from a variety of organisms. Specifically, Parts II and III describe methods for studying progenitor cells in vivo, using nonmammalian and mammalian model systems, respectively. The knowledge of the sophisticated fate-mapping and microscopy techniques described is particularly important, as understanding how progenitor cells interact with their native environment is critical to understanding how we can therapeutically manipulate them. Finally, Part IV (Human Systems) highlights the important strategies for investigating human progenitor cells, including their isolation, characterization, and application in cell-based therapies.

As with all Methods in Molecular Biology series volumes, these chapters are designed to be useful to both novice and expert alike and are written in a detailed, straightforward manner with a helpful Notes section following the protocols. We hope these "tips" will not only aid the reader with the specific protocol in which they are interested but also contribute to their broader understanding of how progenitor cells are studied across multiple systems. Finally, we would like to sincerely thank all of the authors for their contributions and, in particular, Professor John Walker for his help and advice on putting the volume together.

Manchester, UK *Kimberly A. Mace*
London, UK *Kristin M. Braun*

Contents

Preface . . . . *v*
Contributors . . . . *xi*

Part I In Vitro Culture Systems

1 Isolation of Satellite Cells from Single Muscle Fibers from Young, Aged, or Dystrophic Muscles 3
Valentina Di Foggia and Lesley Robson

2 Terminal Differentiation of Human Epidermal Stem Cells on Micro-patterned Substrates 15
John T. Connelly

3 Isolation, Culture, and Potentiality Assessment of Lung Alveolar Stem Cells . . 23
Feride Oeztuerk-Winder and Juan-Jose Ventura

4 Three-Dimensional In Vitro Culture Techniques for Mesenchymal Stem Cells. 31
Fatima A. Saleh, Jessica E. Frith, Jennifer A. Lee, and Paul G. Genever

5 Isolation of Adult Stem Cells and Their Differentiation to Schwann Cells 47
Cristina Mantovani, Giorgio Terenghi, and Susan G. Shawcross

6 Functional Purification of Human and Mouse Mammary Stem Cells 59
Daniela Tosoni, Pier Paolo Di Fiore, and Salvatore Pece

7 Isolation and Expansion of Endothelial Progenitor Cells Derived from Mouse Embryonic Stem Cells. 81
S. Bahram Bahrami, Mandana Veiseh, and Nancy J. Boudreau

Part II Nonmammalian Model Systems

8 Transcriptome Analysis of *Drosophila* Neural Stem Cells 99
Katrina S. Gold and Andrea H. Brand

9 Live Imaging for Studying Asymmetric Cell Division in the *C. elegans* Embryo 111
Alexia Rabilotta, Rana Amini, and Jean-Claude Labbé

10 Tol2-Mediated Gene Transfer and In Ovo Electroporation of the Otic Placode: A Powerful and Versatile Approach for Investigating Embryonic Development and Regeneration of the Chicken Inner Ear 127
Stephen Freeman, Elena Chrysostomou, Koichi Kawakami, Yoshiko Takahashi, and Nicolas Daudet

11 Labeling Primitive Myeloid Progenitor Cells in *Xenopus* 141
Ricardo Costa, Yaoyao Chen, Roberto Paredes, and Enrique Amaya

12 Identification of Oocyte Progenitor Cells in the Zebrafish Ovary. 157
Bruce W. Draper

13 FACS Analysis of the Planarian Stem Cell Compartment as a Tool to Understand Regenerative Mechanisms 167
Belen Tejada Romero, Deborah J. Evans, and A. Aziz Aboobaker

14 Clonal and Lineage Analysis of Melanocyte Stem Cells and Their Progeny in the Zebrafish 181
Robert C. Tryon and Stephen L. Johnson

15 Reconstitution of the Central Nervous System During Salamander Tail Regeneration from the Implanted Neurospheres 197
Levan Mchedlishvili, Vladimir Mazurov, and Elly M. Tanaka

16 Following the Fate of Neural Progenitors by Homotopic/Homochronic Grafts in *Xenopus* Embryos 203
Raphaël Thuret and Nancy Papalopulu

PART III MAMMALIAN MODEL SYSTEMS

17 Analyzing the Angiogenic Potential of Gr-1$^+$CD11b$^+$ Immature Myeloid Cells from Murine Wounds 219
Elahe Mahdipour and Kimberly A. Mace

18 In Vivo Imaging of Hematopoietic Stem Cells in the Bone Marrow Niche . . . 231
Oliver Barrett, Roberta Sottocornola, and Cristina Lo Celso

19 Characterizing the Phenotype of Murine Epidermal Progenitor Cells: Complementary Whole-Mount Visualization and Flow Cytometry Strategies 243
Korinna D. Henseleit, Ann P. Wheeler, Gary Warnes, and Kristin M. Braun

20 Murine Aggregation Chimeras and Wholemount Imaging in Airway Stem Cell Biology 263
Ian R. Rosewell and Adam Giangreco

21 Live Imaging of Primitive Endoderm Precursors in the Mouse Blastocyst 275
Joanna B. Grabarek and Berenika Plusa

PART IV HUMAN SYSTEMS

22 Utilizing DNA Mutations to Trace Epithelial Cell Lineages in Human Tissues 289
Sebastian Zeki, Trevor A. Graham, and Stuart A.C. McDonald

23 NF-Ya Protein Delivery as a Tool for Hematopoietic Progenitor Cell Expansion 303
Alevtina D. Domashenko, Susan Wiener, and Stephen G. Emerson

24 Exploring the Link Between Human Embryonic Stem Cell Organization and Fate Using Tension-Calibrated Extracellular Matrix Functionalized Polyacrylamide Gels 317
Johnathon N. Lakins, Andrew R. Chin, and Valerie M. Weaver

25 Isolation of Circulating Angiogenic Cells 351
Erin E. Vaughan and Timothy O'Brien

26 Methods for Characterization/Manipulation of Human Corneal Stem Cells and their Applications in Regenerative Medicine 357
Francesca Corradini, Beatrice Venturi, Graziella Pellegrini, and Michele De Luca

27 ALDH as a Marker for Enriching Tumorigenic Human Colonic Stem Cells. . . 373
Anitha Shenoy, Elizabeth Butterworth, and Emina H. Huang

28 Protocols for Investigating microRNA Functions in Human Neural Progenitor Cells 387
Sandra Almeida, Celine Delaloy, Lei Liu, and Fen-Biao Gao

Index *403*

Contributors

A. Aziz Aboobaker • *Centre for Genetics and Genomics, Queen's Medical Centre, University of Nottingham, Nottingham, UK*

Sandra Almeida • *Department of Neurology, University of Massachusetts Medical School, Worcester, MA, USA*

Enrique Amaya • *The Healing Foundation Centre, Faculty of Life Sciences, University of Manchester, Manchester, UK*

Rana Amini • *Cell Division and Differentiation Laboratory, Institute of Research in Immunology and Cancer, Université de Montréal, Montreal, QC, Canada*

S. Bahram Bahrami • *Department of Surgery, University of California San Francisco, San Francisco, CA, USA*

Oliver Barrett • *Division of Cell and Molecular Biology, Imperial College, London, UK*

Nancy J. Boudreau • *Department of Surgery, University of California San Francisco, San Francisco, CA, USA*

Andrea H. Brand • *The Gurdon Institute, University of Cambridge, Cambridge, UK*

Kristin M. Braun • *Centre for Cutaneous Research, Blizard Institute, Barts and the London School of Medicine and Dentistry, Queen Mary, University of London, London, UK*

Elizabeth Butterworth • *University of Florida, Gainesville, FL, USA*

Yaoyao Chen • *The Healing Foundation Centre, Faculty of Life Sciences, University of Manchester, Manchester, UK*

Andrew R. Chin • *Center for Bioengineering and Tissue Regeneration, Department of Surgery, University of California, San Francisco, CA, USA*

Elena Chrysostomou • *University College London, UCL Ear Institute, London, UK*

John T. Connelly • *Blizard Institute of Cell and Molecular Science, Barts and The London School of Medicine and Dentistry, London, UK*

Francesca Corradini • *Centre for Regenerative Medicine "Stefano Ferrari", University of Modena and Reggio Emilia, Modena, Italy*

Ricardo Costa • *The Healing Foundation Centre, Faculty of Life Sciences, University of Manchester, Manchester, UK; Departamento de Biologia del Desarrollo Cardiovascular, Centro Nacional de Investigaciones Cardiovasculares, Instituto de Salud Carlos III, Madrid*

Nicolas Daudet • *University College London, UCL Ear Institute, London, UK*

Michele De Luca • *Centre for Regenerative Medicine, "Stefano Ferrari", University of Modena and Reggio Emilia, Modena, Italy*

Celine Delaloy • *The J. David Gladstone Institutes, San Francisco, CA, USA*

PIER PAOLO DI FIORE • *Istituto Europeo di Oncologia, Milan, Italy; IFOM, Fondazione Istituto FIRC di Oncologia Molecolare, Milan, Italy; Dipartimento di Medicina, Chirurgia ed Odontoiatria, Universita' degli Studi di Milano, Milan, Italy*
VALENTINA DI FOGGIA • *Neuroscience and Trauma, Barts and The London School of Medicine and Dentistry, Institute of Cell and Molecular Science, Queen Mary University of London, London, UK*
ALEVTINA D. DOMASHENKO • *Haverford College, Haverford, PA, USA*
BRUCE W. DRAPER • *Department of Molecular and Cellular Biology, University of California Davis, Davis, CA, USA*
STEPHEN G. EMERSON • *Haverford College, Haverford, PA, USA*
DEBORAH J. EVANS • *Centre for Genetics and Genomics, Queen's Medical Centre, University of Nottingham, Nottingham, UK*
STEPHEN FREEMAN • *University College London, UCL Ear Institute, London, UK*
JESSICA E. FRITH • *Department of Biology, University of York, York, UK*
FEN-BIAO GAO • *Department of Neurology, University of Massachusetts Medical School, Worcester, MA, USA*
PAUL G. GENEVER • *Department of Biology, University of York, York, UK*
ADAM GIANGRECO • *Centre for Respiratory Research, University College London, London, UK*
KATRINA S. GOLD • *The Gurdon Institute, University of Cambridge, Cambridge, UK*
JOANNA B. GRABAREK • *Faculty of Life Sciences, University of Manchester, Manchester, UK*
TREVOR A. GRAHAM • *Histopathology Department, Cancer Research UK, London Research Institute, London, UK*
KORINNA D. HENSELEIT • *Centre for Cutaneous Research, Blizard Institute, Barts and the London School of Medicine and Dentistry, Queen Mary, University of London, London, UK*
EMINA H. HUANG • *University of Florida, Gainesville, FL, USA*
STEPHEN L. JOHNSON • *Department of Genetics, Washington School of Medicine, St. Louis, MO, USA*
KOICHI KAWAKAMI • *Division of Molecular and Developmental Biology, National Institute of Genetics, Mishima, Shizuoka, Japan*
JEAN-CLAUDE LABBÉ • *Cell Division and Differentiation Laboratory, Institute of Research in Immunology and Cancer, Université de Montréal, Montreal, QC, Canada; Department of Pathology and Cell Biology, Université de Montréal, Montreal, QC, Canada*
JOHNATHON N. LAKINS • *Department of Surgery, Center for Bioengineering and Tissue Regeneration, University of California, San Francisco, San Francisco, CA, USA*
JENNIFER A. LEE • *Department of Biology, University of York, York, UK*
LEI LIU • *The J. David Gladstone Institutes, San Francisco, CA, USA*
CRISTINA LO CELSO • *Division of Cell and Molecular Biology, Imperial College, London, UK*

KIMBERLY A. MACE • *The Healing Foundation Centre, Faculty of Life Sciences, University of Manchester, Manchester, UK*
ELAHE MAHDIPOUR • *Faculty of Life Sciences,The Healing Foundation Centre, University of Manchester, Manchester, UK*
CRISTINA MANTOVANI • *Blond McIndoe Laboratories, School of Biomedicine, University of Manchester, Manchester, UK*
VLADIMIR MAZUROV • *Max-Planck-Institute of Molecular Cell Biology and Genetics, Dresden, Germany*
STUART A.C. MCDONALD • *Centre for Digestive Diseases, Blizard Institute for Cell and Molecular Science, Barts and the London School of Medicine and Dentistry, Queen Mary, University of London, London, UK*
LEVAN MCHEDLISHVILI • *DFG-Center for Regenerative Therapies Dresden, Cluster of Excellence, University of Technology Dresden, Dresden, Germany*
TIMOTHY O'BRIEN • *Regenerative Medicine Institute, University College Hospital, National University of Ireland, Galway, Ireland*
FERIDE OEZTUERK-WINDER • *The Wellcome Trust Centre for Stem Cell Research, University of Cambridge, Cambridge, UK*
NANCY PAPALOPULU • *Faculty of Life Sciences, University of Manchester, Manchester, UK*
ROBERTO PAREDES • *The Healing Foundation Centre, Faculty of Life Sciences, University of Manchester, Manchester, UK*
SALVATORE PECE • *Istituto Europeo di Oncologia, Milan, Italy; IFOM, Fondazione Istituto FIRC di Oncologia Molecolare, Milan, Italy; Dipartimento di Medicina, Chirurgia ed Odontoiatria, Universita' degli Studi di Milano, Milan, Italy*
GRAZIELLA PELLEGRINI • *Centre for Regenerative Medicine, "Stefano Ferrari", University of Modena and Reggio Emilia, Modena, Italy*
BERENIKA PLUSA • *Faculty of Life Sciences, University of Manchester, Manchester, UK*
ALEXIA RABILOTTA • *Cell Division and Differentiation Laboratory, Institute of Research in Immunology and Cancer, Université de Montréal, Montreal, QC, Canada*
LESLEY ROBSON • *Neuroscience and Trauma, Barts and The London School of Medicine and Dentistry, Institute of Cell and Molecular Science, Queen Mary University of London, London, UK*
IAN R. ROSEWELL • *Centre for Respiratory Research, University College London, London, UK*
FATIMA A. SALEH • *Department of Biology, University of York, York, UK*
SUSAN G. SHAWCROSS • *Blond McIndoe Laboratories, School of Biomedicine, University of Manchester, Manchester, UK*
ANITHA SHENOY • *University of Florida, Gainesville, FL, USA*
ROBERTA SOTTOCORNOLA • *Division of Cell and Molecular Biology, Imperial College, London, UK*
YOSHIKO TAKAHASHI • *Graduate School of Biological Sciences, Nara Institute of Science and Technology (NAIST), Ikoma, Nara, Japan*
ELLY M. TANAKA • *DFG-Center for Regenerative Therapies Dresden, Cluster of Excellence, University of Technology Dresden, Dresden, Germany*

BELEN TEJADA ROMERO • *Centre for Genetics and Genomics, Queen's Medical Centre, University of Nottingham, Nottingham, UK*
GIORGIO TERENGHI • *Blond McIndoe Laboratories, School of Biomedicine, University of Manchester, Manchester, UK*
RAPHAËL THURET • *Faculty of Life Sciences, University of Manchester, Manchester, UK*
DANIELA TOSONI • *Istituto Europeo di Oncologia, Milan, Italy; IFOM, Fondazione Istituto FIRC di Oncologia Molecolare, Milan, Italy*
ROBERT C. TRYON • *Department of Genetics, Washington School of Medicine, St. Louis, MO, USA*
ERIN E. VAUGHAN • *Regenerative Medicine Institute, University College Hospital, National University of Ireland, Galway, Ireland*
MANDANA VEISEH • *Life Science Division, Lawrence Berkeley National Laboratory, Berkeley, CA, USA*
JUAN-JOSE VENTURA • *The Wellcome Trust Centre for Stem Cell Research, University of Cambridge, Cambridge, UK*
BEATRICE VENTURI • *Centre for Regenerative Medicine, "Stefano Ferrari", University of Modena and Reggio Emilia, Modena, Italy*
GARY WARNES • *Flow Cytometry Core Facility, Blizard Institute, Barts and the London School of Medicine and Dentistry, Queen Mary, University of London, London, UK*
VALERIE M. WEAVER • *Department of Surgery and Center for Bioengineering and Tissue Regeneration, University of California, San Francisco, CA, USA; Departments of Anatomy and Bioengineering and Therapeutic Sciences, University of California San Francisco, CA, USA; Eli and Edythe Broad Center of Regeneration Medicine and Stem Cell Research, University of California San Francisco, San Francisco, CA, USA; Helen Diller Family Comprehensive Cancer Center, University of California San Francisco, San Francisco, CA, USA*
SUSAN WIENER • *Haverford College, Haverford, PA, USA*
ANN P. WHEELER • *Blizard Advanced Microscopy Core Facility, Blizard Institute, Barts and the London School of Medicine and Dentistry, Queen Mary, University of London, London, UK*
SEBASTIAN ZEKI • *Centre for Digestive Diseases, Blizard Institute for Cell and Molecular Science, Barts and The London School of Medicine and Dentistry, Queen Mary, University of London, London, UK*

Part I

In Vitro Culture Systems

Chapter 1

Isolation of Satellite Cells from Single Muscle Fibers from Young, Aged, or Dystrophic Muscles

Valentina Di Foggia and Lesley Robson

Abstract

Skeletal muscle contains an identified resident stem cell population called the satellite cells. This cell is responsible for the majority of the postnatal growth and regenerative potential of skeletal muscle. Other cells do contribute to skeletal muscle regeneration and in cultures of minced whole muscle these cells are cultured along with the satellite cells and it is impossible to dissect out their contribution compared to the satellite cells. Therefore, a method to culture pure satellite cells has been developed to study the signaling pathways that control their proliferation and differentiation. In our studies into the role of the resident myogenic stem cells in regeneration, myopathic conditions, and aging, we have optimized the established techniques that already exist to isolate pure satellite cell cultures from single muscle fibers. We have successfully isolated satellite cells from young adults through to 24-month-old muscles and obtained populations of cells that we are studying for the signaling events that regulate their proliferative potential.

Key words: Muscle satellite cell isolation, Myoblast, Myotube, Muscle, Single fibre culture

1. Introduction

Satellite cells are a resident stem cell population of skeletal muscle; they are located under the basal lamina of the muscle fiber and are normally arrested in G_0 of the cell cycle (1, 2). The satellite cells act as reserve population of cells that can proliferate in response to injury and regenerate muscle as well as maintaining the satellite cells number throughout life (3). The overall pool of satellite cells is therefore relatively the same when we are 80 compared to the number we have when we are 20, but their ability to regenerate the muscle becomes compromised (4–8). The biology becomes compromised with increasing age, so that they take longer to return to the cell cycle, do not proliferate as many times, and when they do start to differentiate more of them will follow an adipogenic or fibroblastic lineage rather than myogenic (8–12). These same

Kimberly A. Mace and Kristin M. Braun (eds.), *Progenitor Cells: Methods and Protocols*, Methods in Molecular Biology, vol. 916, DOI 10.1007/978-1-61779-980-8_1,

changes in the satellite cell function are seen in many myopathic conditions such as Duchenne Muscular Dystrophy, where the repeated muscle damage and regenerative cycles appear to cause a premature aging of the satellite cells (13–17). The satellite cells can be identified by the expression of the transcription factor Pax7 which seems to be a universal marker of the population (18, 19), while various subpopulations of satellite cells can be identified by the co-expression with Pax7 with a number of different markers (20).

The early attempts to study the biology of the satellite cells in vitro used established myogenic cell lines such as the C2C12 and primary cultures obtained from dissociated/minced whole muscle, with rounds of purification to obtain almost pure myogenic cultures (21, 22). The use of cell lines and the primary myogenic cells obtained from these mass cultures have several disadvantages especially if the alterations with aging are of interest. Firstly, cell lines are immortalized so that their phenotype at passage 1 is more or less the same as passage 30. There can be selection of particular sub-clones, but they have limited use in understanding the normal aging process. The primary cultures obtained by purification and propagation of eventually pure primary myoblasts require multiple passaging to purify the culture and the addition of growth factors to the media to enhance the growth of the myogenic cells. Both of these can alter the phenotypes of the cells that are purified and which do not represent the starting phenotypes. It also requires a longer time to obtain the pure myogenic culture and is therefore more time consuming. Finally, the satellite cells are derived from multiple muscles that have been pooled together so that the myogenic cells that are obtained are from multiple lineages potentially.

The culture of a single skeletal muscle fiber has the advantage that the cultures of satellite cells are pure satellite cells to begin with, thus eliminating the need to purify the culture from contaminating non-myogenic cells (23–25). In addition, this method only requires a small amount of muscle and can also be more selective for the particular muscle of investigation to study any different phenotypes of the satellite cells derived from different muscles (26, 27). In this culture method, the satellite cells are plated while still attached to their parent muscle fiber. As the culture progresses, the satellite cells migrate from the muscle fiber onto the surrounding substrate. They proliferate and, if given the correct conditions, will differentiate and fuse together to form myotubes. Satellite cells derived from single fibers have been successfully cultured from mouse, rat, goat, and humans using the same basic method that will be outlined below (23, 28–30). Since the introduction of the single muscle fiber culture system, there have been various modifications introduced by individual laboratories. The two most widely used isolation methods use either gentle shaking during the enzymatic digestion or no shaking during the enzymatic digestion stage (23, 25, 31, 32). In this study we describe the single fiber culture method for isolation of satellite cells from young and aged muscles.

2. Materials

2.1. Dissection and Enzymatic Digestion

1. Dissecting instruments, watchmaker forceps, small and spring scissors (Harvard Apparatus Ltd, UK).
2. The appropriate muscles are isolated from mice of varying ages (we have routinely isolated satellite cells from the *mdx* mouse (a model of Duchenne Muscular Dystrophy on a C56Bl6/10 background) and wild-type mice (C57/Bl6/10) aged 2–3 months, 12–18 months, 24 months plus), taking care not to stretch or damage the muscles by holding the muscles by their tendons at all times (see Note 1).
3. Dulbecco's Modified Eagle's Medium (DMEM) high Glucose (4.5 g/L) with l-Glutamine (PAA Laboratories GmbH).
4. Collagenase type I 142 U/ng (Worthington Biochemical Corp. Lakewood NJ) is dissolved at 0.125% in DMEM. Filter the solution through a 0.22 μm filter. Store aliquots at −20°C up to 6 months (see Note 2).
5. Heated incubator with optional shaking facility (Stuart Scientific, UK).
6. 70% ethanol in ddH_2O for surface sterilization.

2.2. Liberation and Purification of Single Fibers

1. DMEM (PAA Laboratories GmbH) supplemented with Penicillin/Streptomycin (Pen/Strep) to produce the following concentrations 100 units per 100 ml/100 μg/100 ml respectively (Sigma-Aldrich, UK).
2. 50×18 mm sterile tissue culture petri dishes (Nunc Thermo Scientific).
3. 1 ml glass disposable Pasteur pipettes of 150 mm diameter (Volac). Pipettes are sterilized by autoclaving prior to use.

2.3. Plating

1. Geltrex™ Reduced Growth Factor Basement Membrane Matrix without phenol red (Invitrogen, UK). Diluted to 10% Geltrex in DMEM supplemented with Pen/Strep: thaw Geltrex on ice and mix it by slowly pipetting up and down without introducing bubbles. Precool DMEM on ice to prevent gelling when Geltrex is added. Store aliquot at −20°C (see Note 3).
2. Plating medium: DMEM supplemented with Pen/Strep, 10% horse Serum (Invitrogen, UK), and 0.5% chick embryo extract (Sera lab, UK). Aliquots of both horse serum and chick embryo extract can be stored at −20°C (see Note 4).
3. 24-multiwell plates (Corning Life sciences).

2.4. Maintaining the Proliferating Satellite Cells

1. Proliferation medium: DMEM supplemented with Pen/Strep, 20% fetal bovine serum, 10% horse serum, 1% chick embryo extract. We use FBS Gold heat inactivated serum

(FBS, PAA Laboratories GmbH); this is a defined FBS so there is no lot-to-lot variation due to the consistent combination of constituents.

2. Differentiation medium: DMEM supplemented with Pen/Strep, 2% FBS (PAA Laboratories GmbH), 10% horse Serum, 0.5% chick embryo extract.
3. Solution of 5% trypsin and ethylenediamine tetra acetic acid (trypsin:EDTA; Sigma-Aldrich, UK). Aliquots can be stored at −20°C.
4. Hanks' Balanced salt solution (Sigma-Aldrich, UK) stored at room temperature.

3. Methods

The single muscle fiber culture method has been validated and is now widely used in various laboratories to examine satellite cell gene expression, myogenesis, regenerative capacity, and activation processes. The method is very powerful as it allows for the maintenance of pure satellite cell cultures without any contaminating fibroblasts or other non-muscle cells. While this is a powerful technique, it is technically difficult to maintain the satellite cells in the optimum conditions to study gene expression in proliferating, differentiating, and quiescent states. As the muscle ages or in certain myopathic conditions the satellite cell behavior changes. It is important to establish the extent of these changes in comparison to satellite cells from younger animals. All methods so far reported for the isolation of satellite cells from single muscle fibers are derived from young animals; in this report we report modifications for the isolation of satellite cells from aged or dystrophic animals where there is a higher proportion of fibrotic material.

3.1. Muscle Isolation and Enzymatic Digestion

1. All animals are euthanized humanely and appropriately according to local regulations.
2. All dissecting instruments are cleaned and sterilized with 70% ethanol before starting.
3. The skin over the appropriate leg muscles is sprayed with 70% ethanol before it is removed; this not only sterilizes the skin and fur but also dampens the fur and helps prevent it from contaminating the muscles. Care should still be taken though not to get too much fur contaminating the muscles during the dissection.
4. The extensor digitorum longus (EDL) muscle is detached at the inferior tendon on the lateral anterior side of the lower leg (Fig. 1a, b arrow) and peeled back with the tibialis anterior and then the superior tendon is cut and the muscle isolated (Fig. 1c).

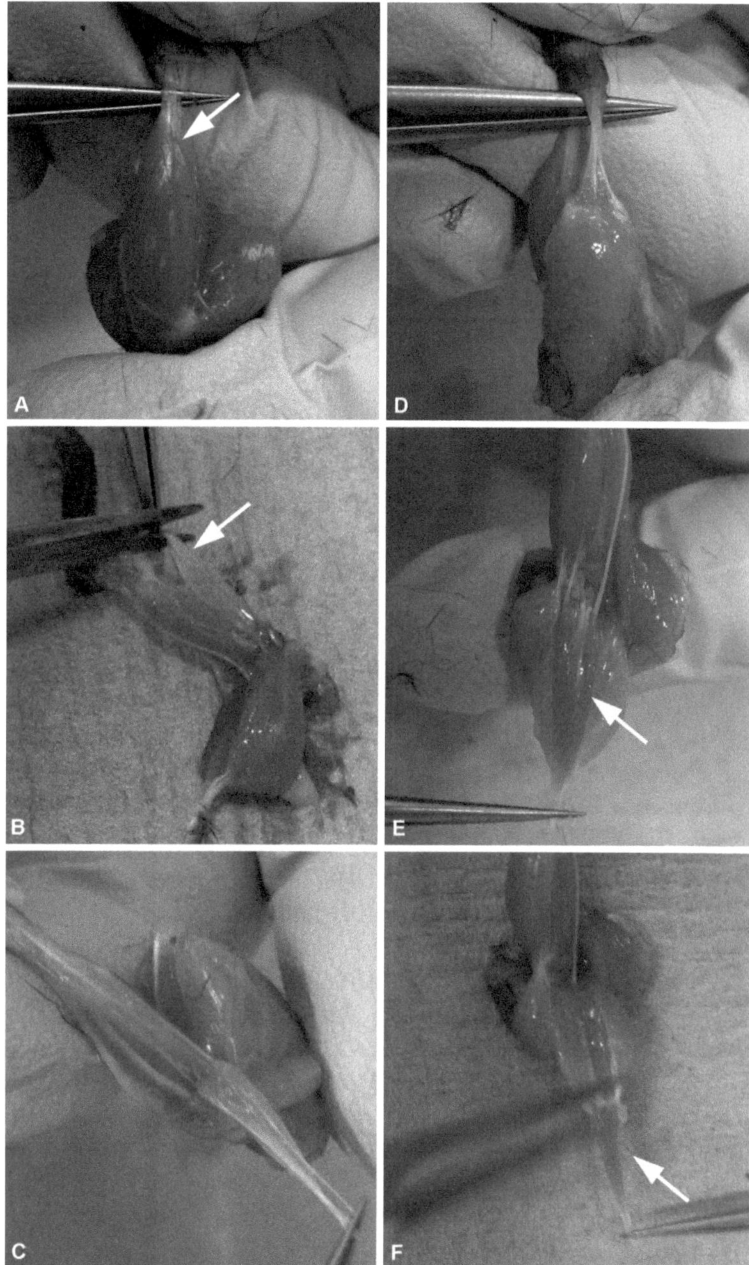

Fig. 1. *Isolation of extensor digitorum longus (EDL) and soleus muscles from a 2 month old C57BL6 mouse.* (**a**) The EDL is detached at the inferior tendon on the lateral anterior side of the lower leg inserting the forceps gently between the tendons of the tibialis anterior and EDL and the bone. Then the muscle is pulled back with the tibialis anterior gently making sure that the muscle is not stretched (**b**). After cutting the superior tendon, the EDL is isolated (**c**). (**d**) The calcaneal tendon (Achilles tendon) is cut freeing the gastrocnemius and soleus muscles at the heel. (**e**) Holding the calcaneal tendon and gently loosening the gastrocnemius and soleus muscles, it becomes possible to see the soleus clearly lying on the interior surface of the gastrocnemius muscle (the soleus has a high proportion of slow type I fibers and appears redder than the gastrocnemius, *white arrow*). (**f**) The soleus is taken gently away from the gastrocnemius always holding it by the calcaneal tendon, once it is separated to the superior tendon that is severed and the muscle can be removed.

The soleus muscle is isolated by cutting the calcaneal tendon (Achilles tendon Fig. 1d) which is the common tendon of the soleus and gastrocnemius. The calcaneal tendon is held by a pair of fine forceps and the muscles gently peeled back to show the soleus lying on the gastrocnemius (Fig. 1e arrow). By separating the soleus and gastrocnemius muscles inferiorly, the soleus is loosened so that the superior tendon of the soleus can be cut and the muscle isolated (Fig. 1f arrow). Other muscles can also be used in this protocol such as the tibialis anterior and gastrocnemius; the most important thing is not to overstretch the muscle during the dissection.

5. The muscles are placed into 50 × 18 mm Petri dishes with 5 ml of Hanks buffered saline. This removes any adhering hairs and then any fat and nerves can be carefully picked off and removed using a dissecting microscope and the tendons trimmed, to reduce the amount on non-muscle contaminating material.
6. The isolated muscles are then incubated individually in the 250 μl of the 0.125% Collagenase type I solution that has been pre-warmed to 37°C. The muscles are incubated in the enzyme for at least 2 h in a heated incubator at 37°C and shaking (150 rev/min).

3.2. Liberation and Purification of Single Muscle Fibers

1. Muscles are transferred carefully into a 50 × 18 mm Petri dish, which contains 5 ml of DMEM supplemented with Pen/Strep.
2. Under a dissecting microscope liberate single fibers by triturating. Use a glass Pasteur pipette with a rubber bulb to triturate the muscle pipetting it up and down very gently; the use of the hand controlled rubber bulb gives better control of the trituration and does not use too much force. The pipette should be pre-wetted by pipetting up and down in the plating media before starting to prevent the muscle sticking to the pipette. During the release of the muscle from the pipette it is possible to see the single fibers being liberated from the muscle (to see these liberated fibers easier use a black background and sidelights). This step is crucial for the yield of the preparation. At this stage, there will be a mixture of healthy and damaged fibers (Fig. 2a).
3. Repeat this step as many times as necessary to liberate all the fibers from the muscle bundles (see Note 5).
4. Make sure that the fibers that are liberated are free of any vascular or connective tissue debris, as this will decrease myogenic

Fig. 2. (continued) A number of satellite cells will still proliferate and will remain undifferentiated; if the culture is left for longer in the differentiation conditions the number of these will decline and greater than 95% of the satellite cells will differentiate after 5 days in differentiation conditions. Magnification of ×10. (**f**) Higher magnification (×20) of myotubes in differentiation conditions show the multinucleated nature of the myotubes and sarcomeric striations indicating full terminal differentiation (*arrow*).

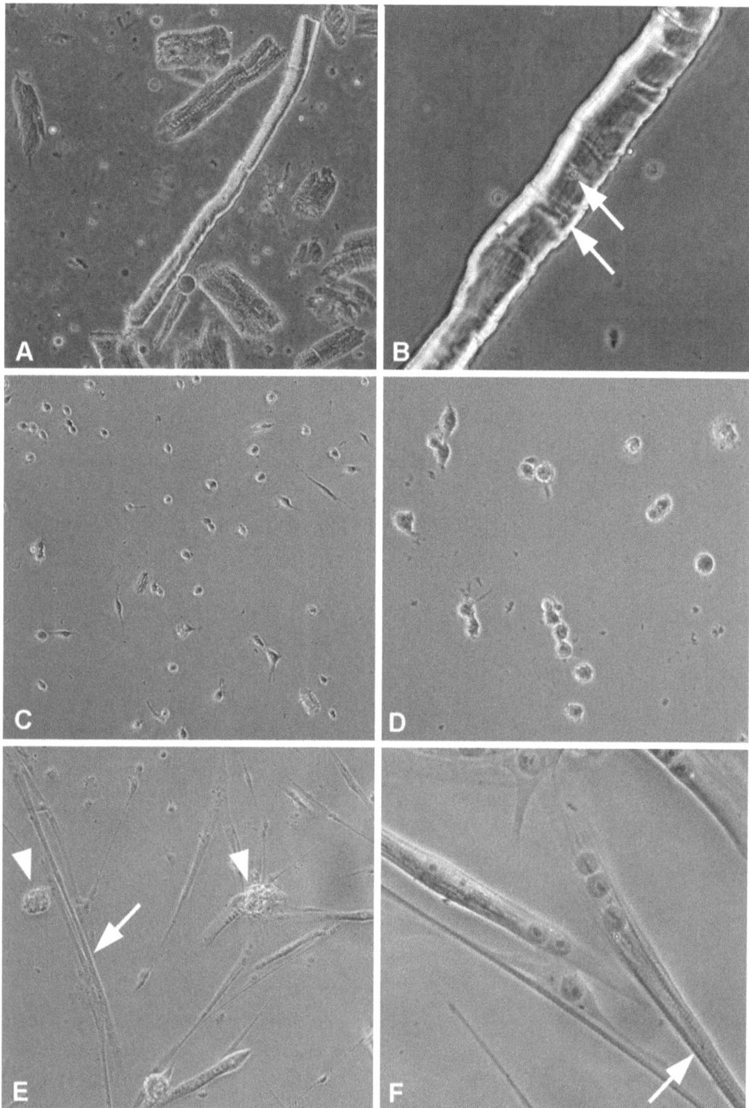

Fig. 2. *Examples of isolated myofibers and satellite cells culture in proliferation and differentiation conditions derived from 2 month old C57BL6 mouse.* (**a**) Example of myofibers obtained after digestion of the muscle in Collagenase and trituration. During the enzymatic digestion and trituration, many of the superficial fibers will be broken, but the deeper fibers are still viable. At this stage, there will be a mixture of both healthy (long and shining) and damaged (short and contracted) fibers. (**b**) During the purification all the healthy fibers will selected to be plated; healthy fibers are transparent and have an even diameter along their length. On a healthy fiber it may be possible to see the satellite cells underneath the basal lamina of the fibers (*white arrows*). (**c**) Pure satellite cells culture after 1 week in proliferation conditions. Satellite cells tend to be very round or short spindle in morphology. Magnification of ×10. (**d**) Higher magnification (×20) of satellite cells after 1 week in proliferation conditions showing their typical morphology. (**e**) Satellite cells after 48 h in differentiation conditions. The lower serum condition in the differentiation medium promotes the fusion between the satellite cells and the formation of new myotubes; the satellite cells will fuse with their neighbors forming long multinucleated myotubes or a more rounded myo-bag type morphology (myotubes; *arrows*, myo-bags *arrow heads*).

purity of the cultures by introducing non-myogenic cells into the culture.

5. Transfer between 20 and 50 identified clean single fibers to a new 50 × 18 mm Petri dish with 3 ml of DMEM supplemented with Pen/Strep. This step helps to increase the purity of the culture (Fig 2b). If more fibers are needed repeat this step as many times as required to obtain more fibers.

3.3. Plating

1. Use 10% Geltrex™ to coat a 24-multiwell plate adding 25 μl of the solution per cm^2 to each well and tilt surface to evenly coat the surface with a thin layer of Geltrex. Incubate the plate with the Geltrex in an incubator at 37°C for 60–90 min with no shaking. After the incubation remove the excess and ungelled Geltrex (excess of the extracellular matrix can be toxic for the cells). We suggest that the plate is prepared just after the muscles have been placed in the Collagenase enzyme solution, as the timing of the relevant incubations is consistent.
2. Pick up a single identified individual muscle fiber in 20 μl of media and place in a single well of the 24-multiwell plate using a sterile p200 tip (if a p20 tip is used, the fiber will remain stuck in the tip; see Note 6).
3. Allow fiber to settle and attach to the Geltrex for 5–10 min in the incubator 37°C, 5% CO_2 (see Note 7).
4. Add 500 μl of plating media and return the plate in the incubator. Do not disturb the cells for the first three days. Fibers are most vulnerable during this period to changes in temperature and CO_2 and any removal from the incubator should be limited to less than 15 mins on the first 3 days (see Note 8).

3.4. Expansion of the Satellite Cells

1. After 3 days remove the plating medium and replace with proliferation medium. Remove the old media carefully with a pipette and not with a vacuum suction system to prevent satellite cells detachment. Add pre-warmed fresh proliferation media gently, by pipetting onto the side of the well and controlling the speed of the ejection of the media.
2. Remove the fiber if it is still attached in the well using a sterile 20-gauge syringe needle. If the fiber is left in the well then any satellite cells will fuse with it.
3. Remove the old media and add fresh proliferation media every 3 days gently.

3.5. Maintenance of Satellite Cells

1. The satellite cells will proliferate and will differentiate even in this high serum media if they are close together, so keep satellite cells below 75% confluent. It is likely that the cells will initially form a small colony of cells around where the fiber adhered (Fig 2c, d). Satellite cells should be small round or short

spindle-shaped cells, fibroblasts tend to be larger and more irregular in shape.

2. When cells start to get close to 75% confluent (even in a small region of the well), remove the media and wash cells with Hanks' Balance Salt solution to remove traces of serum.
3. Add 120 μl of Trypsin/EDTA 10× for a couple of minutes at room temperature; keep a check on the adhesion of the cells under the microscope as the length of time for the satellite cells to detach will depend on the ambient temperature. Over digestion will reduce the viability of the satellite cells.
4. Collect cells in a 1.5 ml eppendorf tube, which contains 500 μl of proliferation media. Add another 120 μl of proliferating media to the well and wash any remaining cells from the well and add to the eppendorf tube (see Note 9).
5. Spin the eppendorf tube at 100 × *g* for 5 min.
6. Remove the supernatant, resuspend the pellet in fresh proliferation media, and plate the cells as required (see Note 10).
7. If differentiated myotubes are required after 12 h switch from the proliferation media to differentiation media. Myotubes will start to form within 48 h of switching the media. A small population of the satellite cells will not terminally differentiate but return to the satellite cells quiescent state and can be reactivated by returning the culture to the proliferating media (Fig. 2e–f).

4. Notes

1. The soleus and extensor digitorum muscles have tendons at each end which makes dissection easier to perform without stretching or damaging the muscle fibers. These muscles also represent a predominately slow and fast muscle fiber phenotype.
2. The Collagenase type I powder is stored at 4°C. It should be taken out of the fridge at least 30 min before making up the solution to stabilize at room temperature. The activity of the Collagenase varies with lot; it is important to pretest a particular lot of enzyme you are planning to use. Worthington offers a free Collagenase sampling service program. Under this program, three different lots of Collagenase can be evaluated and a minimum of 3 g placed on hold. Once an acceptable batch of Collagenase has been identified, it is best to plan ahead as much as possible and reserve enough of that lot of Collagenase to use in all the planned experiments to keep results as consistent as possible.

3. Avoid multiple freeze/thaw cycles of the Geltrex™. Geltrex (and matrigel) is very sensitive and must be used before the expiration date or the fiber adhesion rapidly declines and the yield of satellite cells will be adversely affected as a consequence.
4. The Chick embryo extract can precipitate very easily and should be mixed well before using it.
5. In the case of old or myopathic muscles (characterized by the presence of increased fibrotic tissue) it may be necessary to return the muscle to a fresh 250 μl of Collagenase type I enzyme solution. After the first trituration, which removes the superficial fibers and loosens the muscle, return the muscles to fresh 250 μl of Collagenase type I enzyme solution and incubate again for a further 1 h at 37°C and 150 rev/min and repeat the trituration step to liberate additional fibers. We find that only two rounds of incubation and trituration are required if the Collagenase is of an acceptable quality.
6. When choosing the fibers for plating, choose the shiniest most transparent fibers, these have no contaminating connective tissue, which make fibers less shiny. A fiber that is not contracted and is transparent and rainbow color is sign of a healthy fiber. Fibers that are dull and have an uneven diameter are not healthy and should be discarded.
7. The fibers can be left up to 45 min in the incubator with 20 μl of medium. The settling down is important to allow the fibers to attach so that the satellite cells can leave the fiber and crawl onto the Geltrex surface.
8. It may be difficult to see the satellite cells so it is best just to leave the plates in the incubator undisturbed to avoid any additional stress to the cells. Cell number will increase rapidly when the satellite cells are transferred to the proliferation media. Do not leave the satellite cells longer than 3 days in the plating media, as they will differentiate because there will not be enough serum to keep the cells proliferating.
9. If the colony of satellite cells is small but dense then to avoid any additional lost of cells by having these cells differentiating it is best to disperse the cells in the same well. This can be performed by add the 500 μL of proliferation media to the well after the trypsin EDTA solution has caused the cells to detach this results in the small colony of satellite cells been redistributed in the same well. They will proliferate again and will not be too close to each other to differentiate.
10. To increase satellite cell adhesion you can coat the new dishes that the satellite cells are to be plated onto with 10% Geltrex using the same protocol as above. The number of satellite cells derived from one EDL or one SOL (2–3 months old mouse) is between 100,000 and 250,000 cells from 24 single fibers.

Acknowledgements

The authors would like to thank the other lab members for their help and advice and encouragement especially Dr. X. Zhang. This work is supported by a studentship to V. Di Foggia from the Muscular Dystrophy Campaign.

References

1. Mauro A (1961) Satellite cell of skeletal muscle fibers. J Biophys Biochem Cytol 9:493–495
2. Schultz E, Gibson MC, Champion T (1978) Satellite cells are mitotically quiescent in mature mouse muscle: an EM and radioautographic study. J Exp Zool 206:451–456
3. Cossu G, Biressi S (2005) Satellite cells, myoblasts and other occasional myogenic progenitors: possible origin, phenotypic features and role in muscle regeneration. Semin Cell Dev Biol 16:623–631
4. Gopinath SD, Rando TA (2008) Stem cell review series: aging of the skeletal muscle stem cell niche. Aging Cell 7:590–598
5. Brooks NE, Schuenke MD, Hikida RS (2009) No change in skeletal muscle satellite cells in young and aging rat soleus muscle. J Physiol Sci 59:465–471
6. Barani AE, Durieux AC, Sabido O, Freyssenet D (2003) Age-related changes in the mitotic and metabolic characteristics of muscle-derived cells. J Appl Physiol 95:2089–2098
7. Beccafico S, Puglielli C, Pietrangelo T, Bellomo R, Fano G, Fulle S (2007) Age-dependent effects on functional aspects in human satellite cells. Ann N Y Acad Sci 1100:345–352
8. Brack AS, Conboy MJ, Roy S, Lee M, Kuo CJ, Keller C, Rando TA (2007) Increased Wnt signaling during aging alters muscle stem cell fate and increases fibrosis. Science 317:807–810
9. Carlson ME, Suetta C, Conboy MJ, Aagaard P, Mackey A, Kjaer M, Conboy I (2009) Molecular aging and rejuvenation of human muscle stem cells. EMBO Mol Med 1:381–391
10. Conboy IM, Rando TA (2005) Aging, stem cells and tissue regeneration: lessons from muscle. Cell Cycle 4:407–410
11. Day K, Shefer G, Shearer A, Yablonka-Reuveni Z (2010) The depletion of skeletal muscle satellite cells with age is concomitant with reduced capacity of single progenitors to produce reserve progeny. Dev Biol 340:330–343
12. Vertino AM, Taylor-Jones JM, Longo KA, Bearden ED, Lane TF, McGehee RE Jr, Macdougald OA, Peterson CA (2005) Wnt10b deficiency promotes coexpression of myogenic and adipogenic programs in myoblasts. Mol Biol Cell 16:2039–2048
13. Blau HM, Webster C, Pavlath GK, Chiu CP (1985) Evidence for defective myoblasts in Duchenne muscular dystrophy. Adv Exp Med Biol 182:85–110
14. Blau HM, Webster C, Pavlath GK (1983) Defective myoblasts identified in Duchenne muscular dystrophy. Proc Natl Acad Sci USA 80:4856–4860
15. Irintchev A, Zweyer M, Wernig A (1997) Impaired functional and structural recovery after muscle injury in dystrophic mdx mice. Neuromuscul Disord 7:117–125
16. Schuierer MM, Mann CJ, Bildsoe H, Huxley C, Hughes SM (2005) Analyses of the differentiation potential of satellite cells from myoD-/-, mdx, and PMP22 C22 mice. BMC Musculoskelet Disord 6:15
17. Yablonka-Reuveni Z, Anderson JE (2006) Satellite cells from dystrophic (mdx) mice display accelerated differentiation in primary cultures and in isolated myofibers. Dev Dyn 235: 203–212
18. Seale P, Sabourin LA, Girgis-Gabardo A, Mansouri A, Gruss P, Rudnicki MA (2000) Pax7 is required for the specification of myogenic satellite cells. Cell 102:777–786
19. Oustanina S, Hause G, Braun T (2004) Pax7 directs postnatal renewal and propagation of myogenic satellite cells but not their specification. EMBO J 23:3430–3439
20. Boldrin L, Muntoni F, Morgan JE (2010) Are human and mouse satellite cells really the same? J Histochem Cytochem 58:941–955
21. Rando TA, Blau HM (1994) Primary mouse myoblast purification, characterization, and transplantation for cell-mediated gene therapy. J Cell Biol 125:1275–1287
22. Yaffe D, Saxel O (1977) Serial passaging and differentiation of myogenic cells isolated from dystrophic mouse muscle. Nature 270:725–727

23. Rosenblatt JD, Lunt AI, Parry DJ, Partridge TA (1995) Culturing satellite cells from living single muscle fiber explants. In Vitro Cell Dev Biol Anim 31:773–779
24. Bekoff A, Betz W (1977) Properties of isolated adult rat muscle fibres maintained in tissue culture. J Physiol 271:537–547
25. Bischoff R (1986) Proliferation of muscle satellite cells on intact myofibers in culture. Dev Biol 115:129–139
26. Rosenblatt JD, Parry DJ, Partridge TA (1996) Phenotype of adult mouse muscle myoblasts reflects their fiber type of origin. Differentiation 60:39–45
27. Zammit PS (2008) All muscle satellite cells are equal, but are some more equal than others? J Cell Sci 121:2975–2982
28. Yablonka-Reuveni Z, Rivera AJ (1994) Temporal expression of regulatory and structural muscle proteins during myogenesis of satellite cells on isolated adult rat fibers. Dev Biol 164: 588–603
29. Yamanouchi K, Hosoyama T, Murakami Y, Nishihara M (2007) Myogenic and adipogenic properties of goat skeletal muscle stem cells. J Reprod Dev 53:51–58
30. Bonavaud S, Agbulut O, D'Honneur G, Nizard R, Mouly V, Butler-Browne G (2002) Preparation of isolated human muscle fibers: a technical report. In Vitro Cell Dev Biol Anim 38:66–72
31. Wozniak AC, Pilipowicz O, Yablonka-Reuveni Z, Greenway S, Craven S, Scott E, Anderson JE (2003) C-Met expression and mechanical activation of satellite cells on cultured muscle fibers. J Histochem Cytochem 51:1437–1445
32. Wozniak AC, Anderson JE (2005) Single-fiber isolation and maintenance of satellite cell quiescence. Biochem Cell Biol 83:674–676

Chapter 2

Terminal Differentiation of Human Epidermal Stem Cells on Micro-patterned Substrates

John T. Connelly

Abstract

Extracellular signals play a central role in coordinating the growth and differentiation of epidermal stem cells. This protocol describes a technique for quantitatively examining the influence of extracellular matrix (ECM) interactions on keratinocyte terminal differentiation through the use of micro-patterned, polymer substrates. Circular islands of type I collagen are created, first by micro-contact printing and surface-initiated polymerization of a protein resistant background. The unprotected gold is then coated with collagen by passive adsorption. When human keratinocytes are seeded onto these substrates, limited adhesion on the smallest islands induces terminal differentiation, characterized by increased involucrin, transglutaminase, and periplakin expression, as well as reduced proliferation. This platform provides a robust assay for studying the terminal differentiation of human epidermal stem cells and the regulatory roles of specific cell–matrix interactions in this process.

Key words: Epidermis, Stem cell, Human, Terminal differentiation, Micro-patterned, Microenvironment, Integrin

1. Introduction

The epidermis is a fully stratified epithelial tissue composed of keratinocytes and organized into basal, spinous, granular, and cornified layers. As the outermost layer of the skin it protects our bodies from the external environment and is continuously regenerated throughout life. Normal homeostasis within the epidermis relies on a population of stem cells, which are capable of both self-renewal and terminal differentiation (1). In the human epidermis, a putative population of stem cells resides in the basal layer and expresses high levels of the β1 integrin (2), melanoma chondroitin sulfate proteoglycan (MCSP) (3), and Lrig1 (4). Given the importance of stem cells in epidermal regeneration, understanding the regulatory

Kimberly A. Mace and Kristin M. Braun (eds.), *Progenitor Cells: Methods and Protocols*, Methods in Molecular Biology, vol. 916, DOI 10.1007/978-1-61779-980-8_2, © Springer Science+Business Media, LLC 2012

mechanisms governing their behavior is a key step in developing treatments for conditions such as burns, psoriasis, and cancer.

Robust in vitro assays are essential tools for studying the differentiation of epidermal stem cells. Common in vitro methods for inducing terminal differentiation include elevated Ca^{2+} levels (5), suspension culture (6), and stratification within an organotypic culture model (7). While these techniques promote classical differentiation involving cell cycle exit and expression of differentiation markers, such as involucrin, cornifin, loricrin, and transglutaminase (8), they provide limited control over the external signals that the cells see, and analysis of single cell behavior is challenging.

Recently, the use of engineered materials and extracellular matrices has improved our ability to manipulate the cellular microenvironment, even at the level of individual stem cells. Micro-contact printing for example is one approach for depositing specific adhesion molecules in well-defined shapes and sizes (9), and this strategy has successfully been employed to study cell fate decisions of mesenchymal (10) and epidermal stem cells (11, 12). In the case of epidermal stem cells, restricting the adhesive area and a cell's ability to spread stimulates terminal differentiation, while cells with a larger adhesive area are able to spread and remain undifferentiated. The following protocol describes a technique for creating micro-patterned substrates and quantitatively investigating the role of cell–matrix interactions in the terminal differentiation of human epidermal stem cells.

2. Materials

2.1. Preparation of Micro-Patterned Substrates

1. Glass coverslips (20 × 60 mm) coated sequentially with 1.5 nm chromium and 15 nm gold by vacuum deposition. Gold-coated slides can be stored indefinitely under inert gas (see Note 1).
2. ω-mercaptoundecyl bromo-isobutyrate initiator (13).
3. Oligo(ethylene glycol methyl ether methacrylate), molecular weight 300 (Sigma-Aldrich).
4. $CuBr_2$, CuCl, and bipyridine (Sigma-Aldrich).
5. 100% Ethanol.
6. N_2 gas cylinder.
7. Poly(dimethyl siloxane) (PDMS) and curing agent (Sylgard 184, Dow Corning).
8. Patterned silicon wafers to be used as master molds for casting the PDMS stamps (see Notes 1 and 2).
9. Cotton swabs.
10. 50 ml syringe and 2″, 19-gauge needle.
11. Large glass reaction vessel, flask, or test tube that can be sealed with rubber stopper (Corning).

12. Microscope slide container (e.g., 5-slide open top mailer, Scientific Laboratory Supplies).
13. Dessicator connected to a vacuum source.

2.2. Keratinocyte Culture and Seeding

1. Primary human keratinocytes isolated from adult or neonatal skin.
2. Growth arrested (gamma irradiated or mitomycin C treated) J2 3T3 fibroblasts.
3. FAD medium: 1 part Ham's F-12, 3 parts DMEM and 10^{-4} M adenine, supplemented with 10% fetal bovine serum, 5 μg/ml insulin, 10 ng/ml epidermal growth factor, 0.5 μg/ml hydrocortisone, and 10^{-10} cholera enterotoxin.
4. Versene (Invitrogen).
5. 0.25% Trypsin (No EDTA).
6. 24-well plates.
7. Fine-point forceps with Teflon coating (EMS).
8. Diamond tipped glass cutter (Draper).
9. Sterile 1 mM HCl.
10. Sterile phosphate-buffered saline.
11. Soluble rat tail collagen, Type I (BD Biosciences).

2.3. Immunofluorescence Staining

1. Phosphate-buffered saline (PBS).
2. 4% paraformaldehyde (PFA) in PBS.
3. 0.2% Triton X-100 in PBS.
4. Blocking buffer: 10% FBS and 0.25% fish skin gelatin in PBS.
5. Anti-involucrin mouse monoclonal antibody (SY5) (Abcam, Santa Cruz Biotechnology, or Sigma-Aldrich).
6. Anti-mouse Alexafluor 488 or similar secondary antibody (Invitrogen).
7. 4′,6-diamidino-2-phenylindole (DAPI).
8. Glass slides.
9. Aqueous mounting medium.
10. Inverted fluorescence microscope (e.g., Leica DMI5000).

3. Methods

This procedure employs micro-contact printing and surface-initiated polymerization to create precise patterns of type I collagen. Using patterned PDMS stamps, a chemical initiator is first deposited onto gold-coated substrates, and a controlled polymerization of

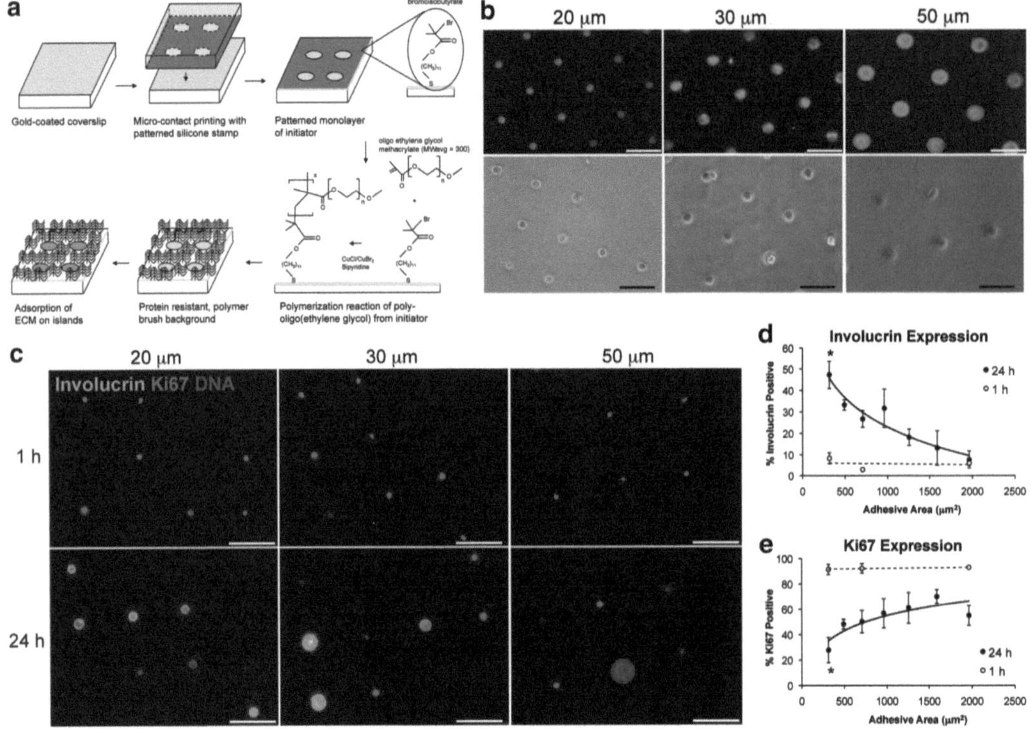

Fig. 1. *Regulation of keratinocyte shape and differentiation on micro-patterned substrates.* Figure is adapted from Connelly et al. 2010 (12). (**a**) Overview of the micro-patterning strategy. (**b**) Immunofluorescence images of type I collagen (*upper panel*) and phase-contrast images of primary human keratinocytes (*lower panel*) on 20, 30, and 50 μm diameter islands. (**c**) Representative immunofluorescence images of involucrin and Ki67 expression on substrates with 20, 30, or 50 μm diameters at 1 h and 24 h after seeding. *Scale bars* = 100 μm. (**d**) Quantification of involucrin positive cells at 1 h and 24 h on substrates with adhesive areas ranging from 314 μm^2 to 1,963 μm^2 (20–50 μm diameters). Data represent mean ± SEM ($n=4$ experiments, $*P=0.0001$ compared to 50 μm). (**e**) Quantification of Ki67 positive cells at 1 h and 24 h. Data represent mean ± SEM ($n=4$ experiments; $*P=0.0472$ compared to 50 μm).

oligo(ethylene glycol) generates a brush-like structure of poly(oligo ethylene glycol methacrylate) (POEGMA). The POEGMA surface provides a highly protein resistant background, surrounding bare gold islands. Passive adsorption of type I collagen onto the exposed gold promotes cell adhesion to the islands (Fig 1a–b). By varying the pattern of the stamp, the size and shape of the islands can easily be manipulated (see Note 2).

3.1. Preparation of Micro-Patterned Substrates

1. Mix ten parts PDMS with one part curing agent. To remove air bubbles place the mixture in a dessicator under vacuum for 30 min.
2. Pour the PDMS into a plastic weigh boat with the master silicon mold face up on the bottom. Continue to de-air for an additional hour.
3. Allow the PDMS to cure overnight at room temperature, followed by 30 min in a 65°C oven. To remove the stamp, cut

away the excess PDMS from the edges and slowly peal it away from the silicon. Be careful not to crack the master as it is extremely brittle.

4. Clean the PDMS stamps by rinsing with 100% EtOH and drying under a stream of N_2. It is essential to keep the patterned surface of the stamp clean. Store in a sealed container, and do not touch the patterned surface.
5. Remove dust and debris from the gold-coated slides with a stream of N_2.
6. Using a cotton swab, coat the patterned surface of the stamp with an even layer of the initiator, rinse with EtOH, and dry with N_2.
7. Bring the coated surface of the stamp into conformal contact with the gold slide for 15 s and remove. Take care not to apply excessive pressure to the stamp and avoid creating air bubbles between the two surfaces.
8. Store the stamped slide under a vacuum until it is ready for polymerization and take care not to touch or scratch the stamped surface.
9. Prepare the monomer and catalyst solution consisting of 1.6 M OEGMA, 3.3 mM $CuBr_2$, and 82 mM bipyridine in a water/ethanol (4:1) solvent mixture. The volume will depend on the number of slides and size of the reaction vessel. Mix and degas for at least 1 h by bubbling N_2 through the solution. This step can be performed while stamping the slides.
10. Before polymerization, add CuCl to the solution for a final concentration of 33 mM and degas for an additional 20 min.
11. Place the slides in a sealed reaction vessel under vacuum, with the needle and syringe inject the monomer solution so that all the slides are completely covered, and allow the reaction to proceed for 15 min.
12. To stop the polymerization, pour off the monomer solution and immerse the slides in deionized water.
13. Rinse the slides thoroughly with EtOH, dry under a stream of N_2, and store in a cool dry place. A sealed container under vacuum or inert gas is preferable and will prolong the stability of the patterned substrates. Under these conditions the patterned substrates will be stable for at least 2 months.

3.2. Keratinocyte Culture and Seeding

1. Cut the patterned slides into 1 cm^2 samples using a diamond tip cutter, and using Teflon-coated forceps under aseptic conditions, place each sample face up into an individual well of a 24-well plate. Take care not to scratch or damage the substrate surface.
2. Sterilize the substrates by immersing in 70% EtOH for 10 min, and rinse twice with sterile PBS.

3. Prepare a 20 μg/ml solution of type I collagen in sterile PBS, add 0.5 ml to each well, and incubate for 1 h at 37°C.
4. After 1 h, rinse the substrates by dilution of the collagen solution in 1 mM HCl. Fill each well with HCl, then aspirate down to the 0.5 ml level, and repeat this process twice. Aspirate off all the HCl and rinse once more with HCl and twice with PBS. Note: Dilution of the collagen in HCl is essential for removing non-adsorbed fibers and preventing the collagen from drying on the POEGMA brushes (14).
5. Substrates can now be set aside until the keratinocytes are ready.
6. Primary human keratinocytes from adult or neonatal skin should be cultured on a confluent layer of growth arrested J2 3T3 cells as previously described (15).
7. Remove the fibroblasts by incubating with Versene at room temperature and continuous tapping on the side of the flask. Rinse twice to remove all remaining fibroblasts.
8. Incubate the keratinocytes in trypsin/Versene (1:4) at 37°C for 5–10 min. Inactivate the trypsin with fully supplemented FAD, transfer to a conical tube, and pellet the cells by centrifugation.
9. Resuspend the cells in FAD and count using a hemocytometer.
10. Dilute the keratinocytes to 5×10^4 cells/ml and seed 0.5 ml (25×10^4 cells) per substrate.
11. Allow the cells to adhere for 1 h at 37°C before rinsing thoroughly (see Note 3). Pipet media over each substrate two to three times, and then exchange the media in each well twice. Ensure that the cells on the substrates do not dry out by replacing the media immediately after aspirating. The gold renders the surface highly hydrophobic and prone to drying, which can quickly kill the cells if exposed to air.
12. Culture the cells for 24 h in a 37°C incubator with 5% CO_2.

3.3. Immunofluorescence Staining

1. Fix the cell-seeded substrates with 4% PFA at room temperature for 10 min and rinse twice with PBS.
2. Permeabilize the cells with 0.2% Triton for 5 min and rinse twice with PBS.
3. Incubate in blocking buffer for 1 h at room temperature.
4. Label the cells with primary antibodies for involucrin (2 μg/ml in blocking buffer) or other differentiation markers for 1 h at room temperature or overnight at 4°C.
5. Rinse three times with PBS. Note: due to the hydrophobicity of the substrates, add PBS to the well immediately after aspirating to avoid drying of the antibody onto the substrate.
6. Label with an appropriate secondary antibody (e.g., anti-mouse IgG Alexafluor 488) and DAPI for 1 h at room temperature.

7. Rinse three times with PBS and mount onto glass slides with mounting medium.
8. Images can be acquired on most inverted fluorescence microscopes, and differentiation is assessed by scoring the percentage of involucrin positive cells.
9. When human keratinocytes are cultured on small islands (20 μm diameter) a significant increase in involucrin expression and decrease in Ki67 expression can be observed after 24 h (Fig. 1c–e).

4. Notes

1. This assay requires several pieces of specialized equipment that are not available in most biology laboratories. Collaboration with micro-fabrication specialists will be necessary for creating the master silicon molds using photolithography, and access to a vacuum deposition system (available in many physics, chemistry, or engineering departments) will be important for coating glass slides with gold.
2. Although the procedure above describes one approach for generating micro-patterns, several similar techniques exist. For example, stamping self-assembled monolayers (9) or ECM proteins directly onto gold or glass surfaces rather than growing polymer brushes may be a simpler procedure, but substrates generated with these techniques are less stable and must be used immediately (16). Other variations include coating with different ECM proteins, patterning shapes other than circular islands (ellipses, squares, triangle), and patterning larger adhesive islands to study the interactions between multiple cells.
3. Given that epidermal stem cells express high levels of β1 integrins (2), allowing the cells to adhere for only 1 h before rinsing enriches for this putative stem cell population (12). Therefore, this procedure has the added advantage of easily isolating and studying a more homogeneous population of epidermal progenitors.

References

1. Potten CS (1981) Cell replacement in epidermis (keratopoiesis) via discrete units of proliferation. Int Rev Cytol 69:271–318
2. Jones PH, Watt FM (1993) Separation of human epidermal stem cells from transit amplifying cells on the basis of differences in integrin function and expression. Cell 73:713–724
3. Legg J, Jensen UB, Broad S, Leigh I, Watt FM (2003) Role of melanoma chondroitin sulphate proteoglycan in patterning stem cells in human interfollicular epidermis. Development 130: 6049–6063
4. Jensen KB, Watt FM (2006) Single-cell expression profiling of human epidermal stem and

transit-amplifying cells: Lrig1 is a regulator of stem cell quiescence. Proc Natl Acad Sci USA 103:11958–11963

5. Hennings H, Michael D, Cheng C, Steinert P, Holbrook K, Yuspa SH (1980) Calcium regulation of growth and differentiation of mouse epidermal cells in culture. Cell 19:245–254
6. Green H (1977) Terminal differentiation of cultured human epidermal cells. Cell 11: 405–416
7. Stark HJ, Baur M, Breitkreutz D, Mirancea N, Fusenig NE (1999) Organotypic keratinocyte cocultures in defined medium with regular epidermal morphogenesis and differentiation. J Invest Dermatol 112:681–691
8. Watt FM (1988) Proliferation and terminal differentiation of human epidermal keratinocytes in culture. Biochem Soc Trans 16:666–668
9. Chen CS, Mrksich M, Huang S, Whitesides GM, Ingber DE (1997) Geometric control of cell life and death. Science 276:1425–1428
10. McBeath R, Pirone DM, Nelson CM, Bhadriraju K, Chen CS (2004) Cell shape, cytoskeletal tension, and RhoA regulate stem cell lineage commitment. Dev Cell 6:483–495
11. Watt FM, Jordan PW, O'Neill CH (1988) Cell shape controls terminal differentiation of human epidermal keratinocytes. Proc Natl Acad Sci USA 85:5576–5580
12. Connelly JT, Gautrot JE, Trappmann B, Tan DW, Donati G, Huck WT, Watt FM (2010) Actin and serum response factor transduce physical cues from the microenvironment to regulate epidermal stem cell fate decisions. Nat Cell Biol 12:711–718
13. Jones DM, Brown AA, Huck WTS (2002) Surface-initiated polymerizations in aqueous media: effect of initiator density. Langmuir 18:1265–1269
14. Gautrot JE, Trappmann B, Oceguera-Yanez F, Connelly J, He X, Watt FM, Huck WT (2010) Exploiting the superior protein resistance of polymer brushes to control single cell adhesion and polarisation at the micron scale. Biomaterials 31:5030–5041
15. Rheinwald JG, Green H (1977) Epidermal growth factor and the multiplication of cultured human epidermal keratinocytes. Nature 265:421–424
16. Ma HW, Hyun JH, Stiller P, Chilkoti A (2004) "Non-fouling" oligo(ethylene glycol)-functionalized polymer brushes synthesized by surface-initiated atom transfer radical polymerization. Adv Mater 16:338–341

Chapter 3

Isolation, Culture, and Potentiality Assessment of Lung Alveolar Stem Cells

Feride Oeztuerk-Winder and Juan-Jose Ventura

Abstract

The cellular and molecular elements involved in the turnover and regeneration of the lung alveolar epithelium remain largely unknown (Kim, Am J Physiol Lung Cell Mol Physiol 293:L1092–1098, 2007). Isolation and characterization of putative stem cells with limited and nonspecific markers have made necessary the use, in parallel, of culture restrictive conditions and optimized reagents to allow selection and long-term expansion of this population.

Key words: Lung alveolar stem cells, Isolation, Cell culture, Differentiation, Self-renewal, MAPK, Flow cytometry

1. Introduction

In order to identify specific target cells that could play a role in alveolar replenishment current techniques have been adapted, together with the development of new methods, to permit studies in the lung to be undertaken. Recently, a discrete population of cells has been proposed to have the potential to differentiate into the major types of cells (Alveolar type-II and Clara cells) in the bronchioalveolar epithelium, thereby acting as putative stem cells (1). Since then many groups have been able to isolate cells with similar profiles that could be considered as bronchioalveolar stem cells (1–3). Culturing and long-term expansion of these cells, in addition to maintaining their differentiation potential, have been a challenge that we have been able to successfully address (Fig. 1). Characterization of any kind of stem cell requires proof of self-renewal and differentiation potential (4). We have developed protocols to assess, both in vitro and in vivo, the potential of lung alveolar stem cells (Figs. 2 and 3). We have efficiently isolated, cultured, and

Kimberly A. Mace and Kristin M. Braun (eds.), *Progenitor Cells: Methods and Protocols*, Methods in Molecular Biology, vol. 916, DOI 10.1007/978-1-61779-980-8_3, © Springer Science+Business Media, LLC 2012

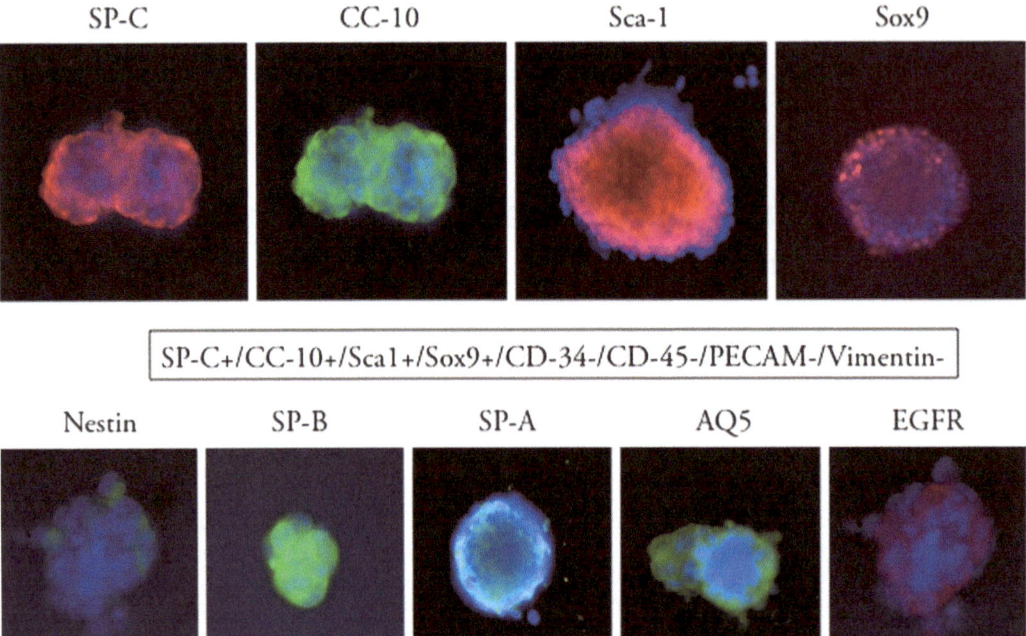

Fig. 1. *Isolation and culturing of lung stem cells.* After disruption and isolation of lung cells, the stem cell population was enriched in serum-free media supplemented with EGF and FGF2. Cells were sorted using the Sca-1 marker and negative selection for endothelial, hematopoietic, and mesenchymal markers. Cells grow forming aggregates that express different lung, SP-C (alveolar-type2), CC-10 (Clara Cell), Aq5 (Alveolar-type1), SP-A,B (lung alveolar), and stem, Sca-1, Sox9, Nestin, markers.

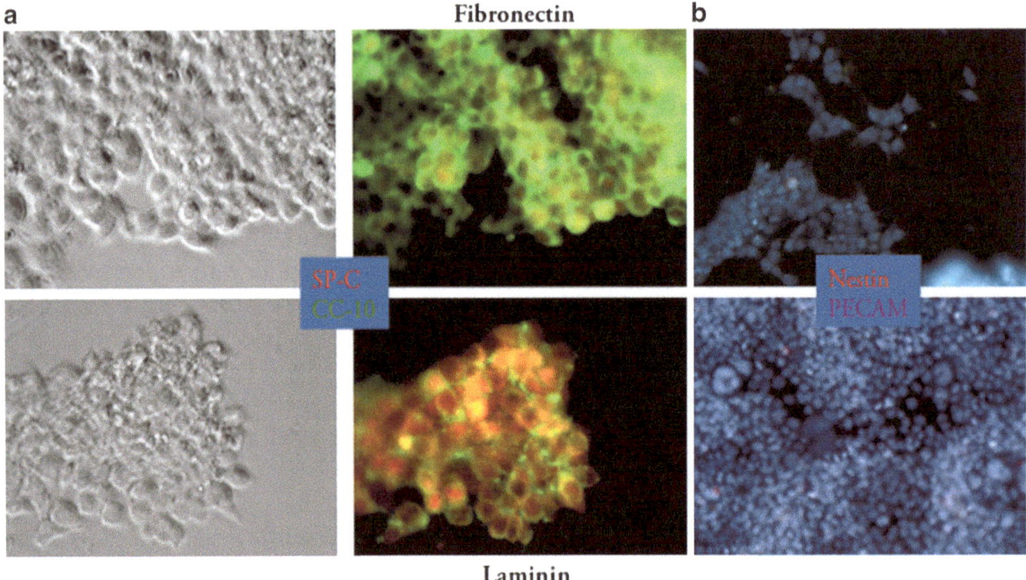

Fig. 2. *In vitro assessment of lung stem cell potential.* (**a**) Lung stem cells culture on matrix proteins form monolayer and change their marker profile. Fibronectin potentiates Clara cell and Laminin Alveolar-type II cell commitment. (**b**) Cells on matrices do not express neural or endothelial markers.

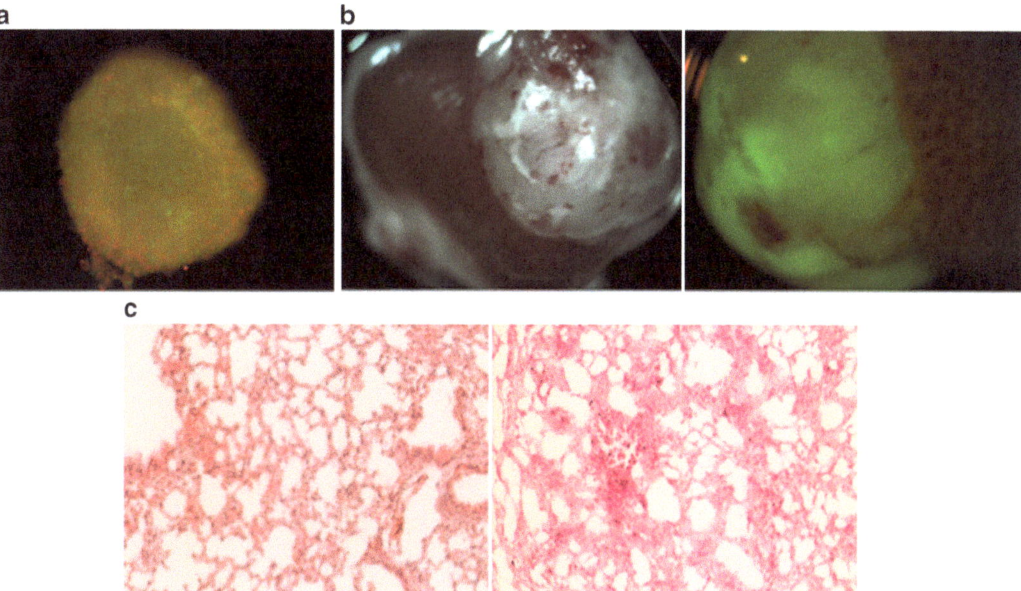

Fig. 3. *In vivo assessment of lung stem cell differentiation potential.* (**a**) Cells in culture form aggregates that express the bronchioalveolar markers CC-10 and SP-C. (**b**) Cells expressing the green fluorescent protein (EGFP) produce tissue when injected under the kidney capsule membrane. (**c**) Lung alveolar stem cells originate an alveolar-like epithelium with bronchiolar and alveoli-like structures.

expanded lung stem cells from not only different mouse strains, but also from human lungs and other species. Our challenge now is to specifically characterize lung alveolar stem cells at different levels of commitment stages using simple molecular markers that could be used for easy detection and isolation of cells from lungs.

2. Materials

2.1. Isolation, Enrichment, and Sorting of Lung Alveolar Stem Cells from Lung Tissue

1. Phosphate-buffered saline (PBS).
2. Collagenase type III (Worthington) at 3 mg/mL in DMEM.
3. Dulbecco's Modified Eagle's Medium (DMEM) (Invitrogen).
4. DNase (Promega) at 1 mg/mL in DMEM.
5. Cell strainers (70 μm or 40 μm mesh size, BD) and 50 mL tubes (Falcon),
6. Antibodies: anti-Sca-1 (PE conjugated, Biolegend), anti-CD-17 (FITC conjugated, Biolegend), anti-CD-105 (FITC conjugated, Biolegend), and anti-CD45 (FITC conjugated, Biolegend).
7. RH-B medium (Stem Cell Science) and Fetal Calf Serum (FCS, Gibco).
8. FGF2 (10 ng/mL, Pepro Tech), EGF (20 ng/mL, Pepro Tech), insulin (5 μg/mL, Pepro Tech).

2.2. Long-term Culture of Lung Alveolar Stem Cells

1. Accutase (PAA).
2. Antibodies: anti-SP-C (Santa Cruz biotech), CC-10 (Santa Cruz biotech).

2.3. In Vitro Assessment of Differentiation Potential

1. Fibronectin (10 μg/mL in PBS, Millipore) or laminin (10 μg/mL in PBS, Sigma).

2.4. Kidney Capsule Engraftments

1. Anesthetic: Isofluorane.
2. EtOH 70% (SIGMA).
3. Surgical material.

2.5. Immuno-fluorescence Microscopy of Lung Alveolar Stem Cells

1. Paraformaldehyde 4% (SIGMA).
2. Hydrophobic pen (Pap Pen, DAKO).
3. Triton X-100 (GIBCO) 0.1% in PBS.
4. DAPI (4,6-diamidino-2-phenylindole).
5. Antifade (Molecular Probes, Eugene, OR).
6. Concave glass slides (Fisher Scientific).
7. Vectastain (Molecular Probes, Eugene, OR).

3. Methods

Isolation of lung alveolar stem cells from lung tissue requires a combination of cell culture and flow cytometry techniques. Due to a lack of specific markers for these cells, we have developed a protocol to enrich stem/progenitor cells, removing most of the contaminating differentiated cells using restriction serum-free media and flow cytometry sorting based on negative and positive selection with different markers. This methodology has allowed us to obtain a homogenous population of alveolar stem/progenitor cells expressing epithelial but not mesenchymal, hematopoietic, or endothelial markers (Fig. 1). The potential of stem cells to differentiate has been assessed using in vitro and in vivo protocols (Figs. 2 and 3).

3.1. Isolation, Enrichment, and Sorting of Lung Alveolar Stem Cells from Lung Tissue

Single-cell suspensions were prepared from lungs of 1 mouse or 3 (or more) mice (6–8 weeks old, see Note 1).

1. Rinse the lungs in sterile phosphate-buffered saline (PBS) followed by removal of the trachea.
2. Mince the lungs very fine with sterile scissors (see Note 2).
3. Resuspend in 0.5–3 mg/mL collagenase type III containing DMEM and incubate for 15–30 min at 37°C in a shaking

(120 rpm) incubator. Alternatively, proceed without enzymatic digestion if the tissue has been very finely minced.

4. Centrifuge the suspension for 5 min at 350 × *g* and then remove the supernatant.
5. Resuspend the pellet with fresh DMEM containing 1 mg/mL DNase and incubate for a further 5–10 min. This step is optional.
6. Centrifuge the suspension for 5 min at 350 × *g* and then remove the supernatant.
7. Resuspend the pellet with PBS followed by filtration through a cell strainer (70 μm) placed on a 50 mL tube, followed by centrifugation for 5 min at 350 × *g*.
8. Resuspend the pellet with PBS followed by filtration through a cell strainer (40 μm) placed on a 50 mL tube, followed by centrifugation for 5 min at 350 × *g*.
9. Determine the viability by trypan blue exclusion and count the cells using a hemocytometer.
10. Proceed immediately with flow cytometry analysis following incubation with anti-Sca-1, anti-CD-17, anti-CD-105, and anti-CD45 antibodies in 100 μl of PBS on ice for 15–20 min *or* seed in 6-well cell culture plates for the further enrichment [5–7 days in RH-B medium, FGF2 (10 ng/mL), EGF (20 ng/mL), with additional insulin (5 μg/mL) and 2% FCS (for the first 2 days) and then without FCS and insulin for the remaining time] prior to flow cytometry analysis.
11. For flow cytometry analysis, wash the cells three times with PBS with 5 min centrifugations at 350 × *g* between washes and then resuspend in 500 μl PBS for flow cytometry analysis. Collect anti-Sca-1 positive and anti-CD31, anti-CD34, anti-CD45 negative cells.
12. Finally resuspend the cells in RH-B culture medium containing 2% FCS, with additional insulin (5 μg/mL), EGF (10 ng/mL), and FGF2 (20 ng/mL) for 2 days and then replace with fresh, serum-free medium without additional insulin, but containing FGF and EGF2. The stem/progenitor cells tend to aggregate when maintained in serum-free media (at 37°C in a 7% humidified CO_2 incubator).

3.2. Long-Term Culture of Lung Alveolar Stem Cells

The culture of cells should form clusters or aggregates when maintained in serum-free media plus FGF2 and EGF (Fig. 3a). Fibroblasts or blood cells cannot proliferate for long under these conditions and will be lost from the culture during serial passaging. Cells can be maintained for more than 80–90 passages, replacing medium every 3 days.

1. To split the cells, obtain a single-cell suspension either by mechanically disrupting the aggregates or treating with accutase (5 min at 37°C). After accutase treatment, wash the cells with PBS and centrifuge at 350 × *g* for 5 min (see Note 3).

3.3. In Vitro Assessment of Differentiation Potential

To test the ability of the lung alveolar stem cells to differentiate, cells can be expanded on fibronectin- and laminin-coated plates in the presence of serum with and without growth factors (Fig. 2).

1. Coat tissue culture dishes (24- or 6-well plates) with 10 μg/mL fibronectin or laminin in PBS and incubate overnight at 37°C.
2. Aspirate solution and wash once with PBS. Air-dry plates in a laminar flow hood.
3. Collect and pellet aggregates by centrifugation at 350 × *g* for 5 min and resuspend in accutase. Incubate for 5 min at 37°C and then centrifuge at 350 × *g* for 5 min.
4. Wash the cells twice with PBS followed by centrifugation at 350 × *g* for 5 min.
5. Count the cells and seed into coated cell culture dishes.
6. Change the medium every 3 days.
7. Analyze differentiation by immunofluorescence or RNA techniques (see Note 4).

3.4. Kidney Capsule Engraftments

To test the ability of the lung alveolar stem cells to differentiate, together with their in vivo self-renewal, we perform kidney capsule engraftments (5). Cells should be disassociated with accutase to generate a single-cell suspension. Different dilutions from one to one million cells can be transplanted (Fig. 3).

1. Anesthetize the mice (6- to 8-week-old) with isofluorane.
2. After the anesthetic has taken effect, shave the right flank of the mouse.
3. Swab the skin of mouse, center-out, and then wipe off with an EtOH swab.
4. Make a small incision in the skin, exposing the peritoneum.
5. Make a small incision in the peritoneum exposing the kidney. Keeping the incision small will help in keeping the kidney raised and exposed.
6. Apply slight pressure to both sides of the incision, raising or "popping" the kidney out.
7. Keep the kidney moist by applying normal saline with a cotton-tipped swab.
8. Using an extended glass Pasteur pipette, make a small scratch on the right flank of the kidney, creating a nick in the kidney capsule, not too deep or too large.
9. Inject the cells slowly, with a very thin capillary glass pipette in a total volume of 20 μl, into the space between the capsule and parenchyma. Some of the media will leak from the hole.

10. Once the grafting procedure has been completed, the kidney is gently eased back into the body cavity; the edges of the body wall and skin are aligned and closed with the aid of suturing. Gather the edges of the skin and clamp with surgical autoclips leaving no gaps in the wound.
11. Monitor the mouse and site of injection weekly for 4–8 weeks.

3.5. Immunofluorescence Microscopy of Lung Alveolar Stem Cells

Lung alveolar stem cells can be stained as aggregates as well as single cells (Fig. 3a).

1. Pellet the aggregates and fix in 4% paraformaldehyde for 10–20 min. The time depends on the aggregate size, the bigger the longer fixation time. Single-cell suspensions should be prepared using accutase. Wash once with PBS and fix with 4% paraformaldehyde for 5–10 min.
2. Wash the aggregates three times with PBS. Single cells can be centrifuged at 350 × *g* for 5 min, whereas it is necessary to wait for the aggregates to sink to the bottom of the tube prior to careful removal of the supernatant. This prevents disruption of the aggregates. Cytospin single cells on to slides at 600 rpm, for 3 min, and draw a hydrophobic circle around the area containing the cells.
3. Resuspend aggregates with 0.1% Triton X-100 in PBS and incubate for 5 min at room temperature. Cover the cells on slides with same buffer for 5 min at room temperature. This step is only required if permeabilization of the cells is required.
4. Remove the supernatant from the aggregates and slides. Wash three times with PBS and block nonspecific binding sites by incubating the aggregates/cells in PBS containing 4% serum from the same species as the secondary antibody for 1 h.
5. Dilute the primary antibody from stock solution in PBS with 1% serum. Apply the primary antibody solution to aggregates as well as cells on the slides in a volume sufficient to cover them.
6. Incubate overnight at 4 °C. Place the slides in a humid chamber.
7. Wash three times with PBS at room temperature (5 min per wash).
8. Gently add the secondary antibody solution (antibody in PBS). Incubate for 1 h at room temperature.
9. Wash three times with PBS.
10. Incubate aggregates with DAPI for 5 min at room temperature or just mount single-cell slides with Vectastain (which contains DAPI) under a coverslip. Place aggregates onto special concave glass slide and mount them with Vectastain which do not contain DAPI or antifade.

4. Notes

1. All animal maintenance and experiments must be performed in accordance with local animal care guidelines and any applicable laws. CD-1® nude mice should be maintained under special pathogen-free conditions.
2. Cell isolation procedures should be performed in a class I sterile hood. All solutions, reagents, media, and equipment used to process and culture lung stem/progenitor cells must be sterile and proper aseptic technique applied.
3. Do not split the cells when they are at low confluency.
4. Periodically analyze an aliquot of the isolated cells by flow cytometry to confirm them as SP-C+/CC10+. The final population in culture should always be SP-C+/CC-10+/Sca1+/CD45−/PECAM−/CD34−. The correct setup and operation of flow cytometers are complex and beyond the scope of this chapter. For general background and theory of flow cytometry, a comprehensive reference/source should be consulted.

References

1. Kim CF, Jackson EL, Woolfenden AE, Lawrence S, Babar I, Vogel S, Crowley D, Bronson RT, Jacks T (2005) Identification of bronchioalveolar stem cells in normal lung and lung cancer. Cell 121:823–835
2. Ventura JJ, Tenbaum S, Perdiguero E, Huth M, Guerra C, Barbacid M, Pasparakis M, Nebreda AR (2007) p38alpha MAP kinase is essential in lung stem and progenitor cell proliferation and differentiation. Nat Genet 39:750–758
3. Teisanu RM, Lagasse E, Whitesides JF, Stripp BR (2009) Prospective isolation of bronchiolar stem cells based upon immunophenotypic and autofluorescence characteristics. Stem Cells 27:612–622
4. Adams GB, Scadden DT (2006) The hematopoietic stem cell in its place. Nat Immunol 7:333–337
5. Eirew P, Stingl J, Raouf A, Turashvili G, Aparicio S, Emerman JT, Eaves CJ (2008) A method for quantifying normal human mammary epithelial stem cells with in vivo regenerative ability. Nat Med 14:1384–1389

Chapter 4

Three-Dimensional In Vitro Culture Techniques for Mesenchymal Stem Cells

Fatima A. Saleh, Jessica E. Frith, Jennifer A. Lee, and Paul G. Genever

Abstract

In recent years there has been a growing interest in culturing adherent cells using three-dimensional (3D) techniques, rather than more conventional 2D culture methods. This interest emerges from the realization that growing cells on plastic surfaces cannot truly re-create 3D in vivo conditions and therefore might be limiting the cells' potential. In addition, adult stem cells exist in specialized microenvironments, or niches, where the spatial organization of different niche elements (such as different cell types, extracellular matrix) contributes significantly to stem cell maintenance, which cannot be represented using 2D in vitro models. We have generated a range of different 3D approaches for the analysis of mesenchymal stem cells (MSCs) using both mono- and co-culture environments.

Key words: MSC Spheroids, Co-culture, Cell tracking, Spinner flask, Rotating wall vessel, hTERT-MSCs

1. Introduction

In vivo, mesenchymal stem cells (MSCs) are sensitive to many aspects of their environment, which together provide the cues that determine cell viability, proliferation, and lineage specification. This includes both soluble signals from growth factors and cytokines (1, 2), signaling between cells, which is dependent upon cell–cell contacts (3), and physical cues such as those provided via extracellular matrix (ECM) molecule composition, presentation, and elasticity (4–8). All these factors form the complex 3D microenvironment which also incorporates appropriate biological signals and external forces such as fluid flow and shear stress that can regulate the MSC activity (9, 10).

Kimberly A. Mace and Kristin M. Braun (eds.), *Progenitor Cells: Methods and Protocols*, Methods in Molecular Biology, vol. 916, DOI 10.1007/978-1-61779-980-8_4, © Springer Science+Business Media, LLC 2012

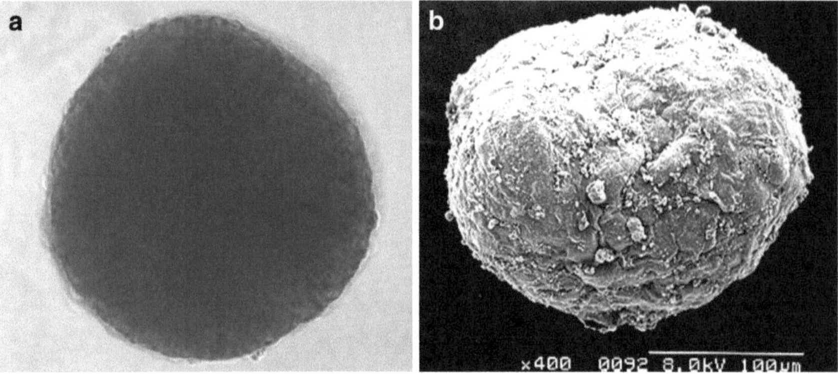

Fig. 1. Light (**a**) and scanning electron (**b**) micrographs of 3D MSCs grown under static culture conditions.

An increasing body of data showing the importance of these factors has led to an understanding that traditional monolayer cultures may not provide the optimal conditions for MSC culture (11, 12).

To mimic this in vivo microenvironment and for better understanding of MSC biology, we have developed static and dynamic 3D spheroid culture systems for the growth and analysis of MSCs (Fig. 1). In line with results using similar techniques for other cell types (13, 14), MSCs cultured as spheroids show many differences to those cultured on tissue culture plastic. These include changes to expression of cell surface antigens, altered gene expression profiles, and enhanced ability to differentiate along the osteogenic and adipogenic lineages (15, 16). Consequently, the 3D MSC system can generate simplified "microtissues" which can be used to monitor differentiation mechanisms in a controlled manner. We have also generated a range of immortalized MSC lines overexpressing human telomerase reverse transcriptase (hTERT) to streamline many of these analyses.

For better reconstruction of the MSC niche, we have co-cultured MSCs with other cell types, such as endothelial cells, to study the effects of heterotypic cell–cell interactions and paracrine signaling on MSC behavior in a putative perivascular niche. These studies aim to identify cues that favor the maintenance of native MSCs and promote the emergence of a more potent phenotype (17).

Here we describe the techniques used to generate 3D MSCs using static and dynamic (co-) culture conditions and methods modified to enable fluorescent cell tracking, viability assays, and analyses of 3D MSCs by flow cytometry as well as the generation of immortalized hTERT–MSCs.

2. Materials

2.1. Cell Growth and Spheroid Formation

1. Basal medium—Dulbecco's Modified Eagle's Medium (DMEM, low glucose) supplemented with 100 units/ml penicillin, 100 μg/ml streptomycin, and 15% batch-tested fetal bovine serum (FBS) (Invitrogen).
2. Endothelial cell growth medium (ECGM) with Supplement Mix (PromoCell).
3. 1× sterile phosphate-buffered saline (PBS, Invitrogen).
4. Solution of trypsin (0.25%) and ethylenediamine tetraacetic acid (EDTA) (1mM) from Invitrogen.
5. Non-adherent 6-well plates. These can be prepared by coating 6-well tissue culture plates, under sterile conditions, with 2 ml of a 1% solution of agarose in PBS. Once set these can be stored at 4°C for up to one week. Alternatively, low-binding tissue culture plates (Corning) may be used.
6. Methyl cellulose (Sigma m-0512, 4,000 cP).
7. Microplate Sterile 96-U Well Polystyrene (Fisher) for static spheroid formation.

2.2. Dynamic 3D Culture

Spinner flasks (Techne) and magnetic stirring platform (Estem) OR rotary cell culture system with slow turning lateral vessel (STLV) from Synthecon.

1. Spinner flasks are thoroughly washed, rinsed, and dried. The surface is then rendered non-adherent using Sigmacote (Sigma). To do this, add 1 ml Sigmacote to each flask and run it around the flask for 5 min. Take care to ensure that both the bottom of the flask and the sides, up to the level at which the media will reach, have been thoroughly coated. Remove excess Sigmacote and leave the flask to air-dry. Once dry the spinner flask should be autoclaved before use.
2. The Synthecon rotating wall vessel (RWV) should be cleaned and dried (NB: do NOT use detergents to clean the vessel as this may damage interior membranes and seals). It can then be loosely assembled and autoclaved in an autoclave bag. Once autoclaved this should only be opened under sterile conditions.
3. The platforms and vessels required for dynamic 3D cultures require a lot of space. Therefore it is advisable to have a separate incubator for these cultures. In addition space will be needed directly next to the incubator for the control boxes.

2.3. Static 3D Culture

For static spheroid generation we use methyl cellulose which is a thickening agent in a non-adherent U-shaped bottom 96-well plate. The preparation of methyl cellulose stock solution is critical;

if the concentration is too low or the solution contains any debris, cells will stick to the walls of the well and several small spheroids will be formed instead of a single spheroid. To prepare methyl cellulose solution:

1. 6 g of methyl cellulose (Sigma M-0512, 4,000 cP) are added to a 250 ml flask containing a magnetic stirrer and autoclaved.
2. 125 ml of pre-warmed (60°C) medium is added and the solution stirred for 20 min. At this stage, some undissolved methylcellulose will remain.
3. Another 125 ml of DMEM (at room temperature) is added to make 250 ml final volume and the entire solution mixed for 2–3 h at 4°C. This should generate a fairly clear viscous solution.
4. The solution is centrifuged at 3,600 × *g* for 2 h at room temperature.
5. The clear final stock solution is aliquoted and stored at 4°C.
6. 5 ml of the stock solution is added to 35 ml of co-culture media (DMEM/15% FBS: ECGM/Supplement Mix/10% FBS) to form 40 ml of co-culture spheroid media (see Note 1).

2.4. Cell Tracking

1. CellTracker™ Green CMFDA (5-chloromethylfluorescein diacetate) (Molecular Probes, Invitrogen, USA).
2. CellTracker™ Red CMTPX (Molecular Probes, Invitrogen, USA).
3. Dimethyl sulfoxide (DMSO) is used to dissolve the lyophilized dyes to make a 10 mM stock solution; e.g., 7.3 μl of DMSO is added to each vial of Red dye. Both aliquots and lyophilized dye are stored in −20°C. Dilute the stock solution to a final working concentration of 0.5 μM in serum-free medium and warm to 37°C just before use.

2.5. Live/Dead Staining in Co-culture Spheroids

1. LIVE/DEAD® Viability/Cytotoxicity Kit: calcein AM and ethidium homodimer-1 (Invitrogen).
2. 4% Paraformaldehyde: dissolve 4 g paraformaldehyde in 100 ml 1× PBS by heating moderately using a stirring hot-plate in a fume hood and then cool to room temperature for use. Adjust the pH to 8.

2.6. Freezing and Cryosectioning of Spheroids for Immunostaining

1. Liquid N_2.
2. Tissue-Tek® O.C.T.™ Compound.
3. Cryovials.
4. Ice-cold acetone.
5. Supersoft slides (Merck).
6. Vectashield (Vector Laboratories).

7. 4′, 6-diamidino-2-phenylindole (DAPI) (5 mg/ml): use 1 in 5,000 dilution.
8. Antibody dilution buffer: 1% bovine serum albumin (BSA) in PBS.

2.7. Harvesting of Spheroids for Flow Cytometry

1. Washing buffer: 0.2% BSA in PBS, 5 mM ethylenediaminetetraacetic acid (EDTA).
2. For flow cytometry, spheroids are dissociated into single suspension using Liberase TL (Roche).
   (a) To reconstitute the lyophilized enzyme add 2 ml of injection-quality sterile water into 5 mg lyophilized enzyme to give 13 Wünsch units/ml target collagenase activity and 2.5 mg/ml target collagenase content. Combined collagenase activity of the collagenase I and II isoforms is measured by the method of Wünsch.
   (b) Place vial on ice to rehydrate the lyophilized enzyme.
   (c) Agitate gently the vial at 2–8°C until enzyme is completely dissolved and then aliquot the stock solution and store at −20°C.
   (d) To prepare the working solution, add 100 μl of stock solution to 1 ml PBS or serum-free medium to give 1.2 Wünsch units/ml collagenase activity.

2.8. Production of hTERT-MSCs

1. pCI-neo-hEST2 (Addgene, courtesy of Robert Weinberg (1)).
2. EcoRI, SalI, XhoI, AatII, and ApaI restriction enzymes (Promega).
3. QAIquick PCR purification kit (Qiagen).
4. pENTR 1A (Invitrogen).
5. Agarose (Invitrogen).
6. QIAquick Gel Extraction Kit (Qiagen).
7. T4 DNA ligase (New England Biolabs).
8. Hi-coli-5A chemically competent cells (Advantagen).
9. LB medium (1% tryptone, 0.5% yeast extract [both Oxoid], 1% NaCl [Fisher Scientific], pH 7.0) and LB agar plates (LB medium, 15 g/l agar) supplemented with either 100 μg/ml ampicillin, 50 μg/ml kanamycin, or 30 μg/ml chloramphenicol.
10. QIAprep Spin Miniprep Kit (Qiagen).
11. ViraPower Lentiviral Gateway Expression Kit (Invitrogen) containing pLenti6/V5-DEST, LR Clonase II, Proteinase K, One Shot Stbl3 chemically competent cells, 293FT cell line, Lipofectamine 2000, Blasticidin.
12. TE buffer, pH 8 (10 mM Tris–HCl, 1 mM EDTA).

13. All tissue culture flasks and plasticware are from Corning.
14. 293FT culture medium (DMEM supplemented with 100 U/ml penicillin, 100 μg/ml streptomycin, 10% fetal bovine serum, 0.1 mM MEM Nonessential Amino Acids (NEAA), 6 mM L-glutamine, 500 μg/ml Geneticin).
15. Lenti-X qRT-PCR Titration Kit (Clontech).
16. Polybrene (Sigma-Aldrich).
17. Hyclone serum (Thermos Scientific).
18. Cloning cylinders, polystyrene, I.D. × H 4.7 mm × 8 mm (Sigma-Aldrich).
19. Silicone grease.

3. Methods

3.1. Spheroid Initiation

MSCs are isolated from human femoral heads following routine hip replacement operation with informed consent (1). To obtain cells for 3D cultures, MSCs are expanded in T175 flasks using standard techniques, passaging upon 75–80% confluence. When the required numbers have been obtained, 3D cultures can be set up.

To initiate spheroid formation, dissociate MSCs from the flask using trypsin/EDTA and collect in basal MSC medium.

1. Count the cells, then centrifuge at $450 \times g$ for 5 min to pellet, and add media to adjust the cell density to 5×10^5 cells/ml.
2. Seed the MSCs into preprepared agarose-coated or low-binding 6-well plates, using 4 ml per well.
3. Transfer the plates to an incubator and culture the MSCs in these high-density, non-adherent conditions for 6 h. During this time the cells will converge into a single large aggregate.
4. Break this single aggregate into smaller clusters by pipetting gently through a 5 ml pipette five times before transferring into dynamic conditions.

3.2. Dynamic Conditions

3.2.1. Spinner Flask Culture

MSC spheroids are cultured at a final density of 2×10^4 cells/ml in volumes ranging between 30 and 150 ml dependent on the number of cells used. Before seeding, the volume of media required for the number of MSCs used should be calculated, e.g., if 1×10^6 MSCs are trypsinized and aggregated, the volume of media used should be 50 ml (see Note 2).

1. Add 75% of the total media volume to the spinner flask. Use the remaining 25% of the media to collect the MSC aggregates from the 6-well plate and transfer to the spinner flask. This should be done through one of the sidearms taking care not to

run the media across the sidearm or to touch the arm with the pipette as this may cause problems with contamination.

2. Once the cells and media are added, the caps on the sidearms should be adjusted so that they are loosely done up. This allows gas exchange throughout the culture period.
3. Place the spinner flasks into the incubator on the magnetic platform and adjust the stirring speed to 30 rpm (see Note 3).
4. For continued culture of MSC spheroids, the medium can be changed by removing all the culture medium, centrifuging the spheroids for 5 min at $450 \times g$, and resuspending them in the original volume fresh culture medium.

3.2.2. Rotating Wall Vessel Culture

1. Human MSC spheroids are cultured at a final density of 2×10^4 cells/ml using the rotary cell culture system with a 110 ml slow turning lateral vessel (STLV) (see Note 4).
2. The preassembled and autoclaved vessel (see Note 5) should be loaded keeping air bubble formation to a minimum. To do this, add culture medium without serum through the syringe port followed by the suspension of MSC aggregates and finally the serum.
3. The vessel should then be topped up with culture medium (containing serum) to ensure that all air bubbles are removed before attachment to the rotator base (see Note 6).
4. The filled vessel can then be attached to the rotator base and placed into an incubator. The stirring speed should be adjusted to 15 rpm (see Note 7). As with the spinner flask cultures, MSC spheroids can be maintained under these conditions with media changes performed by carefully opening the vessel under sterile conditions, removing the media, centrifuging the spheroids for 5 min at $450 \times g$, and resuspending them in the original volume fresh culture medium.

3.3. Co-culture Spheroid Formation

To generate ten spheroids containing 15,000 Human Umbilical Vein Endothelial Cells (HUVECs) (PromoCell) or adult Human Dermal Fibroblasts (HDFa) (Cascade Biologics) and 15,000 MSCs per spheroid:

1. Trypsinise cells using trypsin/EDTA for 5 min.
2. Resuspend the different cell types in specific media and count to have 1.5×10^5 cells/ml of each cell type.
3. Add MSCs to HUVECs and into another universal mix MSCs and HDFa to have cells 1:1 ratio.
4. Centrifuge both universals at $450 \times g$ for 5 min to pellet.
5. Aspirate the media and add the co-culture spheroid media containing 20% methyl cellulose.

6. Distribute the co-culture into the non-adherent U-shaped bottom 96-well plate adding 100 μl per well.
7. Under these conditions, spheroids are generated overnight. Spheroid co-culture media is then exchanged by fresh media twice a week.

3.4. Cell Tracking of Co-culture Spheroids

For tracking cells in co-culture spheroids, HUVECs or HDFa are fluorescently labeled with cell tracker green, whereas MSCs are labeled with cell tracker red prior to co-culturing into MSC/HUVEC spheroids and MSC/HDFa spheroids to serve as controls.

1. HUVEC, HDFa, and MSC cells are trypsinized and counted so that the ratio of HUVECs and HDFa cells to MSC number is 1:1 as described earlier.
2. Centrifuge the cells to pellet them and aspirate the supernatant.
3. Resuspend the cells gently in prewarmed working solution of the dye and incubate for 45 min at 37°C.
4. Centrifuge the cells and replace the probe solution with fresh, prewarmed medium and incubate the cultures for another 30 min at 37°C.
5. Centrifuge and wash the cells with 1× PBS and replace with normal culture medium.
6. Green-labeled HUVECs or HDFa are mixed with red-labeled MSCs in suspension and proceed as in Subheading 3.3 from step 4 (see Note 8).
7. At different time points, co-culture spheroids are then imaged by confocal microscopy to track the movement of labeled cells within a mixed spheroid.

3.5. Live/Dead Cell Viability Assay

1. Aspirate medium from spheroids in 96-well plate and wash with PBS.
2. Add 2 μl calcein AM and 4 μl ethidium homodimer-1 (EthD-1) to 1 ml PBS and vortex thoroughly (gives an 8 μM solution of each label). Calcein AM is converted by live cells to green fluorescent calcein (excitation/emission ~495 nm/~515 nm); while EthD-1 enters cells with damaged membranes and increases fluorescence upon binding to nucleic acids producing bright red fluorescence in dead cells (excitation/emission ~495 nm/~635 nm).
3. Add sufficient live/dead reagent (≥100 ml) to cover the spheroid and incubate in darkness at room temperature for 40 min.
4. Remove excess dye and wash spheroids with PBS prior to fixing with 4% paraformaldehyde for 10 min.
5. Remove the fixative, wash in PBS, and resuspend in a solution of DAPI (1:5,000) and then image using a confocal microscope with an argon laser for green fluorescence and 543 laser for red florescence.

3.6. Freezing Spheroids and Cryosectioning for Immunostaining

1. Aspirate medium from spheroids and wash with PBS.
2. Using pipette tips with the ends cut off; carefully transfer spheroids into 500 μl eppendorf caps (see Note 9).
3. Aspirate off all PBS and add a drop of OCT avoiding air bubbles.
4. Pick up with forceps and snap freeze with liquid nitrogen until OCT is white which takes about 1 min.
5. After freezing, the samples are removed from the liquid nitrogen for immediate use or stored within 1.5-mm cryovials at −80°C until use (see Note 10).
6. For cryosectioning, cool the cryostat to approximately −20°C.
7. Add a small amount of OCT onto a cryostat chuck (at room temperature). Quickly remove the frozen spheroid sample from the lid, place on top of chuck, and overlay with OCT.
8. This block is then put into the cryostat where it must be left to equilibrate before cutting.
9. Cryosections are cut from frozen spheroids at 10 μm using a Microm HM560 cryostat and the sections are collected onto Superfrost slides. Slides can be stored at −20°C until required for immunostaining.
10. For immunostaining, warm slides at room temperature and wash with PBS.
11. Fix the sections for 10 min in 100% ice-cold acetone.
12. After rinsing with PBS, nonspecific binding is blocked by incubation for 30 min at RT with 10% animal serum (of the same origin as the secondary antibody).
13. Incubate the samples with the correct concentration of primary antibody for 1 h at RT or overnight at 4°C (see Note 11).
14. Excess antibody is removed and the slides are washed with 1× PBS (3×5 min).
15. Incubate with the appropriate secondary antibody for 45 min in the dark at RT and then wash with 1× PBS (3×5 min).
16. DAPI is added for 10 min at RT to stain the DNA and identify the nuclei.
17. The samples are then ready to be mounted. Add a drop of mounting medium (Vectashield) onto the microscope slide and then add a coverslip. Nail varnish is used to seal the sample which can be viewed immediately under a fluorescent microscope or can be stored at 4°C in dark for up to a month.

3.7. Flow Cytometry for Spheroids

1. Collect spheroids (at least ten spheroids) into a universal (Sterilin), remove all media, wash twice with PBS.
2. Centrifuge at 450×*g* for 5 min and aspirate supernatant.

3. To obtain a single cell suspension, spheroids are incubated with 400 μl of Liberase TL working solution for 10–30 min (depending on the size of the spheroid) on an orbital shaker at 37°C with repeated pipetting every 5 min to ensure that a single cell suspension resulted.
4. When no cell aggregates are visible, spheroid-derived cells are collected by centrifugation at 450 × *g* for 5 min.
5. Resuspend cells in washing buffer to give 1×10^6 cells in 100 μl per sample.
6. Incubate samples with the primary antibody or IgG control at 4°C for 30 min followed by washing with washing buffer and centrifugation at 450 × *g* for 5 min.
7. Add the secondary antibody at the correct concentration and incubate for 4°C for 30 min.
8. Wash samples with the washing buffer and centrifuge. Resuspend in 400 μl buffer prior to being analyzed on a Dako Cytomation CyAN Flow Cytometer (see Note 12).

3.8. Production of hTERT Lentiviral Vector

1. This protocol uses the ViraPower Lentiviral Gateway Expression Kit with pLenti6/V5-DEST from Invitrogen.
2. The pCI-neo-hEST2 stab culture is propagated by streaking on LB/ampicillin plates and incubating overnight at 37°C. Four single cell colonies are inoculated in 5 ml LB/ampicillin liquid cultures and incubated overnight at 37°C. The plasmid is then isolated using a QIAprep Spin Miniprep Kit according to the manufacturer's instructions.
3. The hTERT DNA sequence is isolated from pCI-neo-hEST2 by performing a sequential restriction digest with EcoRI and SalI. 1.5 μg pCI-neo-hEST2 is digested with 5 units EcoRI, in the presence of Buffer H, in a 20 μl reaction volume at 37°C for 1 h. The linearized pCI-neo-hEST2 is purified using a QAIquick PCR Purification Kit according to the manufacturer's instructions and eluted into 20 μl dH_2O. The 20 μl linearized pCI-neo-hEST2 is then digested with 5 units SalI, in the presence of Buffer D, at 37°C for 1 h.
4. 1.5 μg pENTR 1A is digested with 5 units EcoRI and 5 units XhoI, in the presence of Buffer H, in a 20 μl reaction volume at 37°C for 1 h.
5. The digested pCI-neo-hEST2 and pENTR 1A plasmids are run down a 1% agarose gel. The 3,450 bp hTERT gene and the 2,279 bp pENTR 1A plasmid backbone are excised from the gel using a scalpel blade, purified using a QIAquick Gel Extraction Kit according to the manufacturer's instructions, and eluted into 20 μl dH_2O. 2 μl of each DNA fragment are then run down a 1% Agarose gel to determine the plasmid concentrations.

6. 450 ng hTERT sequence DNA is ligated into 100 ng pENTR 1A backbone (in a 1 M vector:3 M insert ratio) with 400 units T4 DNA ligase, in 1× T4 DNA Ligase Reaction Buffer, in a 10 μl reaction volume at 16°C overnight (a negative control ligation without the insert is also performed).
7. The ligated plasmid is transformed into Hi-coli-5A chemically competent cells according to the manufacturer's instructions, then spread on LB/kanamycin plates, and incubated overnight at 37°C. Five colonies are isolated and propagated in 5 ml LB/kanamyin liquid cultures overnight at 37°C. The plasmids are then isolated using a QIAprep Spin Miniprep Kit according to the manufacturer's instructions.
8. To check the integrity of pENTR 1A-hTERT a restriction digest is carried out with 500 ng plasmid DNA, 5 units AatII, in the presence of Buffer J, in a 20 μl reaction volume at 37°C for 1 h. The reaction is then run down a 1% Agarose gel. The correct product sizes are 568 bp, 2,283 bp, and 2,827 bp.
9. An LR recombination reaction is performed with 150 ng pENTR 1A-hTERT and 150 ng pLent6/V5-DEST using 2 μl LR Clonase II in a reaction volume of 8 μl TE buffer and incubated at 25°C for 1 h. 2 μg Proteinase K is then added to the reaction and incubated at 37°C for 10 min.
10. 3 μl of the ligation reaction is transformed into One Shot Stbl3 chemically competent cells following the manufacturer's instructions, then spread on LB/ampicilin plates, and incubated overnight at 37°C. To check the integrity of pLenti6/V5-DEST-hTERT the bacterial stock is also streaked on LB/chloramphenicol plates and incubated overnight at 37°C. True expression clones are chloramphenicol sensitive.
11. Single cell colonies of pLenti6/V5-DEST-hTERT from the LB/ampicillin plates are propagated in 5 ml liquid cultures of LB/ampicillin medium overnight at 37°C. Plasmid DNA is isolated using a QIAprep Spin Miniprep Kit according to the manufacturer's instructions. Plasmid integrity is also confirmed by restriction digest of 500 ng plasmid DNA, 5 units ApaI, in the presence of Buffer A, in a 20 μl reaction volume at 37°C for 1 h. The reaction is then run down a 1% agarose gel. The correct product sizes are 1,246 bp, 3,028 bp, and 6,284 bp.

3.9. Production and Titering of Lentiviral Stocks

1. pLenti6/V5-DEST-hTERT lentiviral stocks are produced using 293FT cells according to the manufacturer's instructions. Lentiviral stocks of a pLenti6/V5-DEST-EGFP positive control are also produced. The day after transfection with Lipofectamine 2000 the medium is removed and replaced with 5 ml fresh 293FT culture medium to concentrate the viral particles. The medium is applied to the cells really carefully to prevent the cells detaching from the plate.

Table 1
Transduction efficiency of pLenti6/V5-DEST-EGFP at various MOIs

MOI	Transduced cells (%)
1	0.00
10	0.03
1×10^2	0.21
1×10^3	2.17
1×10^4	20.47
1×10^5	69.20
3.36×10^5 (neat viral stock)	92.04

2. 48 h post-transfection lentivirus-containing medium is harvested and centrifuged at 1900×*g* at 4°C for 15 min to pellet cell debris. The lentiviral stocks are then aliquoted and stored at −80°C.
3. We found that the antibiotic selection protocol in the manufacturer's instructions does not give satisfactory results. Therefore, lentiviral stocks are titered using the Lenti-X qRT-PCR Titration Kit according to the manufacturer's instructions. Using this kit we get titers in the range 1×10^3–2×10^4 viral copies/ml.
4. To determine the appropriate multiplicity of infection (MOI) MSCs are plated in 6-well plates at a density of 5×10^3 cells/cm^2 in MSC medium. The next day the pLenti6/V5-DEST-EGFP viral stock is added to the MSCs at MOIs of 1, 10, 1×10^2, 1×10^3, 1×10^4, 1×10^5, and 3.36×10^5 (neat viral stock) in a total volume of 1 ml MSC medium. Polybrene is then added to the medium to give a final concentration of 6 μg/ml. The next day the lentivirus-containing medium is discarded and 2 ml/well fresh medium is added to the MSCs.
5. 72 h after lentiviral transduction the MSCs are trypsinized and the percentage of EFGP fluorescent cells is analyzed by flow cytometry to determine the transduction efficiency (results in Table 1).

3.10. Production of hTERT-MSCs

1. An MOI of 1×10^4, which gives a 20% transduction efficiency, is used to transduce MSCs with pLenti6/V5-DEST-hTERT as this is the MOI which is most likely to produce a single genomic integration (19).
2. MSCs are plated out in a 6-well plate at a density of 5×10^3 cells/cm^2 in MSC medium. The next day the MSCs are

transduced with the pLenti6/V5-DEST-hTERT viral stock (or a mock control) at an MOI of 1×10^4 in a volume of 1 ml MSC medium. Polybrene is added to the medium to give a final concentration of 6 μg/ml. The next day the lentivirus-containing medium is discarded and replaced with 2 ml/well fresh MSC medium.

3. The next day the MSCs are passaged into T25 tissue culture flasks and Blasticidin is added to the medium at a final concentration of 2 μg/ml to select for transduced cells. The medium is replaced every 3/4 days with fresh Blasticidin-containing MSC medium for 12 days or until no living cells remain in the mock transduced flask.
4. To select single cell lines the transduced MSCs are trypsinized and plated out at 10 cells/cm^2 in 10 cm plates in triplicate. Use MSC medium containing 20% Hyclone serum for plating out the cells; then replace the medium with fresh MSC medium containing 15% Hyclone serum every 3/4 days for 14 days or until nice sized single cell colonies are visible.
5. Single cell colonies are isolated using cloning cylinders, silicone grease, and trypsin-EDTA and then transferred to wells of 24-well plates. When the cells reach 70% confluence passage to 6-well plates, then 25 cm^2 flasks, and finally 75 cm^2 flasks.
6. Count cells at each passage to determine population doublings and therefore hTERT activity.
7. Southern blotting can also be used to confirm a single genomic integration of the hTERT gene, comparative genomic hybridization can be performed to confirm the absence of chromosomal abnormalities, and telomerase activity can be analyzed using a telomerase activity kit such as the TRAPEZE Telomerase Detection Kit (Millipore).
8. hTERT-MSCs can then be used in 3D culture models to determine the effects of gene knockdown, using shRNAs such as the BLOCK-iT Lentiviral RNAi Expression System (Invitrogen), and gene knockout (20).

4. Notes

1. Due to the high viscosity of methylcellulose stock solution, the use of a pipette is avoided. It is preferred to pour the solution into a 50 ml falcon tube without touching its walls and then add the media to make the working solution.
2. In addition to the use of basal MSC media, osteogenic or adipogenic medium may also be used within the dynamic 3D cultures to promote differentiation. These should be added to the culture vessel at the point of spheroid formation.

3. 30 rpm was determined to be the optimal stirring speed. Increasing stirring speeds decrease the size of the resulting spheroids but can also affect cell viability (15).
4. Unlike spinner flask cultures, the volume used for RWV cultures is fixed because the vessel cannot contain air bubbles and must be filled completely. For a typical 110 ml vessel, 2.2×10^6 cells are required although other chamber sizes can be obtained from Synthecon.
5. The culture vessel should be cleaned between each experiment. Note that it is not recommended to use detergents to clean the vessel and that this may result in damage to the inner membrane or rubber seals. The cleaned and dried vessel should be loosely assembled and placed into an autoclave bag with all ports open. This can then be autoclaved normally and should then only be removed from the autoclave bag under sterile conditions and at the time the next experiment is set up.
6. Removal of all the air bubbles is the hardest part of the RWV setup. We have found that the best method is to take the vessel once it is almost filled with media, cells, and serum and then to use a syringe to top it up with complete media. The vessel can be tipped from side to side to direct any air towards the exit port as more media is added through the syringe port.
7. To achieve low gravity conditions the rotation speed should be set so that the speed of sedimentation of the spheroids is balanced by the upwards rotation of the vessel (so that the spheroids neither fall to the bottom or are moved around the vessel). This will need to be monitored regularly and changed as required. For smaller aggregates it may not be possible to achieve this. We found that the lowest rotation speed of 15 rpm, although not producing low gravity conditions, was optimum for the MSC spheroids.
8. HUVEC/MSC co-culture spheroids are more fragile than HDFa/MSC spheroids and smaller in size.
9. Spheroids generally show a stable cellular organization which is unlikely to be disrupted by pipetting. However, cutting the ends of the pipette tip before harvesting the spheroids will prevent their disruption especially at early time points.
10. Never allow the frozen spheroids samples to thaw and always carry them across in dry ice.
11. An isotype (IgG) control and a secondary antibody control should be included in each experiment.
12. Dead samples and doublets are eliminated from the population and samples are gated against the IgG control. 10,000 cells are counted for each sample. Data are analyzed using Cytomation Summit version 3.1 software.

Acknowledgments

This work was supported by BBSRC, Smith and Nephew (FAS, JEF), and the Dr. Hadwen Trust for Humane Research, the UK's leading medical charity funding exclusively nonanimal research techniques to replace animal experiments (JAL).

References

1. Etheridge SL, Spencer GJ, Heath DJ, Genever PG (2004) Expression profiling and functional analysis of wnt signaling mechanisms in mesenchymal stem cells. Stem Cells 22:849–860
2. Huang Z, Ren PG, Ma T, Smith RL, Goodman SB (2010) Modulating osteogenesis of mesenchymal stem cells by modifying growth factor availability. Cytokine 51:305–10
3. Pittenger MF, Mackay AM, Beck SC, Jaiswal RK, Douglas R, Mosca JD et al (1999) Multilineage potential of adult human mesenchymal stem cells. Science 284:143–147
4. Mauney JR, Kirker-Head C, Abrahamson L, Gronowicz G, Volloch V, Kaplan DL (2006) Matrix-mediated retention of in vitro osteogenic differentiation potential and in vivo bone-forming capacity by human adult bone marrow-derived mesenchymal stem cells during ex vivo expansion. J Biomed Mater Res A 79:464–475
5. Mauney JR, Volloch V, Kaplan DL (2005) Matrix-mediated retention of adipogenic differentiation potential by human adult bone marrow-derived mesenchymal stem cells during ex vivo expansion. Biomaterials 26:6167–6175
6. Kundu AK, Putnam AJ (2006) Vitronectin and collagen I differentially regulate osteogenesis in mesenchymal stem cells. Biochem Biophys Res Commun 347:347–357
7. Engler AJ, Sen S, Sweeney HL, Discher DE (2006) Matrix elasticity directs stem cell lineage specification. Cell 126:677–689
8. Rowlands AS, George PA, Cooper-White JJ (2008) Directing osteogenic and myogenic differentiation of MSCs: interplay of stiffness and adhesive ligand presentation. Am J Physiol Cell Physiol 295:C1037–1044
9. Meinel L, Karageorgiou V, Fajardo R, Snyder B, Shinde-Patil V, Zichner L et al (2004) Bone tissue engineering using human mesenchymal stem cells: effects of scaffold material and medium flow. Ann Biomed Eng 32:112–122
10. Datta N, Pham QP, Sharma U, Sikavitsas VI, Jansen JA, Mikos AG (2006) In vitro generated extracellular matrix and fluid shear stress synergistically enhance 3D osteoblastic differentiation. Proc Natl Acad Sci USA 103:2488–2493
11. Baxter MA, Wynn RF, Jowitt SN, Wraith JE, Fairbairn LJ, Bellantuono I (2004) Study of telomere length reveals rapid aging of human marrow stromal cells following in vitro expansion. Stem Cells 22:675–682
12. Reiser J, Zhang XY, Hemenway CS, Mondal D, Pradhan L, La Russa VF (2005) Potential of mesenchymal stem cells in gene therapy approaches for inherited and acquired diseases. Expert Opin Biol Ther 5:1571–1584
13. Campos LS (2004) Neurospheres: insights into neural stem cell biology. J Neurosci Res 78:761–769
14. Bates RC, Edwards NS, Yates JD (2000) Spheroids and cell survival. Crit Rev Oncol Hematol 36:61–74
15. Frith JE, Thomson B, Genever PG (2010) Dynamic three-dimensional culture methods enhance mesenchymal stem cell properties and increase therapeutic potential. Tissue Eng Part C Methods 16:735–749
16. Wang W, Itaka K, Ohba S, Nishiyama N, Chung UI, Yamasaki Y et al (2009) 3D spheroid culture system on micropatterned substrates for improved differentiation efficiency of multipotent mesenchymal stem cells. Biomaterials 30:2705–2715
17. Saleh FA, Whyte M, Ashton P, Genever PG (2011) Regulation of mesenchymal stem cell activity by endothelial cells. Stem Cells Dev 20:391–403
18. Meyerson M, Counter CM, Eaton EN, Ellisen LW, Steiner P, Caddle SD, Ziaugra L, Beijersbergen RL, Davidoff MJ, Liu Q, Bacchetti S, Haber DA, Weinberg RA (1997) hEST2, the putative human telomerase catalytic subunit gene, is up-regulated in tumor cells and during immortalization. Cell 90:785–95
19. Fehse B, Kustikova OS, Bubenheim M, Baum C (2004) Pois(s)on–it's a question of dose. Gene Ther 11:879–81
20. Liizumi S, Nomura Y, So S, Uegaki K, Aoki K, Shibahara K, Adachi N, Koyama H (2006) Simple one-week method to construct gene-targeting vectors: application to production of human knockout cell lines. Biotechniques 41:311–316

Chapter 5

Isolation of Adult Stem Cells and Their Differentiation to Schwann Cells

Cristina Mantovani, Giorgio Terenghi, and Susan G. Shawcross

Abstract

Peripheral nerve injuries are an economic burden for society in general and despite advanced microsurgical reconstruction of the damaged nerves the functional result is unsatisfactory with poor sensory recovery and reduced motor functions (Wiberg and Terenghi, Surg Technol Int 11:303–310, 2003). In the treatment of nerve injuries transplantation of a nerve graft is often necessary, especially in nerve gap injuries.

Schwann cells (SC) are the key facilitators of peripheral nerve regeneration and are responsible for the formation and maintenance of the myelin sheath around axons in peripheral nerve fibers. They are essential for nerve regeneration after nerve injuries as they produce extracellular matrix molecules, integrins, and trophic factors providing guidance and trophic support for regenerating axons (Wiberg and Terenghi, Surg Technol Int 11:303–310, 2003; Bunge, J Neurol 242:S19–21, 1994; Ide, Neurosci Res 25:101–121, 1996; Mahanthappa et al. J Neurosci 16:4673–4683, 1996). However, the use of ex vivo cultured SC within conduits is limited in its clinical application because of the concomitant donor site morbidity and the slow growth of these cells in vitro (Tohill et al. Tissue Eng 10:1359–1367, 2004).

Mesenchymal stem cells (MSC or bone marrow stromal cells) and adipose-derived stem cells (ASC) are easily accessible non-hematopoietic stem cells that have proven essential for research purposes due to their plasticity and ability to differentiate into several functional cell types. This alternative source of cells is relatively simple to isolate and expand in culture. We have demonstrated that MSC and ASC can trans-differentiate along a SC lineage with functional properties and growth factor synthesis activities similar to those of native SC and could provide nerve fiber support and guidance during nerve regeneration.

Key words: Schwann cells, Differentiated mesenchymal stem cells, Adipose-derived stem cells

1. Introduction

In the last few decades a large group of publications described the use and the application of stem cells in a diverse range of injuries and diseases. The aim of our research is to utilize mesenchymal stem cells for repair and regeneration of damaged peripheral nerves (1–5). Stem cells can be identified as either embryonic or adult

Kimberly A. Mace and Kristin M. Braun (eds.), *Progenitor Cells: Methods and Protocols*, Methods in Molecular Biology, vol. 916, DOI 10.1007/978-1-61779-980-8_5,

stem cells. In the hierarchy of embryonic stem cells (ES), cells isolated from the fertilized oocyte are defined as *totipotent stem cells*; ES taken from the blastocyst are called *pluripotent* as these cells appear to be forming the three germ layers during embryogenesis. Fully developed adult tissues and organs contain niches of multipotent stem cells; these cells have been isolated from a wide range of adult tissues such as brain, heart, lungs, kidney, and spleen. However, the most well-characterized source of adult stem cells is the bone marrow. The bone marrow contains a mixed population of cells, including *hematopoietic stem cells* (HSC) and a subset of non-hematopoietic stem cells commonly called *marrow stromal cells* or *mesenchymal stem cells* (MSC).

The characterization of stem cells can lead to confusion as there is no universally accepted definition of the term "stem cell" and no unified theory describing their origin, plasticity, and function in the adult organism (6). The currently accepted characteristics of a stem cell are that the cells must be (1) undifferentiated (that is, lacking a tissue-specific differentiation markers), (2) capable of proliferation, (3) self-renewable, (4) able to produce a large number of differentiated functional progeny, and (5) able to regenerate tissue following injury (7).

MSC originate from the mesoderm germ layer; they give rise to connective tissue, skeletal muscle cells, and cells of the vascular system. Nowadays, there are still many unanswered questions about the true identity of the MSC, including location, origins, and multipotential capacity. Although isolation of MSC from many different tissues, such as adipose tissue, liver, amniotic fluid, umbilical cord blood, and dental pulp, has been described, the bone marrow remains the principal source of MSC with most potential and clinical application. Based upon recent knowledge, it is estimated that MSC constitute between approximately 0.001 and 0.01% of the nucleated cells isolated from the bone marrow (8, 9).

The advantage of MSC as therapeutic tools is that they can be easily isolated from the bone marrow and expanded in vitro, used in allogeneic transplantation, and show paracrine-mediated effects and migratory behavior to site of injury. There is evidence that MSC are capable of neuronal antigen expression in vitro (10, 11) and in vivo (12, 13). They have been shown to differentiate into astrocytes following direct transplantation into the rodent brain (14). Recent studies described remyelination of spinal cord lesions and showed that local delivery of MSC at the site of spinal cord injury was associated with the formation of neurofilament bundles at the interface between scar tissue and graft (15, 16). It is not clear what mechanisms govern the in vivo differentiation and migration of MSC within zones of injury; however, it is likely that the local milieu of growth factors, cytokines, and local stem cells has some influence.

Adipose-derived stem cells (ASC) are isolated from the stromal vascular fraction (SVF) of homogenized adipose tissue. ASC can be easily isolated from liposuction waste and exhibit the potential for chondrogenic, osteogenic, adipogenic, and myogenic differentiation (17, 18) and some aspects of neurogenesis (19). Although ASC show some similarities to MSC, they have a number of distinct features in terms of cells surface markers, differentiation potential, and abundance in the body. Up to 300-fold more stem cells can be harvested from 100 g of adipose tissue compared to 100 ml of bone marrow aspirate (8, 20). ASC are generally defined as CD34-positive CD31-negative ($CD34^{+}$ $CD31^{-}$). Many factors can influence the cellular composition of ASC cultures, such as species of origin, donor age, tissue location, isolation procedures, culture conditions, and cell storage. Moreover, the choice of experimental methods and reagents may also affect the outcome of any given study concerning ASC expression profile, differentiation potential, and therapeutic capacity. ASC and MSC share more than 90% of phenotypic markers; however, differences in surface protein expression have been reported (17). Furthermore, ASC are easier to culture for long periods and showed faster growth rates than MSC (21). Finally, ASC differentiated into a SC-like phenotype were recently shown to improve axonal regeneration across gaps repaired with fibrin conduits seeded with these differentiated ASC cells (22). Taken together these observations suggest that ASC are ideal candidates for tissue engineering-based injury repair and for future clinical applications.

Standard techniques such as histological, immunohistological, biochemical, and mechanical assays have been used to characterize the differentiation of both animal and human MSC. Cell lineage analysis has shown that the differentiation of cells is stimulated by cell-specific transcription factors that act as gene expression switches (23).

In vitro osteogenic differentiation can be induced using ascorbic acid, β-glycerophosphate, and dexamethasone: the differentiation of the cells is observed as increased expression of alkaline phosphatase (AP) and calcium accumulation with time (24).

In vitro chondrogenic differentiation is stimulated by transforming growth factor β (TGF-β), which results in an induction of protein kinases and is observed as an increase in the proteoglycan extracellular matrix (25).

In vitro adipogenic differentiation is promoted by the addition to the MSC cell culture of dexamethasone, indomethacin, and isobutyl methyl xanthine, which inhibits the enzymatic conversion of cyclic AMP to 5′ AMP by phopshodiesterase. This results in the upregulation of protein kinase A, which in turn upregulates hormone-sensitive lipase; this lipase converts triacylglycerides to glycerol and free fatty acids and is observed as an accumulation of lipid-rich vesicles (8).

2. Materials

2.1. Mesenchymal Stem Cell Isolation and Expansion

1. α-MEM (M8042; Sigma-Aldrich UK).
2. L-glutamine 200 mM (M11-004; PAA, UK).
3. Fetal Bovine Serum (FBS) (10270-106; Invitrogen Life Technologies, UK: Origin: South American—EU Approved).
4. Penicillin–Streptomycin (P11-010; PAA).
5. Hank's balanced salt solution (HBSS) (H9394; Sigma-Aldrich).
6. Trypsin, 0.25% (1×) with EDTA•4Na, liquid (25200-056; Invitrogen Life Technologies).
7. *Chlorhexidine in Spirit*, 70% (D549, Williams Medical Supplies).
8. A class II microbiological safety cabinet for cell culture and a vertical laminar flow cabinet for tissue dissection: both should be equipped with UV light for decontamination purposes.
9. A static water bath with temperature control.
10. Centrifuge U-320/U-320R (swing out rotor 45°, 138 radius) BOECO-Germany.
11. Sterile 15 ml Falcon-type conical bottom polypropylene centrifuge tubes (for example, 62.554.002; Sarstedt, UK).
12. Incubator with temperature and gas composition controls.
13. Inverted microscope with phase-contrast ability (for example, the Olympus IX51).
14. Sterile 10 ml syringes with 21-gauge needles.
15. Pipettes—sterile Pasteur-type, sterile serological, and Gilson-type autopipettes.
16. 70 μm mesh filters (1520012; BD-Falcon).
17. 75 cm^2 vent-cap tissue culture flasks (Corning 430641 or similar).
18. Sterile surgical scissors, surgical forceps, scalpel, bone nibblers, and forceps (Fine Science Tools, Germany).
19. Dissecting stereomicroscope.

2.2. Adipose-Derived Stem Cells Isolation and Expansion

1. α-MEM with (M8042; Sigma-Aldrich).
2. L-Glutamine 200 mM (M11-004; PAA).
3. Fetal bovine serum (FBS) (10270-106; Invitrogen: Origin: South American—EU Approved).
4. Penicillin–Streptomycin (P11-010; PAA).
5. HBSS (H9394; Sigma-Aldrich).
6. Trypsin, 0.25% (1×) with EDTA•4Na, liquid (25200-056; Invitrogen).

7. Sterile serological pipettes (5, 10 and 25 ml).
8. Sterile 15 ml and 50 ml Falcon-type conical bottom polypropylene centrifuge tubes (for example, 62.554.002 (15 ml) and 62.547.004 (50 ml); Sarstedt).
9. Sterile razor blades.
10. Sterile plastic Petri dish, 75 mm diameter.
11. Type I Collagenase (LS004197; Worthington Biochemical Products, UK).
12. Some 0.22 μm low protein-binding sterilization/filter units (for example, SLGP033RS Millex GP 0.22 μm unit; Millipore, UK).
13. 75 cm^2 vent-cap tissue culture flasks (Corning 430641 or similar).
14. Water bath with shaker facility.

2.3. MSC and ASC Differentiation to SC-Like Cell

1. Sigmacote® (SL2-25 ml; Sigma-Aldrich). *Important:* see siliconizing method (see Subheading 3.1) before proceeding with preparation of growth factor solutions.
2. Liquid nitrogen for snap-freezing growth factor/reagent aliquots.
3. Stem cell growth medium (SCGM): 45 ml α-MEM, 5 ml FBS, and 0.5 ml penicillin-streptomycin solution (10,000 U/ml and 10,000 μg/ml, respectively) and 0.5 ml of L-glutamine (2 mM).
4. Harvest buffer preparation: 10 ml HBSS medium containing 100 μl of penicillin-streptomycin; store on ice during the harvesting process.
5. Collagenase solution preparation: this should be freshly prepared just before starting the isolation process. Dissolve 30 mg collagenase type I in 15 ml of HBSS and filter sterilize using a 0.22 μm Millex® filter unit.
6. β-mercaptoethanol (βME) (M3148; Sigma): the final concentration of βME is 1 mM in the cell growth media without growth factors. Add 3.905 μl in 50 ml of SCGM.
7. All-*trans*-retinoic acid (RA) (R2625-100MG; Sigma-Aldrich) preparation: To 50 mg All-*trans*-RA solid, as supplied by Sigma-Aldrich, add 1.43 ml dimethyl sulfoxide (DMSO) to obtain a stock solution of 35 mg/ml (w/v). Dilute 2 μl stock in 20 ml of medium to a final concentration of 3.5 μg/ml. Dilute this stock solution of RA into SCGM to obtain a 350 ng/ml solution.
8. Basic Fibroblast Growth Factor (bFGF) (RCYT-218B; Sera Laboratories International (SLI, UK) is supplied lyophilized at 50 μg per vial and the final concentration of bFGF in differentiation medium is 10 ng/ml. First prepare the 5 mM Tris

(pH 7.6) solution by dissolving 0.6057 g Tris base (mwt 121.14) in total a volume of 1,000 ml (adjusted to pH 7.6). Next, dissolve the supplied 50 μg bFGF in 0.5 ml of 5 mM Tris (pH 7.6) solution to make a stock solution of concentration 0.1 mg/ml. Add 5 μl of bFGF stock solution to 50 ml SCGM to obtain a concentration of 10 ng/ml. Aliquot into small volumes, snap freeze in liquid nitrogen, and store at –40°C.

9. Platelet-Derived Growth Factor (PDGF) (RCYT-568B; SLI) is supplied as a lyophilized powder of 10 μg per vial and the final concentration of PDGF in differentiation medium is 5 ng/ml. Prepare a stock of 10 mM acetic acid by adding 60.24 μl acetic acid (99.5%) to 100 ml distilled water. Dissolve the 10 μg PDGF powder, as supplied, in 100 μl 10 mM acetic acid, which will produce a final concentration of 0.1 mg/ml. Aliquot into small volumes, snap freeze in liquid nitrogen, and store at –80°C. Add 2.5 μl of the 0.1 mg/ml PDGF stock to 50 ml SCGM to make a stock concentration of 5 ng/ml.
10. Forskolin (catalogue number F3917, SIGMA-Aldrich) is supplied as a powder at 10 mg per vial. Dissolve 10 mg Forskolin, as supplied, in 2.436 ml in DMSO to produce a 10 mM stock solution, Add 70 μl of the 10 mM stock solution to 50 ml SCGM. Aliquot into small volumes, snap freeze, and store at –80°C.
11. Neuregulin NRG1-1 (377-HB; R & D Systems, Abingdon, UK) is used to differentiate stem cells to a Schwann cell phenotype. NRG1-1 is supplied as a lyophilized powder at 50 μg per vial. The NRG-1 should be reconstituted in sterile PBS containing at least 0.1% (v/v) human or bovine serum albumin and to a concentration of 100 μg/ml. Store the reagent at –20°C. Add 10 μl stock solution to 50 ml SCGM.
12. Stem cell differentiation medium (SCDM) is prepared as follows: 50 ml SCGM plus 70 μl forskolin solution, 5 μl bFGF, 2.5 μl PDGF-AA solution, 5 μl or 10 μl NRG1-1 solution (for bone marrow or adipose stem cells, respectively). Preparation of the growth factors and forskolin stocks as detailed above (see Note 6).

3. Methods

3.1. Siliconizing Plasticware

Treat plasticware (tips and centrifuge tubes) with a solution of silicone to reduce or eliminate cell and growth factor attachment: proteins adhere to plastics and in the case of growth factors this reduces the amount available to the cultured cells. From our experience, we recommend using Sigmacote®. Start with clean plasticware and apply the Sigmacote® at about 5 ml Sigmacote® per 50 ml volume of the item to be coated; thoroughly coat the entire inner surface of the tube(s) by swirling the Sigmacote® around within the tube(s). Drain excess Sigmacote® from the plasticware and allow it to air-dry for about 10 h: the excess solution can be retained and reused for future applications—store in a dark-colored, capped bottle or tube at 4°C. Thoroughly rinse the plasticware with sterile distilled water; then leave to air-dry in a class II tissue culture cabinet. The coated plasticware should be stored at 4°C and used within a couple of months of coating. Do not autoclave the plasticware after the coating with Sigmacote®.

3.2. Isolation and Culture of Mesenchymal Stem Cells from Rat Bone Marrow

1. Bone marrow is harvested from the long bones of one adult Sprague-Dawley (or other strain of laboratory *Rattus norvegicus*) rat. Following terminal anesthesia with CO_2, the hindquarters of the rat are shaved and washed with chlorhexidine solution. After incising the skin, the muscles are split along the lines of the femur and tibia and the bones dissected free of muscle. The bones are then dislocated at the hip and ankle joints and placed in chilled HBSS with 1% (v/v) penicillin–streptomycin. The bones are then transferred to the tissue culture laboratory for marrow harvesting.
2. Harvest the BM in a class II microbiological safety cabinet or laminar flow cabinet using aseptic technique. Cut the proximal and distal ends from the tibia and femur—just below the end of marrow cavity—using a pair of sterile, sharp bone nibblers. Insert a 21-gauge needle attached to a 10 ml syringe containing SCGM into the spongy bone. Flush out the marrow plug through the cut end of the bone with 5 ml of complete medium and collect it in a sterile 15 ml tube on ice. This procedure is repeated for each bone and the resulting cell suspension is twice triturated through a 21-gauge needle (see Note 1).
3. Filter the cell suspension through a 70 μm mesh filter to remove any bone fragments and other debris.
4. Centrifuge the cell suspension at 600 × *g* for 5 min, then gently aspirate the supernatant, and resuspend the pellet in 10 ml of fresh SCGM.
5. Seed the BM cells in a 75 cm^2 vent-cap tissue culture flask and incubate at 37°C in 5% CO_2 in a humidified chamber. The flasks must remain undisturbed for 24 h (see Note 2).
6. Next day: the cells present at this stage are a mixture of adherent marrow stem cells, non-adherent haematopoietic cells, dead cells, and debris. Remove the non-adherent, free-floating cells (hematopoietic cell lineage) by washing the culture twice with 5 ml HBSS. Add fresh SCGM. Repeat this step for the first 3 days of culture (see Note 3).

7. On the third day of culture the adherent cells appear isolated when viewed under a phase-contrast microscope. With proliferation, the number and size of the cell colonies appear to increase gradually on days 4–6. During the following days the culture becomes more confluent and reaches 65–70% confluence (=passage 0).
8. When the cells reach confluence, they should be trypsinized and counted as described below. Aspirate the medium and wash the cells with HBSS to remove all traces of FBS as this interferes with the trypsinization process. Add 3 ml of 0.25% Trypsin–EDTA (ready to use from the supplier) to the cells and incubate for 5 min at 37°C (incubator). Neutralize the trypsin by adding 7 ml of SCGM and transfer the cell suspension into a new 15 ml tube. Centrifuge at 600 × *g* for 5 min. After centrifugation, resuspend the pellet in 10 ml of fresh medium and count the cells using a hemocytometer slide. Replate the MSC at a concentration of 3.75×10^5 cells per 75 cm^2 flask. The cells can also be stored in liquid nitrogen at each passage stage to build up cell stocks (see Note 4).
9. Change the culture medium (SCGM) every 3 days (10 ml per 75 cm^2 flask). Typically, cell confluence is reached in 7 days. Figure 1a shows uMSC in culture.

3.3. Adipose-Derived Stem Cell Harvesting

1. Dissect out the visceral fat encasing the stomach and intestines of adult rats and chop to a fine consistency using a sterile razor blade in a clean, sterile Petri dish.
2. Transfer the tissue into a sterile 50 ml tube containing 15 ml freshly made collagenase type I solution and place in a static water bath at 37°C for 1–2 h.
3. Neutralize the enzymes by the addition of an equal volume of SCGM and centrifuge the solution at 600 × *g* for 5 min—at this stage an upper layer of floating adipose cells should be observed and a pellet of cells constituting the stromal fraction, which contains the stem cells, will have formed. Carefully aspirate and discard the upper layer and the remaining medium to leave behind the pellet of stromal cells.
4. Resuspend the pellet in 10 ml SCGM and pass the suspension through a 70 μm filter to remove any large pieces of undissociated tissue; then transfer the filtrate to a fresh 15 ml tube.
5. Centrifuge cell suspension at 600 × *g* for 5 min and resuspend the resulting pellet in 10 ml SCGM before transferring into 75 cm^2 tissue culture flasks. Maintain cultures at sub-confluent levels in a 37°C incubator with 5% CO_2; trypsinize and split as required (see Note 5). Figure 1b shows uASC in culture.

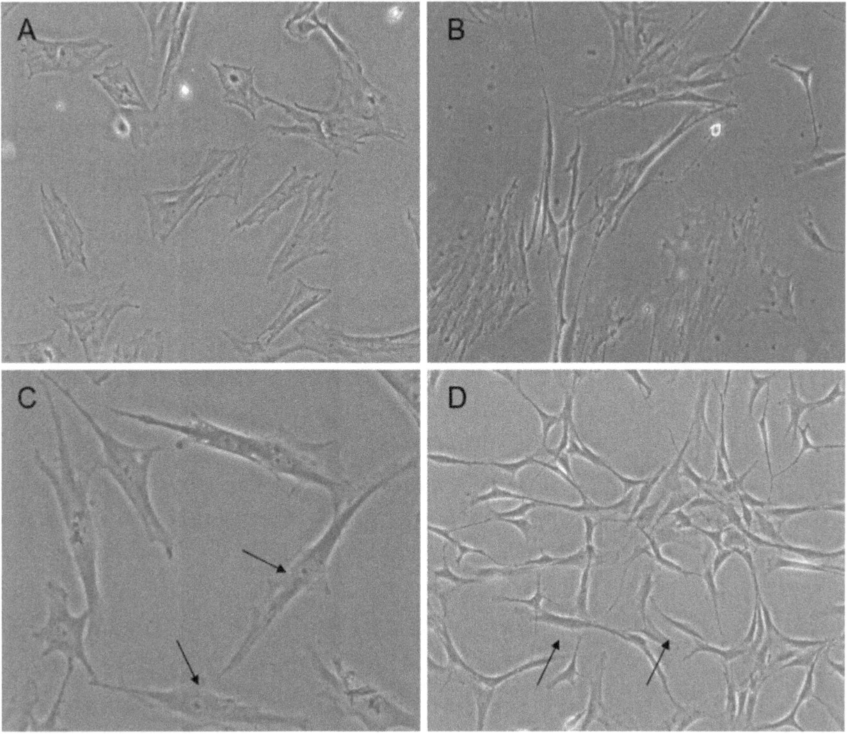

Fig. 1. *Morphology of the undifferentiated stem cells and the differentiated Schwann cell-like cells.* Phase-contrast images of various types of cell in culture were obtained using an Olympus IX51 inverted microscope and Olympus DP12 camera. (**a**) uMSC at passage 1 and (**b**) uASC at passage 2 display star-like morphology. Cultures of MSC (**c**) and ASC (**d**) undergoing differentiation and displaying the spindle-shaped (*arrows*) morphology similar to that of Schwann cells: the dMSC were at passage 5 and the dASC at passage 3. Magnification: Images **a** through **c** were taken at ×20 magnification and image **d** at ×10 magnification.

3.4. MSC and ASC Differentiation Procedure

1. Prior to differentiation of the MSC and ASC it is advisable to freeze cells from a few flasks to create stocks for future analysis and experiments.
2. Aspirate the SCGM from sub-confluent cultures of passage 2–3 undifferentiated MSC or ASC and replace with SCGM supplemented with 0.8 μl βME per 10 ml of medium and incubate for 24 h at 37°C in 5% CO_2.
3. Next day, wash the cells with HBSS and replace with fresh SCGM containing 350 ng/ml all-*trans*-retinoic acid and incubate at 37°C with 5% CO_2.
4. Following 3 days of incubation, wash the cells with HBSS and replace with 10 ml stem cell *differentiation* medium (SC*DM*).
5. Incubate the cells (3) for 2 weeks and with medium (SCDM) changes approximately every 72 h (see Note 7). Figure 1c, d shows dMSC and dASC, respectively, in culture.

4. Notes

1. Avoid bubble formation. The cells may become trapped within the bubbles and lost from the preparation.
2. Ensure that the incubation conditions are exactly as described in the method. Hypoxia, high CO_2 level and low humidity are highly detrimental to MSC growth and survival.
3. At this stage remove the culture medium very gently because rapid aspiration may cause the MSC to lift and detach from the bottom surface of the flask with the consequent loss of the adherent (MSC) cells. Bacterial contamination is always a possibility at this stage so particular vigilance is required.
4. The *time* and *temperature* of the trypsin incubation are very important. A longer incubation in this solution may cause cell death. It is important to prepare the fresh medium in advance so that it is ready to stop action of the trypsin.
5. A similar protocol for isolation of mouse adipose-derived stem cells from inguinal fat pads has been reported (26).
6. It is helpful to treat the polypropylene tubes in which the stock growth factors are stored and used with Sigmacote® (SL2; Sigma) to minimize the adhesion of the growth factors to plasticware.
7. Prepare only the required quantity of differentiation medium: upon storage at 4°C there is a loss of growth factor activity.

References

1. Wiberg M, Terenghi G (2003) Will it be possible to produce peripheral nerves? Surg Technol Int 11:303–10
2. Bunge RP (1994) The role of the Schwann cell in trophic support and regeneration. J Neurol 242(1 Suppl 1):S19–21
3. Ide C (1996) Peripheral nerve regeneration. Neurosci Res 25:101–121
4. Mahanthappa NK, Anton ES, Matthew WD (1996) Glial growth factor 2, a soluble neuregulin, directly increases Schwann cell motility and indirectly promotes neurite outgrowth. J Neurosci 16:4673–4683
5. Tohill MP, Mann DJ, Mantovani CM et al (2004) Green fluorescent protein is a stable morphological marker for schwann cell transplants in bioengineered nerve conduits. Tissue Eng 10:1359–1367
6. Tohill M, Terenghi G (2004) Stem-cell plasticity and therapy for injuries of the peripheral nervous system. Biotechnol Appl Biochem 40:17–24
7. Loeffler M, Bratke T, Paulus U et al (1997) Clonality and life cycles of intestinal crypts explained by a state dependent stochastic model of epithelial stem cell organization. J Theor Biol 186:41–54
8. Pittenger MF, Mackay Am Beck SC et al (1999) Multilineage potential of adult human mesenchymal stem cells. Science 284:143–147
9. Wexler SA, Donaldson C, Denning-Kendall P et al (2003) Adult bone marrow is a rich source of human mesenchymal 'stem' cells but umbilical cord and mobilized adult blood are not. Br J Haematol 121:368–378
10. Dezawa M, Takahashi I, Esaki M et al (2001) Sciatic nerve regeneration in rats induced by transplantation of in vitro differentiated bone-marrow stromal cells. Eur J Neurosci 14: 1771–1776
11. Kim S, Honmou O, Kato K et al (2006) Neural differentiation potential of peripheral blood- and bone-marrow-derived precursor cells. Brain Res 1123:27–33

12. Kopen GC, Prockop DJ, Phinney DG (1999) Marrow stromal cells migrate throughout forebrain and cerebellum, and they differentiate into astrocytes after injection into neonatal mouse brains. Proc Natl Acad Sci USA 96: 10711–10716
13. Mezey E, Chandross KJ (2000) Bone marrow: a possible alternative source of cells in the adult nervous system. Eur J Pharmacol 405: 297–302
14. Azizi SA, Stokes D, Augelli BJ et al (1998) Engraftment and migration of human bone marrow stromal cells implanted in the brains of albino rats–similarities to astrocyte grafts. Proc Natl Acad Sci USA 95:3908–3913
15. Hofstetter CP, Schwarz EJ, Hess D et al (2002) Marrow stromal cells form guiding strands in the injured spinal cord and promote recovery. Proc Natl Acad Sci USA 99: 2199–2204
16. Akiyama Y, Shirasugi N, Aramaki O et al (2002) Intratracheal delivery of a single major histocompatibility complex class I peptide induced prolonged survival of fully allogeneic cardiac grafts and generated regulatory cells. Hum Immunol 63:888–892
17. Gimble JM, Katz AJ, Bunnell BA (2007) Adipose-derived stem cells for regenerative medicine. Circ Res 100:1249–1260
18. Strem BM, Hicik KC, Zhu M et al (2005) Multipotential differentiation of adipose tissue-derived stem cells. Keio J Med 54:132–141
19. Kingham PJ, Kalbermatten DF, Mahay D et al (2007) Adipose-derived stem cells differentiate into a Schwann cell phenotype and promote neurite outgrowth in vitro. Exp Neurol 207: 267–274
20. Aust L, Devlin B, Foster SJ et al (2004) Yield of human adipose-derived adult stem cells from liposuction aspirates. Cytotherapy 6:7–14
21. Locke MJ, Windsor Dunbar PR (2009) Human adipose-derived stem cells: isolation, characterization and applications in surgery. ANZ J Surg 79:235–244
22. di Summa PG, Kingham PJ, Raffoul W et al (2009) Adipose-derived stem cells enhance peripheral nerve regeneration. J Plast Reconstr Aesthet Surg 63:1544–1552
23. Zheng H, Guo Z, Ma Q et al (2004) Cbfa1/osf2 transduced bone marrow stromal cells facilitate bone formation in vitro and in vivo. Calcif Tissue Int 74:194–203
24. Gronthos S, Zannettino AC, Hay SJ et al (2003) Molecular and cellular characterisation of highly purified stromal stem cells derived from human bone marrow. J Cell Sci 116:1827–1835
25. Mackay AM, Beck SC, Murphy JM et al (1998) Chondrogenic differentiation of cultured human mesenchymal stem cells from marrow. Tissue Eng 4:415–428
26. Taha MF, Hedayati V (2010) Isolation, identification and multipotential differentiation of mouse adipose tissue-derived stem cells. Tissue Cell 42:211–216

Chapter 6

Functional Purification of Human and Mouse Mammary Stem Cells

Daniela Tosoni, Pier Paolo Di Fiore, and Salvatore Pece

Abstract

Normal and tumor stem cells are present in rare quantities in tissues and this has historically represented a major hurdle to in-depth investigations of their biology. In the case of the mammary gland, the relative promiscuity of the immunophenotypical markers described in several studies for the isolation of human and mouse mammary stem cells limits their usefulness, in particular when highly purified mammary stem cell fractions are required for an in-depth molecular and functional characterization (Stingl et al. Nature 439:993–997, 2006; Shackleton et al. Nature 439:84–88, 2006; Liao et al. Cancer Res 67:8131–8138, 2007; Eirew et al. Nat Med 14:1384–1389, 2008; Raouf et al. Cell Stem Cell 3:109–118, 2008; Lim et al. Nat Med 15:907–913, 2009). In fact, most so-called stem cell markers are not selectively expressed by mammary stem cells, but are instead also expressed by terminally differentiated luminal and/or myoepithelial cells or by bipotent progenitors within the mammary gland (Stingl et al. Nature 439:993–997, 2006; Eirew et al. Nat Med 14:1384–1389, 2008; Raouf et al. Cell Stem Cell 3:109–118, 2008; Stingl et al. Differentiation 63:201–213, 1998; Jones et al. Cancer Res 64:3037–3045, 2004). Here, we describe a new methodology that does not require the use of immunophenotypical markers to obtain highly pure populations of mammary stem cells. This approach exploits two functional properties of mammary stem cells: (1) their quiescent or slowly proliferative phenotype, as compared to their progeny; and (2) their ability to survive and proliferate in anchorage-independent conditions, giving rise to clonal spheroids, commonly known as "mammospheres" (Dontu et al. Genes Dev 17:1253–1270, 2003; Pece et al. Cell 140:62–73, 2010; Cicalese et al. Cell 138:1083–1095, 2009). In the context of mammospheres, stem cells, which perform one or two rounds of division and then reenter quiescence, are identified based on their ability to retain a lipophilic fluorescent dye, PKH26, that is by contrast progressively lost by dilution in the actively proliferating progeny of precursors (Pece et al. Cell 140:62–73, 2010; Cicalese et al. Cell 138:1083–1095, 2009). Following mammosphere dissociation, the differential degree of PKH26 epifluorescence displayed by stem cells compared to precursor cells is exploited for their purification by FACS sorting. As a result, the scarcely represented PKH26-labeled mammary stem cells are purified to near homogeneity and can be used for further molecular and biological studies.

Key words: Normal mammary stem cells, Mammospheres, PKH26 dye, Breast cancer stem cells, Quiescence.

Kimberly A. Mace and Kristin M. Braun (eds.), *Progenitor Cells: Methods and Protocols*, Methods in Molecular Biology, vol. 916, DOI 10.1007/978-1-61779-980-8_6, © Springer Science+Business Media, LLC 2012

1. Introduction

The methodological approach described herein can be used for the isolation of pure populations of mammary stem cells from both human and mouse mammary glands (1, 2). The overall strategy that we have devised for the prospective isolation of mammary stem cells relies on two key defining stem cell properties, namely, their relative quiescence and their ability to survive in anchorage-independent conditions. In brief, the critical steps in our procedure involve (1) the use of a lipophilic fluorescent dye, PKH26, to label dissociated bulk mammary epithelial cells. This dye has been used to label relatively quiescent cells within a proliferating population (3, 4). In our method, PKH26 is selectively retained in the stem cells present in the mammary gland, but not in their proliferating progeny; (2) the mammosphere culture method, i.e., the generation of clonal three-dimensional spheroids in suspension growth conditions. This method relies on the specific ability of stem cells to self-renew in suspension by withstanding apoptosis due to anchorage detachment (a process termed *anoikis*) (1, 2, 5); (3) FACS sorting of cells obtained from the dissociation of mammospheres to purify PKH26-labeled mammary stem cells from the bulk of progenitors (Fig. 1).

Our strategy, therefore, does not rely on the use of immunophenotypical markers and it circumvents the limitations posed by the promiscuous distribution of such markers among the different cell types that make up the normal mammary gland. Another important advantage of our methodology is that it is also suitable for the isolation of cancer stem cells (1, 2), which, given the heterogeneity of breast cancer, would otherwise require the extremely complex task of identifying universal cancer stem cell-specific immunophenotypical markers. Finally, our approach allows not only for the purification of nearly pure stem cell populations, but also of nearly pure undifferentiated progenitors. These latter populations can be more easily, therefore, compared to differentiated cells in primary cultures (not covered here, but see refs. (6, 7) for methodologies to prepare primary cultures from normal and cancer tissues of various origins).

As well as describing the general protocols, we provide detailed descriptions of the procedures specific to the purification of stem cells from human and mouse mammary glands in two distinct sections (see Subheadings 3.4 and 3.5), since important differences exist between these respective protocols.

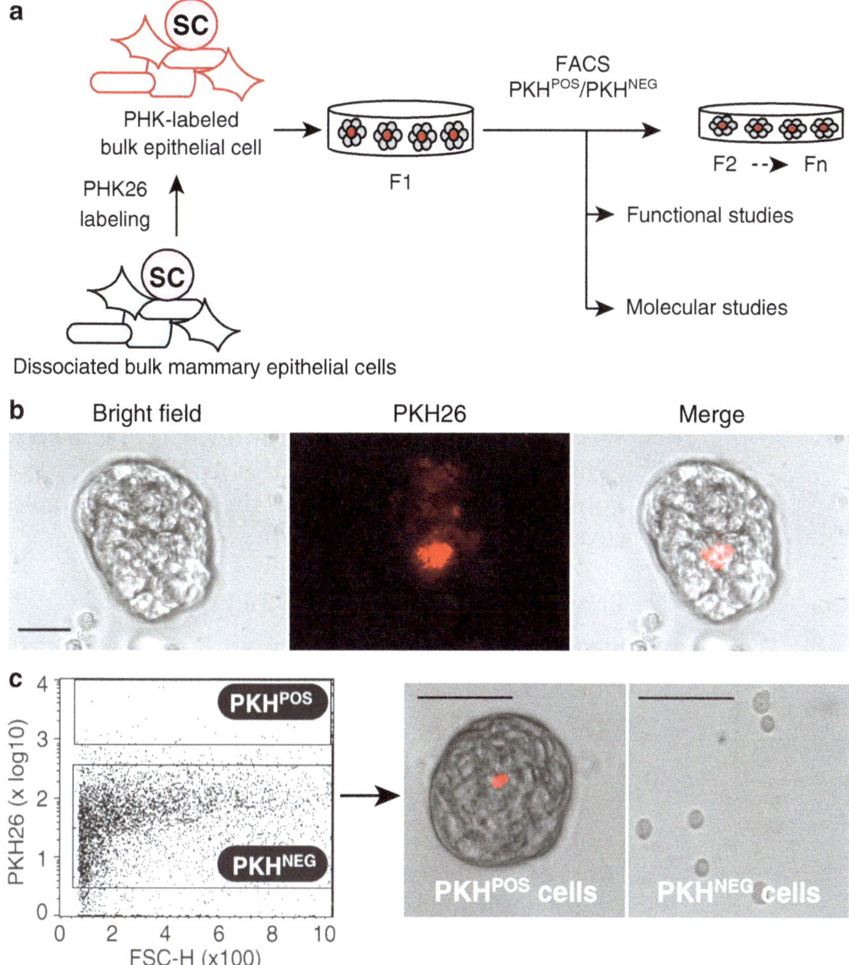

Fig. 1. *Functional purification of normal mammary stem cells*. (**a**) Schematic representation of the procedure used to generate PKH26-labeled mammospheres. Cells are freshly isolated from the mammary gland to yield the "dissociated bulk mammary epithelial cells (*bottom*). This population is then labeled with the PKH26 dye (PKH-labeled bulk epithelial cells, *top*) and plated in suspension culture to allow for the formation of primary generation mammospheres (F1). Cells from mammosphere dissociation can be replated in suspension to yield multiple generations of mammospheres (mammosphere generations are indicated as F1 through Fn). (**b**) A typical mammosphere derived from PKH26-labeled epithelial cells. *Bar*, 100 μm. (**c**) *Left*, a typical FACS profile of a PKH26-labeled mammosphere population with gated populations; *right*, suspension cultures of FACS-sorted $PKH26^{POS}$ or $PKH26^{NEG}$ cells showing that only $PKH26^{POS}$ cells display mammosphere-forming ability, as expected of true stem cells. *Bar*, 100 μm (adapted from ref. (1) with permission from Elsevier).

2. Materials

2.1. Preparation of Low Adhesion Plates for Suspension Culture Using Poly-HEMA

1. Prepare a stock concentration (12%) Poly-HEMA solution by dissolving 10 g Poly-HEMA (2-hydroxyethylmethylmethacrylate) (Sigma) in 83.3 mL 95% ethanol overnight at 55°C using a rotating shaker.
2. From this stock concentration, prepare a 1:10 dilution in 95% ethanol to yield the Poly-HEMA working solution at a final

concentration of 1.2%. Filter sterilize through a 0.2-μm vacuum filter unit. Both the stock and the working Poly-HEMA solutions can be stored at room temperature for several months.

3. Prepare low adhesion cell culture plates using the following amounts of 1.2% Poly-HEMA solution: 400 μL/well for 6-well plates (BD Falcon™ 6-well Multiwell Plate); 250 μL/well for 12-well plates (BD Falcon™ 12-well Multiwell Plate); 150 μL/well for 24-well plates (24-well plates (BD Falcon™ 24-well Multiwell Plate). Coat plates by uniformly distributing the indicated volumes of working Poly-HEMA solution onto wells under a sterile hood.
4. Let the plates air dry under sterile conditions (see Note 1).
5. When the plates have completely dried, repeat the coating procedure (see step 3) to ensure that no breaks are present in the Poly-HEMA polymer film. Let the plates dry completely, then cover with lids, and store at 4°C.

2.2. Isolation and Culture of Primary Mammary Epithelial Cells in Adhesion and in Suspension Conditions

1. Breast tissue (see Subheading 3.1).
2. Sterile tweezers, scissors, and scalpel.
3. Dulbecco's Phosphate-Buffered Saline (D-PBS).
4. 100-mm tissue culture Petri dishes.
5. Enzyme Digestion Mixture (EDM): prepare a solution of DMEM (Lonza) + HAM's nutrient mixture F12 medium (Gibco) (1:1 ratio) supplemented with 1 μg/mL insulin (Roche), 1 μg/mL Hydrocortisone (Sigma), 100 U/mL Penicillin, 100 U/mL Streptomycin, 2 mM L-Glutamine; 200 U/mL Collagenase type 1A (Sigma), 100 U/mL Hyaluronidase (Sigma). Filter sterilize through a 0.2-μm vacuum filter unit and store at 4°C until needed (see Note 2). Add 10 ng/mL EGF (Peprotech) immediately before use. Prewarm the EDM at 37°C in a water bath before adding to sample tissues. The EDM will be used to digest the tissue.
6. Mammary Epithelial Cell Medium (MECM): prepare a solution of DMEM + HAM's nutrient mixture F-12 medium (1:1 ratio) supplemented with 10 nM Triiodothyronine (Sigma), 10 mM Hepes, 50 μM Ascorbic acid (Sigma), 10 nM β-Estradiol (Sigma), 1 μg/mL Insulin, 1 μg/mL Hydrocortisone, 0.1 mM Ethanolamine (Sigma), 10 μg/mL Transferrin (Sigma), 2 mM L-Glutamine, 100 U/mL Penicillin, 100 U/mL Streptomycin, 15 nM Sodium Selenite (Sigma), 50 ng/mL Cholera toxin (Sigma), 1% Fetal Bovine Serum (FBS) (Gibco), 35 μg/mL Bovine Pituitary Extract (BPE) (Gibco), 100 μg/mL Gentamicin. Sterilize the MECM through a 0.2-μm vacuum filter unit and store at 4°C until needed. Prepare complete MECM by freshly adding 10 ng/mL EGF immediately

before use. This medium will be used to cultivate mammary epithelial cells in adhesion.

7. Mammary Epithelial Stem Cell Medium (MESCM): prepare using MEBM Basal Medium supplemented with 100 U/mL Penicillin, 100 U/mL Streptomycin, 2 mM L-Glutamine, 5 μg/mL Insulin, 0.5 μg/mL Hydrocortisone, 1 U/mL Heparin (Wockhardt). Filter sterilize through a 0.2-μm vacuum filter unit and store at 4°C until needed. Prepare complete MESCM by adding 20 ng/mL EGF, 20 ng/mL FGF (Peprotech), and 2% B-27 Supplement (Gibco) immediately before use. Filter sterilize through a 0.2-μm syringe filter prior to adding complete MESCM to cells. This medium will be used to cultivate cells in suspension.
8. ACK buffer (Bio Whittaker).
9. 35-mm tissue culture dishes.

2.3. Dissociation of Organoids and Preparation of Single Epithelial Cells for Mammosphere Culture

1. Organoids (see Subheading 3.1).
2. 2.5% trypsin–0.2 mg/mL EDTA (Bio Whittaker).
3. Fetal bovine serum.
4. D-PBS.
5. DMEM with 2 mM L-Glutamine.

2.4. PKH26 Labeling and Mammosphere Culture Method

1. Single cell suspension of mammary epithelial cells.
2. Complete MESCM.
3. Poly-HEMA-treated 6-well plates.
4. D-PBS.
5. PKH26 Red Fluorescent Cell Linker Kit (PKH26GL, Sigma-Aldrich).
6. Ethanol.
7. FBS.
8. Complete MECM.

2.5. Generation of Human Mammospheres in Suspension Culture Conditions

1. Complete MESCM.
2. Poly-HEMA-coated 6-well plate.

2.6. Isolation of Primary Normal and Tumor Mammary Epithelial Cells from Mouse Breast Tissues

1. Mice.
2. D-PBS.
3. 100-mm tissue culture Petri dishes.
4. Sterile tweezers, scissors, and scalpel.
5. EDM.

6. Pasteur pipette.
7. DMEM with 2 mM L-Glutamine.
8. 100-, 70-, and 40-μm cell strainers.
9. Sterile tubes to fit cell strainers (e.g., 50 mL Falcon tubes).
10. Filcon 10-μm syringe filters (BD Bioscence).
11. 0.2% sodium chloride solution (in sterile distilled water).
12. 1.6% sodium chloride solution (in sterile distilled water).

2.7. Generation of Mouse Mammospheres in Suspension Culture Conditions

1. Dissociated mouse mammary cells.
2. Complete MESCM.
3. Poly-HEMA-coated 6-well plates.

2.8. PKH26 Labeling Procedure for Mouse Mammary Epithelial Cells

1. Cell suspension in DMEM plus 2 mM L-Glutamine.
2. 50-mL conical tubes.
3. D-PBS.
4. PKH26 Red Fluorescent Cell Linker Kit (PKH26GL, Sigma-Aldrich).
5. Ethanol.
6. FBS.
7. Complete MESCM.
8. Poly-HEMA-coated dishes.
9. 35-mm tissue culture dishes.
10. Trypan blue.
11. Poly-HEMA-coated 6-well plates.

2.9. FACS Sorting Analysis to Purify Human and Mouse Mammary Stem Cells from Their Precursors

1. PKH26-labeled mammary cells and PKH26^{NEG} cells.
2. 15-mL conical polypropylene tubes.
3. D-PBS.
4. 0.25% trypsin–0.2 mg/mL EDTA (BioWittaker).
5. FBS.
6. L-15 Leibovitz Medium (Sigma).
7. 40-μm filter cell strainers.
8. 5-mL polystyrene round bottom tubes (12 × 75 mm).
9. Flow cytometer capable of sorting cells, e.g., FACS Vantage SE flow cytometer (Becton & Dickinson).
10. Complete MESCM.
11. Plastic sealed bags.
12. Trypan blue.

3. Methods

The methodology combining PKH26 labeling and mammosphere culture presents a number of possible pitfalls mostly due to the tendency of epithelial cells to form aggregates when plated in suspension culture conditions, particularly in the case of the mouse mammary gland. Formation of aggregates is a major source of misinterpretation, resulting in poorly reproducible results. Therefore, prior to proceeding with a detailed description of the procedure, it is useful to introduce a series of relevant parameters, whose measurement is instrumental to monitor the quality of experimental execution and to assess the outcome of the different biological assays.

The first important parameter is the (mammo)sphere-forming efficiency (SFE), which expresses the percentage of mammospheres/number of epithelial cells seeded in suspension growing conditions (1, 2, 5). The SFE is a function of the stem cell content, be they normal or tumor stem cells, within a given epithelial cell population, whether this is derived from freshly dissociated mammary tissue or from dissociated mammospheres. The SFE of freshly isolated normal mammary epithelial cells should be, under the experimental conditions herein described, ~0.01% and ~0.05% for the human and the mouse normal mammary glands, respectively (i.e., 1 stem cell every ~10,000 human or every ~2,000 mouse total mammary primary epithelial cells) (1, 2). The mammospheres obtained from freshly dissociated epithelial cells from mammary tissue are called primary mammospheres (F1 generation). Mammospheres can be dissociated and the resulting epithelial cell suspension replated to generate the next generation of mammospheres (F2 to F*n*) (Fig. 1a). The SFE of cells obtained from the dissociation of F1 mammospheres and replated in suspension to yield F2 mammospheres is ~0.1% and 0.4% for human and mouse tissue, respectively (1, 2). Of note, normal mammospheres can be serially propagated for at least 4–5 generations (1, 2). During these passages the clonogenic ability of stem cells decreases exponentially, with the SFE at every generation being ~23% of that measured in the preceding generation (1, 2).

For both the human and the mouse mammary gland, ~1 stem cell is present per mammosphere and each mammosphere contains ~350–400 cells (1, 2). This knowledge is at the basis of the gating strategy in the FACS sorting analysis, which is typically set to purify the most epifluorescent 0.2–0.4% of the total cell population (PKH^{POS} stem cells) derived from dissociation of PKH26-labeled mammospheres (Fig. 1c). The rest of the mammosphere population (gated at 10^1–10^2 fluorescence units) represents the fraction of PKH^{NEG} progenitors (Fig. 1c). If the combined PKH26 labeling/mammosphere culture procedure is properly executed, the purified $PKH26^{POS}$ stem cell fraction should be ~90% pure (see *detailed calculation in ref.* (1)).

3.1. Isolation of Primary Normal and Tumor Mammary Epithelial Cells from Human Breast Tissues: Tissue Digestion and Preparation of Organoids

Human breast tissues can be obtained from patients undergoing surgery for reductive mammoplasty or for the removal of breast cancer. Human tissue specimens are potential sources of blood-borne pathogens and therefore must be handled using the appropriate biosafety precautions for biohazardous materials. Tissues can be stored at 4°C in sterile saline solution or PBS for up to 48 h in 50-mL sterile conical tubes without any loss of cell viability. Avoid the use of FBS. All the steps required for the isolation of primary epithelial cells from biopsy specimens are executed under a sterile tissue culture hood.

1. Transfer the sample into a 100-mm tissue culture Petri dish.
2. Remove the skin and as much as possible of the yellow adipose tissue using sterile tweezers, scissors, and scalpel.
3. Mince the sample into ~1–2 mm^3 pieces with a sterile scalpel or scissors. The sample is ready for subsequent enzymatic digestion when tissue pieces pipette easily with a 10-mL pipette.
4. Resuspend the tissue pieces in 10 mL pre-warmed EDM and incubate at 37°C in a 5% CO_2 humidified incubator until all large tissue fragments are digested. Pipette the sample up and down several times (~5–7) every 20–30 min to aid tissue dissociation. Digestion times should be adjusted according to sample size (see Note 3).
5. Transfer the suspension in a 15-mL conical polypropylene tube and centrifuge at 80 × *g* for 10 min at room temperature (see Note 4).
6. Carefully decant off the supernatant and retain the pellet that contains organoids.
7. Wash the pellet with sterile D-PBS by gently pipetting up and down two times with a 5-mL pipette.
8. Centrifuge at 400 × *g* at room temperature for 5 min.
9. Carefully aspirate and discard the supernatant without disturbing the pellet.
10. Resuspend the pellet containing the organoids in a 15-mL conical tube using 0.5–1 mL (the exact volume depends on the pellet size) of ACK buffer to lyse red blood cells.
11. Gently pipette up and down continuously for 1 min with a 2-mL pipette.
12. Stop the lysis reaction by adding 10 mL of D-PBS and then centrifuge at 400 × *g* for 5 min at room temperature (see Note 5).
13. Carefully pour off the supernatant without disturbing the pellet and wash the pellet again with 10 mL of D-PBS.
14. Centrifuge at 400 × *g* for 5 min at room temperature and then discard the supernatant (see Note 6).

15. Resuspend the pellet containing organoids in 3 mL of complete MECM and plate the suspension in a 35-mm tissue culture Petri dish (see Note 7).
16. Place the dish at 37°C in a 5% CO_2 humidified incubator to allow organoids to adhere to the cell culture dish. This normally takes ~3–5 h.
17. Incubate the primary organoid culture for 24 h (see Note 8).

3.2. Dissociation of Organoids and Preparation of Single Epithelial Cells for Mammosphere Culture

Since each mammosphere is clonally derived from one stem cell and the frequency of stem cells is one stem cell every ~10^4 normal human mammary epithelial cells, it is important to use enough starting material to ensure the recovery of a sufficient amount of stem cells for subsequent studies (see Note 9). The preparation of a homogeneous single cell suspension is the prerequisite for an efficient PKH26 labeling.

1. Pipette 500 μL of a pre-warmed 2.5% trypsin–0.2 mg/mL EDTA solution pre-warmed at 37°C into the 35-mm tissue culture dish containing adherent organoids.
2. Transfer the dish to 37°C in a 5% CO_2 humidified incubator and allow organoids to dissociate for 10 min to yield a single cell suspension.
3. Stop the digestion reaction by adding 500 μL of FBS.
4. Transfer the single cell suspension into a fresh 15-mL conical tube.
5. Rinse the 35-mm tissue culture dish three times with 2 mL D-PBS, aspirate, and transfer the rinsing buffer to the 15-mL tube containing the single cell suspension.
6. Centrifuge at 400 × *g* for 5 min at room temperature to pellet cells.
7. Resuspend the cells in 1 mL DMEM plus 2 mM L-Glutamine and then determine the cell count using a hemocytometer.
8. Centrifuge at 400 × *g* for 5 min at room temperature.
9. Carefully decant off the supernatant and retain the pellet.
10. Rinse the pellet with 10 mL D-PBS and then centrifuge again at 400 × *g* for 5 min at room temperature.
11. Carefully decant off the supernatant and resuspend the pellet in D-PBS to yield a single cell suspension at a final concentration of 10^6 cells/mL.

3.3. PKH26 Labeling and Mammosphere Culture Method

Since the PKH26 labeling procedure is not a saturation reaction but rather a function of both dye and cell concentrations, it is essential to carefully adjust the amount of dye to be incorporated into cells (see Note 10). For the human mammary gland (see Note 11), a final PKH26 concentration of 10^{-7} M in PBS is suitable for staining up to 10^6 primary human mammary epithelial cells/mL in 1 mL volume (see Note 12). Perform all steps at RT.

1. Start from the single cell suspension of epithelial cells (10^6 cells/mL D-PBS) (from Subheading 3.2, step 11).
2. Save 10% of this material (~10^5 cells) and plate these cells in suspension conditions in complete MESCM at a concentration of ~3×10^4 cells/mL in Poly-HEMA-treated 6-well plates to permit mammosphere formation (see Note 13).
3. Place the remaining single cell suspension in a conical 15-mL polypropylene tube.
4. Add 10 mL D-PBS (see Note 14).
5. Centrifuge the cells at $400 \times g$ for 5 min.
6. Carefully aspirate and discard the supernatant without disturbing the pellet.
7. Resuspend the cells in 500 μL of D-PBS with gentle pipetting to ensure complete dispersion. Do not vortex.
8. Immediately prior to staining, prepare a 1:10 dilution in 100% ethanol of the 10^{-3} M stock PKH26 concentration to yield a 10^{-4} M PKH26 dye dilution in ethanol.
9. Prepare a 2×10^{-7} M PKH26 solution by adding 1 μL of the 10^{-4} M PKH26 solution into 500 μL of D-PBS in polypropylene tubes (see Note 15).
10. Rapidly add 500 μL of the 2×10^{-7} M PKH26 solution to 500 μL of cell suspension and immediately mix the sample by pipetting up and down several times (see Note 16).
11. Incubate at room temperature for 5 min.
12. Stop the staining reaction by adding 1 mL of FBS. Incubate 1 min at room temperature.
13. Add 10 mL of complete MECM.
14. Centrifuge the cells at $400 \times g$ for 5 min at room temperature.
15. Gently decant off the supernatant.
16. Resuspend the pellet using 10 mL of D-PBS, centrifuge at $400 \times g$ for 5 min at room temperature, and discard the supernatant. Repeat twice (see Note 17).

3.4. Generation of Human Mammospheres in Suspension Culture Conditions

A typical example of a normal mammosphere derived from PHK26-labeled human normal mammary epithelial cells grown in suspension conditions is depicted in Fig. 1b (see Note 18).

1. At the end of the PKH26 labeling procedure (see Subheading 3.3, step 16), resuspend the cells in complete MESCM at a concentration of 3×10^4 cells/mL.
2. Plate 3 mL of this cell suspension (~9×10^4 cells) onto 1 well of a Poly-HEMA-coated 6-well plate to allow for mammosphere formation (see Note 19).

3. Plate the remaining preparation in suspension culture conditions using the appropriate type and number of Poly-HEMA-coated plates (see Note 20).
4. Allow mammosphere formation to proceed by incubating the cell culture suspensions for 7–10 days at 37°C in a 5% CO_2 humidified incubator.

3.5. Isolation of Primary Normal and Tumor Mammary Epithelial Cells from Mouse Breast Tissues

For the preparation of reasonable amounts of purified normal mouse mammary stem cells, we recommend starting from at least 20 animals (age range, 4–6 weeks).

1. Start from pooling freshly dissected inguinal and thoracic mammary glands into several 50-mL conical tubes, each containing 30 mL of D-PBS (see Note 21).
2. Transfer the pooled mammary tissue from each tube onto a 10-cm tissue culture dish using sterile tweezers.
3. Mince the sample with a sterile scalpel or scissors into ~1–2 mm^3 pieces. The sample is ready for subsequent enzymatic digestion when tissue pieces pipette easily with a 10-mL pipette.
4. Add 10 mL of pre-warmed EDM into each 10-cm dish and incubate at 37°C in a 5% CO_2 humidified incubator until the large tissue fragments are dissociated. This step typically takes 4 h during which time the sample should be pipetted up and down several times every 30 min using a 10-mL pipette. During the last hour, we suggest fitting a 1,000-μL pipette tip on a 10-mL pipette to aid tissue dissociation into single cells.
5. Transfer the digested tissue from each dish into a fresh 15-mL conical polypropylene tube.
6. Centrifuge at 80 × *g* for 5 min at room temperature (see Note 22).
7. Use a handheld Pasteur pipette to remove the upper fat layer (see Note 23).
8. Carefully aspirate the supernatant without disturbing the pellet.
9. Add 10 mL DMEM plus 2 mM L-Glutamine into each 15-mL tube and resuspend the tissue pellet by gently pipetting with a 10-mL pipette. To yield a homogeneous single cell suspension of mouse mammary epithelial cells, the material is filtered sequentially using membrane syringe filters of decreasing pore sizes.
10. Using a 10-mL pipette, aspirate the digested material already resuspended in 10 mL DMEM plus 2 mM L-Glutamine (from step 9).
11. Sieve the suspension through a 100-μm cell strainer fitted on a 50-mL sterile tube to remove undigested cell clumps and separate the single cells (see Note 24).

12. Rinse the filter with 10 mL of DMEM plus 2 mM L-Glutamine.
13. Using a 10-mL pipette, sieve the suspension through a 70-μm cell strainer fitted on a fresh 50-mL sterile tube.
14. Rinse the filter with 10 mL of DMEM plus 2 mM L-Glutamine.
15. Collect and sieve the suspension through a 40-μm cell strainer fitted on a fresh 50-mL sterile tube.
16. Rinse the filter with 10 mL of DMEM plus 2 mM L-Glutamine.
17. Collect and pass the cell suspension through a Filcon 10-μm syringe filter using a 50-mL syringe (see Note 25). Repeat this step twice.
18. Centrifuge the resulting single cell suspension at 400 × *g* for 5 min at room temperature. Carefully decant off the supernatant without disturbing the pellet.
19. Add 5 mL sterile D-PBS to each tube to resuspend the pellet (see Note 26).
20. Pool single cell suspensions from different tubes into a fresh 50-mL conical tube.
21. Centrifuge at 400 × *g* for 5 min at room temperature and carefully discard the supernatant.
22. Lyse erythrocytes by resuspending the pellet with 3 mL of a 0.2% sodium chloride solution (in sterile distilled water) pipetting continuously for 45 s.
23. Immediately add 3 mL of a 1.6% sodium chloride solution to stop the lysis reaction.
24. Add 30 mL of DMEM plus 2 mM L-Glutamine.
25. Centrifuge at 400 × *g* for 5 min at room temperature. Carefully discard the supernatant and retain the pellet.
26. Resuspend the pellet in 10 mL of DMEM plus 2 mM L-Glutamine.
27. Count cells in a hemocytometer using the vital dye Trypan blue (Sigma-Aldrich). At this stage, the cell suspension typically contains ~2×10^6 cells/mL.

3.6. Generation of Mouse Mammospheres in Suspension Culture Conditions

There is a substantial difference in the procedure for obtaining mammary stem cells and mammospheres from mouse tissue compared to the procedures used for human tissue. Unless an immunophenotypical purification is performed to specifically enrich the epithelial portion of the mouse mammary tissue, the single cell suspension obtained from freshly dissociated mouse mammary tissue also contains cellular contaminants, mostly in the form of leukocytes. The short-term adhesion step (see Subheading 3.1, step 17 and Note 8), which is used to enrich epithelial cells after dissociation of the human mammary gland, cannot be performed for the mouse mammary tissue since it heavily affects the ability of mouse mammary stem cells to grow in suspension.

1. Start from the cell suspension obtained at the end of the tissue dissociation procedure (see Subheading 3.5, step 27).
2. Take an aliquot of medium containing 10^6 cells (~500 μL) and transfer to a fresh 15-mL tube.
3. Centrifuge at 400 × *g* for 5 min at room temperature.
4. Discard the supernatant without disturbing the pellet and resuspend the cells in complete MESCM at a concentration of 10^5 cells/mL.
5. Plate 3 mL/well of the cell suspension in poly-Hema-coated 6-well plates to allow for primary mammosphere formation for 7 days (see Note 27).

3.7. PKH26 Labeling Procedure for Mouse Mammary Epithelial Cells

The PKH26 labeling procedure described in this section has been optimized for the mouse mammary gland. All the steps are performed at room temperature.

1. Start from the 10 mL cell suspension in DMEM plus 2 mM L-Glutamine (see Subheading 3.6, steps 1–2).
2. Divide the cell suspension into two fresh 50-mL conical tubes (each containing ~10^7 cells in 5 mL medium).
3. Add 10 mL D-PBS and centrifuge at 400 × *g* for 5 min at room temperature.
4. Carefully decant off the supernatant and resuspend the cells in 5 mL of D-PBS. Gently pipette to ensure complete dispersion. Do not vortex.
5. Immediately prior to staining, dilute 4 μL of the PKH26 stock concentration (10^{-3} M in 100% ethanol) in 5 mL of D-PBS in 50-mL polypropylene tubes to yield a working PKH26 solution of 4×10^{-7} M in D-PBS.
6. Rapidly mix 5 mL of working PKH26 solution with 5 mL of cell suspension (the final PKH26 concentration becomes 2×10^{-7} M). Immediately mix the sample by pipetting up and down several times (see also Note 16).
7. Incubate at room temperature for 5 min.
8. Stop the staining reaction by adding 30 mL of MESCM.
9. Centrifuge the cells at 400 × *g* for 5 min at room temperature.
10. Gently decant off the supernatant.
11. Resuspend the pellet using 10 mL of MESCM, then centrifuge again at 400 × *g* for 5 min at room temperature. Carefully decant off the supernatant and retain the pellet containing PKH26-labeled cells (see also Note 17).
12. Resuspend the cells in complete MESCM at a final concentration of 10^5 cells/mL.
13. Plate the suspended cells in poly-Hema-coated dishes to allow for primary mammosphere formation.

14. After 7 days of culture, collect the primary mammospheres in 50-mL polypropylene tubes.
15. Centrifuge at 400 × *g* for 5 min at room temperature. Discard the supernatant and retain the pellet containing primary mammospheres (see also Note 27).
16. Resuspend the pellet in 500 μL of complete MESCM.
17. Mechanically dissociate the mammospheres by pipetting up and down several times with a 1-mL pipette (~200 times) to yield a single cell suspension.
18. Transfer the material to a 35-mm tissue culture dish and check the efficiency of cell dissociation under an inverted microscope.
19. Count the cells with the vital dye Trypan blue.
20. Plate 5×10^3 cells/mL in suspension using poly-Hema-coated 6-well plates to allow for secondary mammosphere formation (see Note 28).

3.8. FACS Sorting Analysis to Purify Human and Mouse Mammary Stem Cells from Their Precursors

The starting material for FACS sorting is epithelial cells obtained from the dissociation of PKH26-labeled mammospheres. Cells from unlabeled mammospheres are used as negative controls. Cells to be subjected to FACS sorting are derived from primary mammospheres in the case of human mammary tissue, and from secondary mammospheres in the case of mouse mammary tissue.

3.8.1. Preparation of a Single Cell Suspension for FACS Analysis from Human Mammospheres

1. Harvest mammospheres after 7–10 days of culture and transfer into a 15-mL conical polypropylene tube (see Note 29).
2. Centrifuge at 80 × *g* for 10 min at room temperature to yield a loose pellet (see Note 30).
3. Gently decant off the supernatant.
4. Resuspend the pellet with 10 mL of sterile D-PBS with gentle pipetting.
5. Centrifuge at 400 × *g* for 5 min at room temperature and then pour off the supernatant.
6. Resupend the pellet in 300 μL of a 0.25% trypsin–0.2 mg/mL EDTA solution to dissociate mammospheres.
7. Incubate for 15 min at 37°C in a 5% CO_2 humidified incubator pipetting every 5 min with a 1-mL pipette.
8. Stop the digestion reaction by adding 300 μL of FBS for 3 min.
9. Add 10 mL of sterile D-PBS and centrifuge at 400 × *g* for 5 min at room temperature to pellet dissociated cells.
10. Resuspend the pellet with 10 mL of D-PBS (see Note 31).
11. Centrifuge again at 400 × *g* for 5 min at room temperature and discard the supernatant.

12. Resuspend the pellet in 3 mL of L-15 Leibovitz Medium pre-warmed at 37°C (see Note 32).
13. Filter the cell suspension through a 40-μm filter cell strainer and collect the material into in a 5-mL polystyrene round bottom tube (12 × 75 mm) for subsequent FACS analysis.

3.8.2. Preparation of a Single Cell Suspension for FACS Analysis from Mouse Mammospheres

1. Harvest mammospheres after 7–10 days of culture and transfer into a 50-mL conical polypropylene tube (see Note 29).
2. Centrifuge at 80 × *g* for 10 min at room temperature to yield a loose pellet (see Note 30).
3. Carefully decant off the supernatant and retain the pellet.
4. Resuspend the pellet in 250 μL of L-15 Leibovitz Medium (see Note 32).
5. Mechanically dissociate the mammospheres by pipetting several times (~200 times) to yield a single cell suspension.
6. Transfer the material to a 35-mm tissue culture dish and check the efficiency of cell dissociation under an inverted microscope.
7. Add 2.75 mL of L-15 Leibovitz Medium.
8. Filter the cell suspension through a 40-μm cell strainer into a 5-mL polystyrene round bottom tube (12 × 75 mm).
9. Proceed to FACS sorting (see Subheading 3.8.3.).

3.8.3. FACS Sorting

Cells can be sorted based on the differential intensity of PKH26 epifluorescence using a FACS Vantage SE flow cytometer (Becton & Dickinson), or any other appropriate cell sorter, equipped with an argon ion laser tuned to 488 nm excitation wavelength at 120 mW (see Note 33). The PKH26 fluorescence is detected with a band-pass 575/26 nm optical filter (FL2 channel). Single cell suspensions obtained from mammosphere dissociation are sorted using a 70-μm ceramic nozzle (BD Biosciences), a sheath pressure of 20 lb per square inch (PSI), and an average acquisition rate of 800 events per second ("gentle FACS") (see Note 34). All the FACS sorting steps should be performed under sterile conditions. Based on the expected SFE of epithelial cells derived from mammosphere dissociation (see also Subheading 3, *initial paragraph*), mammary stem cells are purified as the top 0.2–0.4% most intensively epifluorescent cells (PKH^{POS} cells, gated at 10^3–10^4 fluorescence units) as opposed to a "dull" population of progenitor cells (PKH^{NEG} cells, gated at 10^1–10^2 fluorescence units) (Fig. 1c) (see Note 35). The CellQuest acquisition and analysis software is used to quantify the fluorescence signal intensities and forward light scattering, and to set logical electronic gating parameters designed for sorting simultaneously PKH^{POS} and PKH^{NEG} cells.

1. Physically sort an equal amount of PKH^{POS} and PKH^{NEG} cells (see Note 36) into designated sterile microcentrifuge tubes (1.5-mL tube) prefilled with 50 μL of complete MESCM (see Note 37).
2. Carefully remove the collection tubes from the sorting chamber and seal each tube with a sterile cap.
3. Using a system (plastic sealed bags, for instance) with a controlled 5% CO_2 atmosphere, immediately transfer the tubes under a sterile cell culture hood.
4. Add an appropriate amount of complete MESCM to each collection tube and mix gently to yield a homogeneous cell suspension (see Note 38).
5. Remove a small aliquot (~10 μL) of each PKH26 fraction to carefully assess the number and viability of cells using the vital dye Trypan blue (1:1 vol/vol) in a hemocytometer (see Note 39).
6. Centrifuge cells at 400 × *g* for 5 min at room temperature and carefully decant off the supernatant without disturbing the pellet.
7. At this stage, the cell pellet can be frozen or resuspended in MESCM for subsequent functional and molecular analysis (1).

4. Notes

1. For the alcohol to thoroughly evaporate off, it takes ~1 h for 6-well plates, ~1.30 h for 12-well plates, and ~2 h for 24-well plates (the more the surface area the faster the drying process). Do not force alcohol evaporation by blowing air onto plates.
2. EDM is stored at 4°C without EGF. It is best to prepare small aliquots (~100 mL) of EDM to be used for no longer than 10–15 days, since the enzymatic activity dramatically declines over time.
3. 4–5 h of digestion pipetting up and down every 20–30 min are usually enough for complete digestion of 1–2 cm^3 biopsy specimens. Decrease digestion times according to sample size. Inappropriately prolonged digestion will unavoidably affect cell viability rather than increase cell yield.
4. This step of differential centrifugation will allow to separate single stromal cells, mostly fibroblasts and endothelial cells, which will remain in the supernatant (8, 9), from large clusters of mammary tissue commonly referred as to "organoids". Organoids correspond to the epithelial portion of the mammary

gland and are therefore composed of luminal epithelial and myoepithelial cells. Stromal cells can be eliminated if not needed for other uses.

5. Inappropriately prolonged lysis reaction dramatically affects epithelial cell viability.
6. At this step, the pellet contains organoids that can be frozen at −80°C for prolonged storage. To freeze down organoids, resuspend the pellet in 900 μL of MECM + 100 μL of DMSO.
7. To distribute the organoids evenly over the bottom of the Petri dish, gently move the plate in a star pattern on a level surface. Swirling will cause the organoids to accumulate excessively in the center of the well.
8. This short-term adhesion step is crucial to remove the majority of non-epithelial cellular contaminants, mostly leukocytes, without affecting stem cell properties (1, 10).
9. For instance, if ~100 normal mammary stem cells are required for a given biological assay, it will be necessary to start from at least 10^6 primary human mammary epithelial cells derived from dissociation of short-term adherent organoids. If large quantities of stem cells are needed, human normal mammary gland specimens can be pooled.
10. PKH26 dye is provided as an ethanol solution (stock PKH26 concentration is 10^{-3} M in 100% ethanol) and therefore an excess PKH26 dye can affect cell viability.
11. The PKH26 labeling procedure adopted for the mouse mammary gland substantially deviates from the one used for human mammary epithelial cells (see Subheading 3.7).
12. With these labeling conditions, a 5% loss in cell viability is a very frequent, and almost unavoidable, occurrence. We suggest using no less than $3–5 \times 10^5$ primary human mammary epithelial cells for the PKH26 labeling reaction. When higher amount of cells are to be stained, PKH26 concentrations and reaction volumes should be adjusted accordingly to avoid insufficient PKH26 staining and cell aggregation. For instance, for 3×10^6 primary human mammary epithelial cells, use a final reaction volume of 3 mL and a threefold higher final PKH26 concentration (3×10^{-7} M in PBS).
13. After 7 days, these unlabeled mammospheres will be dissociated and the resulting epithelial cell population will be used as a negative control in the FACS sorting. The presence of a negative control population in each experiment, whether for normal or tumor samples, allows the comparison of the distribution of PKH26 across different experiments and samples. This is important, for instance, to compare the PKH26 distribution in a population of cells obtained from PKH26-labeled tumor

mammospheres compared to their normal counterpart from the same patient.

14. This washing step allows the removal of any residual contamination of serum proteins and lipids that, by binding the lipophilic PKH26 dye, would reduce efficiency of the labeling procedure.
15. This 2×10^{-7} M PKH26 solution represents a 2× PKH26 dye concentration in D-PBS to be used for cell staining. In fact, adding the PKH26 ethanol solution directly to cells would unavoidably result in a dramatic loss of cell viability and also in a nonuniform membrane staining.
16. Rapid and homogeneous mixing is critical to yield a uniform cell labeling since membrane incorporation of PKH26 is nearly instantaneous.
17. These washing steps are crucial to remove any residual contamination of PKH26 dye from the cell suspension.
18. In most of cases, the PKH26-labeled stem cell is centrally located in the mammosphere. The presence of several PKH26-positive cells centrally located or peripherally distributed in the context of a mammosphere suggests the presence of a cellular aggregate rather than of a functional structure. If the PKH26 labeling procedure has been performed correctly the bulk mammosphere population should display no evident, or at least a barely detectable, PKH26 staining. By contrast, in the case of mammary tumor tissues, more than one PKH26-labeled cell can be visualized in tumor mammospheres (1).
19. The limited amount of cells plated in this well will easily permit counting the number of the first generation (F1) mammospheres. Based on the expected SFE (numbers of mammospheres/number of plated cells × 100) of 0.01% associated with freshly isolated human mammary epithelial cells, the number of F1 mammospheres detected in this control well after 7 days of suspension culture should be ~9. This will also allow monitoring of the formation of cellular aggregates, which might sometimes be difficult to distinguish from true mammospheres and may therefore affect the correct interpretation of results. Of note, in case of tumor tissues, the SFE of the F1 generation may vary according to the intrinsic tumor characteristics (1).
20. For instance, the rest of the single cell suspension can be plated in several wells of a 6-well plate or in 1 (or more, if needed) 10-cm cell culture dish.
21. Pool the inguinal and thoracic mammary glands from no more than five mice per 50-mL conical tube.

22. This is a differential centrifugation step also described for the human mammary gland (see Note 4 and Subheading 3.1, step 5). In the case of the mouse mammary tissue, a layer of fat tissue is typically present in the upper part of the tube at the end of the centrifugation.
23. Do not use a Pasteur pipette hooked up to a vacuum flask.
24. Use a separate filter for each pool of mammary glands from five mice. If needed, the plunger of a 10-mL syringe can be used to press the suspension through the filter.
25. For this step, it is best to proceed as follows: (1) remove the plunger of a 50-mL syringe; (2) mount a 10-μM syringe filter on the syringe; (3) fit the syringe filter on a fresh 50-mL conical tube; (4) pour off the cell suspension into the syringe; (5) use the plunger to press the suspension through the filter.
26. Perform steps 1–10 separately for each pool of mammary glands obtained from five animals. At the end of this centrifugation step, the pellets from the different tubes are pooled together. The pellet typically contains ~2×10^7 cells.
27. After 7 days, together with primary mammospheres, there will also be present cellular aggregates and floating single cells. From this material, second-generation mammospheres will be derived. For the mouse mammary gland, this step is crucial to obtain a pure mammosphere population and eliminate cellular aggregates (see Note 28). The epithelial population obtained from the dissociation of unlabeled second-generation mammospheres will be used as a negative control in the FACS sorting (see Subheading 3.7, step 20).
28. Second-generation mammospheres are no longer contaminated by cellular aggregates and single floating cells, and are therefore suitable for subsequent FACS analysis.
29. From this step onwards, PKH26-labeled and unlabeled mammospheres are processed in parallel.
30. This differential centrifugation step allows the separation of mammospheres from single epithelial cells that have undergone apoptosis due to the anchorage-independent growth conditions.
31. We suggest transferring the 10 mL suspension to a Poly-HEMA-coated dish to check the efficiency of dissociation under an inverted microscope.
32. The specific formulation of this medium is suitable to maintain physiological pH control in CO_2-free systems. It is therefore used to help maintain cell viability in the different steps needed to prepare the cells for FACS analysis. The use of D-PBS might determine a loss of up to 30% of the material. Another crucial step is the transfer of samples to the FACS sorter, for which it

is best to use systems with a controlled 5% CO_2 atmosphere (the simplest way is to use sealed plastic bags pre-equilibrated in a 5% CO_2 humidified incubator).

33. FACS sorting is the step at which a number of factors may concur in determining loss of material and cell viability. The duration of the FACS sorting procedure should not exceed 1–1.5 h and throughout this time the material should be kept in a controlled 5% CO_2 atmosphere (see also Note 32).
34. These sorting parameters are the optimized conditions for human and mouse normal mammary stem cells. A higher sheath pressure and an acquisition velocity are not well tolerated by the fragile mammary stem cells and result in loss of viability in subsequent functional analyses.
35. The negative control is the population of epithelial cells derived from dissociation of unlabeled mammospheres.
36. The amount of $PKH26^{POS}$ cells is the limiting step at this stage. The excess $PKH26^{NEG}$ cells can be collected in separate tubes for different purposes.
37. The use of microcentrifuge tubes prefilled with MESCM allows the deflected microdroplets to land directly in the medium, rather than hit the dry wall of the collection tube.
38. Given the limited amount of microdroplets generated during the sorting procedure, it is typically sufficient to add 50 μL of MESCM to yield a final cell suspension of ~100 μL.
39. When possible, it is best to count cells, given the frequent discrepancy between the number of events recorded by the cell sorter and the actual number of cells present in each fraction. An accurate determination of the number and viability of purified cells is crucial for the assessment of the SFE associated with a given cell subpopulation.

Acknowledgments

The authors thank all members of the laboratory, past and present, especially Drs. Daniele Galvagno, Silvia Zecchini, and Simona Ronzoni for contributing to the development of the protocols and procedures described in this chapter. Also, we thank Dr. Pascale Romano for critical comments on the manuscript. Our research is supported by grants from the Associazione Italiana per la Ricerca sul Cancro (AIRC); the Italian Ministry of Education-University-Research (MIUR); the Italian Ministry of Health; the European Community (FP6 and FP7); the CARIPLO foundation; the Ferrari Foundation; the Monzino Foundation; the G. Vollaro Foundation; the European Research Council (ERC).

References

1. Pece S, Tosoni D, Confalonieri S, Mazzarol G, Vecchi M, Ronzoni S, Bernard L, Viale G, Pelicci PG, Di Fiore PP (2010) Biological and molecular heterogeneity of breast cancers correlates with their cancer stem cell content. Cell 140:62–73
2. Cicalese A, Bonizzi G, Pasi CE, Faretta M, Ronzoni S, Giulini B, Brisken C, Minucci S, Di Fiore PP, Pelicci PG (2009) The tumor suppressor p53 regulates polarity of self-renewing divisions in mammary stem cells. Cell 138:1083–1095
3. Lanzkron SM, Collector MI, Sharkis SJ (1999) Hematopoietic stem cell tracking in vivo: a comparison of short-term and long-term repopulating cells. Blood 93:1916–1921
4. Huang S, Law P, Francis K, Palsson BO, Ho AD (1999) Symmetry of initial cell divisions among primitive hematopoietic progenitors is independent of ontogenic age and regulatory molecules. Blood 94:2595–2604
5. Dontu G, Abdallah WM, Foley JM, Jackson KW, Clarke MF, Kawamura MJ, Wicha MS (2003) In vitro propagation and transcriptional profiling of human mammary stem/progenitor cells. Genes Dev 17:1253–1270
6. Pece S, Serresi M, Santolini E, Capra M, Hulleman E, Galimberti V, Zurrida S, Maisonneuve P, Viale G, Di Fiore PP (2004) Loss of negative regulation by Numb over Notch is relevant to human breast carcinogenesis. J Cell Biol 167:215–221
7. Westhoff B, Colaluca IN, D'Ario G, Donzelli M, Tosoni D, Volorio S, Pelosi G, Spaggiari L, Mazzarol G, Viale G, Pece S, Di Fiore PP (2009) Alterations of the Notch pathway in lung cancer. Proc Natl Acad Sci USA 106: 22293–22298
8. Eirew P, Stingl J, Raouf A, Turashvili G, Aparicio S, Emerman JT, Eaves CJ (2008) A method for quantifying normal human mammary epithelial stem cells with in vivo regenerative ability. Nat Med 14:1384–1389
9. Ginestier C, Hur MH, Charafe-Jauffret E, Monville F, Dutcher J, Brown M, Jacquemier J, Viens P, Kleer CG, Liu S, Schott A, Hayes D, Birnbaum D, Wicha MS, Dontu G (2007) ALDH1 is a marker of normal and malignant human mammary stem cells and a predictor of poor clinical outcome. Cell Stem Cell 1: 555–567
10. Raouf A, Zhao Y, To K, Stingl J, Delaney A, Barbara M, Iscove N, Jones S, McKinney S, Emerman J, Aparicio S, Marra M, Eaves C (2008) Transcriptome analysis of the normal human mammary cell commitment and differentiation process. Cell Stem Cell 3: 109–118

Chapter 7

Isolation and Expansion of Endothelial Progenitor Cells Derived from Mouse Embryonic Stem Cells

S. Bahram Bahrami, Mandana Veiseh, and Nancy J. Boudreau

Abstract

The unlimited differentiation and proliferation capacity of embryonic stem cells represents a great resource for regenerative medicine. Here, we describe a method for differentiating, isolating, and expanding endothelial cells (ECs) from mouse embryonic stem cells (mESCs). First, mESCs are expanded on a mouse embryonic fibroblast (mEF) feeder layer and partially differentiated into embryoid bodies (EBs) by growing the cells in an ultra-low attachment plate for up to 5 days. The EBs are then differentiated along the endothelial lineage using endothelial growth medium supplemented with 40 ng/mL vascular endothelial growth factor (VEGF). The differentiated endothelial population expresses both Fetal Liver Kinase 1 (Flk-1) and VE-Cadherin on the cell surface which can be further purified using a fluorescence-activated cell sorting (FACS) system and subsequently expanded on 0.1 % gelatin-coated plates. The differentiated cells can be analyzed by real-time PCR and flow cytometry to confirm enrichment of EC-specific genes and proteins.

Key words: Embryonic stem cells, Differentiation, FACS, Endothelial cells

1. Introduction

Embryonic stem cells (ESCs), which are derived from the inner cell mass of blastocysts, are capable of self-renewal and differentiation along different cell lineages, a property known as pluripotence (1). Because ESC differentiation can be directed into particular lineages with specialized functional properties for tissue repair and replacement, they are considered to be an excellent resource for regenerative medicine (2). Three major obstacles associated with using ESCs for regenerative medicine are: (1) precise and controlled differentiation of ESCs toward a well-defined lineage; (2) isolation of homogeneous populations of stably differentiated cells that are fully functional; and (3) retaining the expansion potential.

Kimberly A. Mace and Kristin M. Braun (eds.), *Progenitor Cells: Methods and Protocols*, Methods in Molecular Biology, vol. 916, DOI 10.1007/978-1-61779-980-8_7, © Springer Science+Business Media, LLC 2012

A common approach is to pre-differentiate ESCs into three-dimensional cell aggregates known as embryoid bodies (EBs). These bodies contain cells from all three germ layers, mesoderm, ectoderm, and endoderm, and can self-renew and differentiate into different cell types. Embryoid bodies can be produced in three main ways: (1) hanging drop; (2) methylcellulose hydrogel; and (3) suspension culture. The most convenient and efficient of these is suspension culture of ESCs in non-adherent plates (3, 4). This approach produces large amounts of EBs in a short time, which is advantageous for high-throughput drug screening and tissue engineering. However, an important consideration is embryoid body size, which may affect the outcome and properties of differentiated cells. For example, uncontrolled overgrowth of EBs may result in cavity formation due to apoptosis, after which EBs eventually become cystic and contain fluid.

Methods for developing differentiated endothelial cells from EBs are well established and the corresponding gene expression patterns are well characterized (4–6). Embryonic stem cells have been directly differentiated toward specific lineages without first making EBs by using conditioned medium for endothelial differentiation of ESC in collagen IV plates (7, 8).

Large-scale changes in gene expression accompany the initial differentiation of ESCs into EBs, and subsequent large-scale changes in gene expression are linked to lineage-specific differentiation of EBs along mesenchymal, epithelial, neural, or hematopoietic lineages (9, 10, 11). Embryoid bodies can be directed to give rise to hemangioblasts, which subsequently undergo further differentiation into either hematopoietic or endothelial cells. Hemangioblasts have been widely used to study the expression of transcription factors that control EC lineage and recapitulate many aspects of vascular development in vivo (12). The expression of vascular endothelial growth factor receptor-2 (VEGF-R2), also known as Flk-1, in mice, is a mesodermal indicator and the earliest functional marker for hemangioblasts (13).

After EBs are formed and exposed to endothelial differentiation medium containing VEGF, a heterogeneous cell population, including mesenchymal, hematopoietic, and epithelial cells, emerges. Only a subset of these cells will differentiate toward endothelial cells (less than 2 % with our current method). Therefore, once differentiation has been induced the endothelial cell population must be isolated and purified for further expansion and analysis.

Herein, we describe our methods for endothelial differentiation of mouse ESCs (mESCs) and FACS-based isolation and expansion of the resulting cell populations that express the endothelial-specific markers Flk-1 and VE-Cadherin on their surface.

2. Materials

2.1. Cells

1. *mEF cells*: Untreated *mEF cells*-C57BL/6 Cat. No. SCRC-1008, (ATTC Manassas, VA) OR primary mouse embryo fibroblasts, hygro-resistant strain C57BL/6, passage 3 (Millipore, Billerica, MA).
2. *mESCs*: mESC were gifts from the laboratory of Robert Blelloch, Eli and Edythe Broad Center of Regeneration Medicine and Stem Cell Research (University of California at San Francisco). The cells were made by targeted insertion of EGFP into the *Rosa26* locus in C57BL/6 embryonic stem cells.

2.2. Cell Culture Media

1. *mESCs and mEF media*: Dulbecco's Modified Eagle's Serum (DMEM) supplemented with 15 % fetal bovine serum (FBS) (HyClone), 2 mM L-glutamine, 0.1 mM MEM nonessential amino acids (Invitrogen), 2.5×10^{-5} M 2-Mercaptoethanol (Sigma 99 %), 1 mM MEM sodium pyruvate (Invitrogen), 100 U/mL penicillin, 100 μg/mL streptomycin, and 1,000 U/mL Esgro LIF (Chemicon, Billerica, MA) (Only for mESCs).
2. *mEF inactivation medium*: Mitomycin C (Sigma): Mix 2 mg Mitomycin C powder with 4 mL Dulbecco's Phosphate-Buffered Saline (D-PBS) (see Note 1).
3. *Freezing medium*: DMEM 60 %—FBS 20 %—Dimethyl sulfoxide (DMSO) 20 %: Add 6 mL DMSO (Sigma), 6 mL Hyclone FBS, 18 mL DMEM into the 50 mL falcon tube and mix properly (see Note 2).
4. *Endothelial differentiation medium*: Endothelial Cell Growth Medium-2 (EGM®-2, Lonza) supplemented with 40 ng/mL recombinant human vascular endothelial growth factor (rhVEGF-R&D Systems): Add 100 μL of D-PBS to 10 μg of VEGF powder. Mix the solution by pipetting up and down. Add 40 μL of VEGF solution to 100 mL of EGM-2 (see Note 3).

2.3. Solution and Culture Plates

1. *Gelatin coating solution and gelatinized plate*: 0.1 % Gelatin Solution ES Cell Qualified (Chemicon/Millipore, Embryomax): Add 3 mL of gelatin per 10 cm dish or 1 mL per 6 cm dish, and coat the entire plate by swirling the plate. Allow the plate to sit for 30 min; then aspirate excess gelatin. Culture the cells immediately before the dish becomes dry.
2. *mEF cell lysate solution*: 0.25 % trypsin with Ethylenediaminetetraacetic acid (EDTA) (1×) (Invitrogen).
3. *mESCs cell lysate solution*: Dissolve 0.5 g collagenase IV in 500 mL of Knockout D-MEM (Invitrogen) (see Note 4).

4. *Poly (2-hydroxyethyl methacrylate) (Poly-HEMA) solution*: Add 1 L of 100 % ethanol and a magnetic stirrer to an autoclaved 1 L bottle. Put the bottle on the stirrer and slowly add 25 g of Poly-HEMA (Sigma) preferably over a course of 20–30 min to avoid clumping. Stir the Poly-HEMA solution overnight.
5. *Ultra-low attachment plates*: Add 3 mL of Poly-HEMA solution per 10 cm dish or 1 mL per 6 cm dish. Coat the entire plate by gently swirling the Poly-HEMA solution over the surface. Heat the coated plates in a 45 °C oven for a minimum of 24 h. Place the plates under a UV light in a tissue culture hood with lids on. Wash the plates three times with D-PBS before use (see Note 5).

2.4. Fluorescence-Activated Cell Sorting

1. *Harvesting solution*: Use cell dissociation medium (Sigma) or 10 mM (EDTA) in D-PBS.
2. *Washing buffer*: dissolve 0.2 g sodium azide (NaN_3, Sigma) in 200 mL D-PBS and prepare 0.09 % NaN_3-PBS solution and supplement it with 5 % D-FBS.
3. *Blocking buffer*: supplement the 0.09 % NaN_3-PBS solution with 10 % FBS.
4. Phycoerythrin (PE) conjugated anti-mouse Flk-1 (VEGF-R2, eBioscience, Inc., San Diego, CA).
5. Allophycocyanin (APC) conjugated anti-mouse VE-Cadherin (CD144, eBioscience).
6. PE conjugated rat IgG2a isotype control (eBioscience).
7. APC conjugated rat IgG1 isotype control (eBioscience).

3. Methods

One method to expand mESCs is to culture them on mEF feeder layers that have been inactivated with Mitomycin C or via irradiation and with the differentiation inhibitory factor, Leukemia inhibitory factor (LIF) (14). When mEF cells are inactivated, the mESCs can attach, form clumps, and use the growth factors produced from feeder layers, while mEF cell proliferation is inhibited. Once expanded LIF and stromal contact are both withdrawn, mESCs can be grown in ultra-low attachment plates and partially differentiated to form EBs containing all germ layers. The EBs are subsequently differentiated toward the endothelial cell lineage by growing in gelatinized plates and exposing the cells to differentiation medium containing VEGF. The differentiated ECs can be selected via cell sorting with endothelial-specific cell surface antigens.

3.1. mEF Feeder Layer

3.1.1. Plating mEF Cells

1. Thaw a frozen vial of mEF cells at 37 °C until all ice is completely melted.
2. Transfer the contents to 10 mL of mEF medium in a 15 mL falcon tube and spin down for 5 min at 125×*g*, at room temperature.
3. Aspirate the supernatant and resuspend the cells in 1 mL mEF medium.
4. Determine cell viability using trypan blue stain and count the cells using a hemocytometer.
5. Plate 1×10^6 cells with >85 % viability on a 10 cm 0.1 % gelatinized plate and add 10 mL mEF medium.
6. Incubate mEF cultures at 37 °C in 5 % CO_2 and change medium every 2–3 days.

3.1.2. Splitting mEF Cells

1. Wash confluent mEF cultures twice with D-PBS.
2. Add 2 mL of 0.25 % trypsin per 10 cm plate and incubate for 4–5 min at 37 °C in a 5 % CO_2 incubator.
3. Tap the side of the plate to loosen the cells; then wash cells using 5 mL mEF medium.
4. Transfer the plate contents (cells + medium) to a 15 mL tube and spin down for 5 min at 125×*g* at room temperature.
5. Aspirate the medium and resuspend the cells in the pellet in 3 mL mEF medium.
6. Split cells at a 1:3 ratio by plating 1 mL volume of the above suspension to each of 3×10 cm gelatinized plates and add an additional 10 mL of the mEF medium to each plate.
7. Feed the mEF plate with mEF medium every 2–3 days.

3.1.3. Inactivation of mEF Cells

1. Grow the mEF cells until they become 75–85 % confluent.
2. Aspirate the mEF medium and wash the cells with D-PBS.
3. Mix 200 μL of 0.5 mg/mL Mitomycin C in 10 mL MEF medium (10 μg/mL) and transfer it to each 10 cm mEF plate.
4. Incubate the mEF plates at 37 °C and 5 % CO_2 for 2–3 h.
5. Wash the plates three times with 7–10 mL of D-PBS.
6. Add 1 mL of 0.25 % trypsin per 10 cm plate and incubate for 4–5 min at 37 °C in a 5 % CO_2 incubator.
7. Tap the side of the plates to loosen the cells; then wash cells in each plate using 3 mL mEF medium.
8. Transfer the plate contents (cells + medium) to a 15 mL tube and spin down for 5 min at 125×*g* at room temperature.
9. Aspirate the medium and resuspend the pellet in 1 mL mEF medium.

10. Culture the cells in a gelatinized plate at a density of $2–3 \times 10^6$ cells per 10 cm plate.
11. Add 10 mL mEF medium to each plate; then incubate them at 37 °C and 5 % CO_2.
12. Feed the mEF plates with mEF medium every 2–3 days (see Notes 6 and 7).

3.1.4. Freezing the mEF Cells

1. Wash cells twice with D-PBS.
2. Add 2 mL of 0.25 % trypsin per 10 cm plate and incubate for 4–5 min at 37 °C.
3. Tap the plate sides to loosen cells and wash them using 10 mL mEF medium, then transfer cells to a 15 mL tube, and spin them for 5 min at $125 \times g$.
4. Aspirate the trypsin and medium, and resuspend the cells using 0.5 mL mEF medium.
5. Transfer cells to cryo-vial/freezing vials and add 0.5 mL cold freezing medium per vial; then put the vials on ice.
6. Place vials in −80 °C freezer overnight; after 24 h, place them in liquid nitrogen.

3.2. Culturing mESCs

3.2.1. Plating mESCs on Feeder Layer

1. Thaw a frozen vial of mESC at 37 °C until ice is completely melted.
2. Transfer contents to a 15 mL falcon tube with 10 mL of mESC medium and spin down for 5 min at $125 \times g$ at room temperature.
3. Aspirate the medium and resuspend the cells in 1 mL mESC medium.
4. Count the cell numbers and determine cell viability using trypan blue stain and a hemocytometer.
5. Plate 1×10^6 cells on a 10 cm plate of inactivated mEF feeder layer and add 10 mL of mESC medium.
6. Incubate in 37 °C and 5 % CO_2 and feed mESC every day with mESC medium (see Note 8).

3.2.2. Splitting mESCs

This method is for splitting the mESCs once they have been cultured on the mEF feeder layer. The mESCs will begin to form colonies on the mEF feeder layer that can eventually merge. Because mESCs only maintain their undifferentiated state when colonies are not merged, cells must be passaged before colonies come in contact with each other.

1. Wash the mESC plates twice with D-PBS.
2. Add 5 mL 0.1 % collagenase IV per 10 cm plate and incubate for 5–10 min at 37 °C at 5 % CO_2.

3. Tap plate sides to loosen cells and wash the cells with 10 mL of mESC medium (collagenase IV primarily detaches the mESCs colonies; however some mEFs may also become detached during this treatment).
4. Transfer contents to a 15 mL tube and spin down for 5 min at 125 × *g* at room temperature.
5. Aspirate the supernatant; then resuspend the cells in 4 mL mESC medium.
6. Split the cells at a 1:4 ratio by plating 1 mL volume of the above suspension to each of 4 × 10 cm inactivated new mEF feeder layers and add 10 mL of mESC medium into each plate.
7. Feed the plates with mESC medium every day. Colonies merge together, usually between 3 and 4 days.

3.2.3. Freezing the mESCs

1. Wash cells twice with D-PBS.
2. Add 5 mL 0.1 % collagenase IV per10 cm plate and incubate for 5–10 min at 37 °C and 5 % CO_2.
3. Tap the plate sides to loosen cells and wash using 10 mL mESC medium, then transfer cells to a 15 mL tube, and spin for 5 min at 125 × *g*.
4. Aspirate the supernatant and resuspend the cells with 0.5 mL mESC medium.
5. Transfer cells to cryo-vial/freezing tubes and add 0.5 mL cold freezing medium per vial; then put them on ice.
6. Place vials in −80 °C freezer overnight; after 24 h place them in liquid nitrogen.

3.3. Preparing the Cells for Differentiation

After treating the mESCs and mEFs cells with 0.1 % collagenase IV (see Subheading 3.2.2), mESCs will primarily detach; however, some mEF cells may also detach from the plate. As only mESCs will be used for EB formation, mESCs can be separated from detached mEF cells by allowing the mEF cells to differentially adhere onto new 10 cm culture dishes.

Subculture all cells (mESC + mEF) onto a new 10 cm plate using mESC medium and incubate at 37 °C and 5 % CO_2 for 0.5–1 h. The mEF cells will attach during this time and mESCs will stay in suspension. After 0.5–1 h, transfer the non-adherent mESCs into a 15 mL falcon tube and triturate the cells using a 5 mL pipette to break down the colonies. These cells will be used to prepare EBs.

3.4. EB Formation in Suspension Culture

1. Transfer the mESC suspensions isolated by differential adhesion (see Subheading 3.3) to a 15 mL tube and spin down for 5 min at 125 × *g*.
2. Aspirate the mESC medium and gently resuspend the cells in 1 mL mEF medium.

3. Add 10 mL of mEF medium to a Poly-HEMA coated plate (ultra-low attachment plate) and transfer the suspended cells to this plate. Let the cells sit for 2 days before feeding with fresh medium.
4. Use a 10 mL tissue culture pipette to transfer suspended EBs from each 10 cm plate to a 15 mL falcon tube and let the EBs settle to bottom of tube by gravity for 10–15 min at 37 °C and 5 % CO_2.
5. Aspirate used medium down to the 2 mL mark; then add 8–10 mL fresh mEF medium.
6. Use a 10 mL tissue culture pipette to transfer contents back into the original 10 cm plates and incubate them at 37 °C and 5 % CO_2.
7. Change the medium every other day and grow the EBs for up to 5 days.
8. Switch the medium from mEF medium to EC differentiation medium a day before transferring the EBs to gelatinized plate (see Subheading 3.4) and treat the EBs with EC differentiation medium in Poly-HEMA plate overnight. The following day, transfer the EBs to gelatinized plates for use in EC differentiation (see Subheading 3.5).

3.5. EC Differentiation of mESCs

The size of EBs is an important factor that can affect properties of differentiated cells. The mean size of EBs after 5–6 days of culture in ultra-low attachment plates described above should be around 100–250 μm (approximately to 10–25 cells wide). Smaller EBs might not contain all three germ layers, whereas overgrown EBs will start undergoing apoptosis (Fig. 1).

1. Transfer the optimally sized (100–250 μm) EBs from each 10 cm Poly-HEMA plate to a 15 mL falcon tube and let the EBs settle to the bottom of the tube for 10–15 min at 37 °C and 5 % CO_2.
2. Aspirate used medium down to the 2 mL mark.
3. Plate the EBs onto 0.1 % gelatinized plates and add 10 mL fresh EC differentiation medium.
4. Replace the medium every other day and culture for 7 days.
5. EBs will attach to 0.1 % gelatinized plates usually overnight and adherent cells will grow out of the EBs after 1–2 days.

Differentiation starts when EBs form (while in suspension culture, see Subheading 3.4) and will continue after the EBs adhere to the 0.1 % gelatinized plates. The cells adhering and growing out from the attached EBs have an extended morphology at early time points (day 1–2) in gelatinized plate (Fig. 2A) and become more rounded at later time points (day 7) (Fig. 2D).

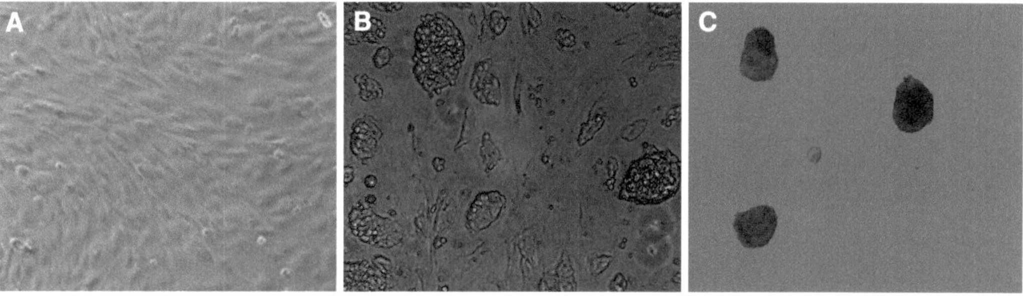

Fig. 1. Photomicrographs of (**A**) mouse embryonic fibroblasts (mEF) as feeder layer, (**B**) mouse embryonic stem cells (mESCs) cultured on the feeder layer, (**C**) embryoid bodies cultured on ultra-low attachment plates.

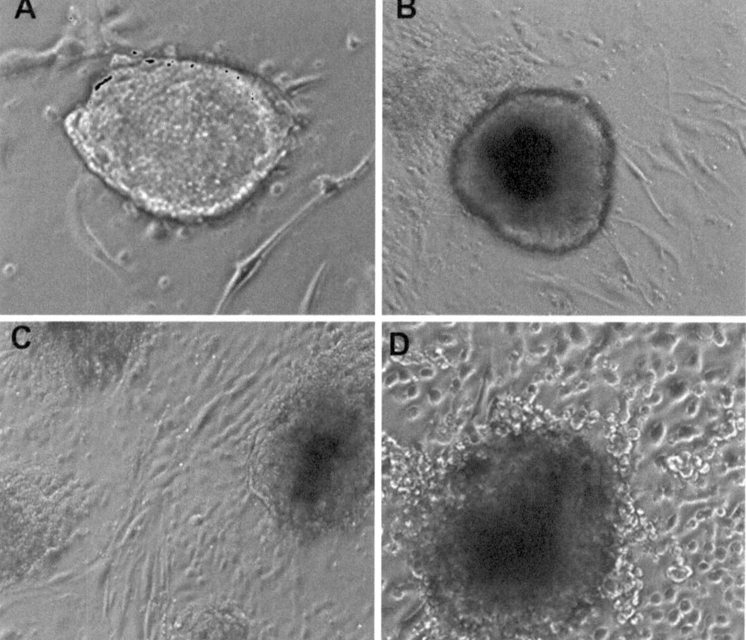

Fig. 2. Emergence of differentiating endothelial cells from mESCs cultured on gelatin-coated plates at various time points: (**A**) day 1, (**B**) day 3, (**C**) day 5, and (**D**) day 7.

3.6. Cell Sorting

3.6.1. FACS

In the early stages of differentiation, the culture conditions to promote endothelial differentiation described above give rise to a heterogeneous population with less than 2 % of the cells being pure endothelial lineage. Thus, to study the characteristics and fate of purified endothelial progenitors, this small subpopulation of cells must be isolated from the mixed cultures. FACS is a specialized type of flow cytometry for separating heterogeneous mixtures of biological cells into multiple fractions, one cell at a time, based on light scattering and fluorescent characteristics of each cell. Lasers are used to excite intrinsic or extrinsic fluorescence of cells and the fluorescence intensity is measured from cells or particles through sensitive photomultiplier tubes (15, 16).

The combination of flow cytometry and single-cell sorting is a powerful way to identify and isolate cells with particular characteristics, for instance, based upon markers expressed on the cell surface during differentiation. Control setup and data validation can sometimes be complex, especially when cells have a high fluorescence background or are transfected with a fluorescence reporter before being labeled. Nevertheless, compared to other techniques, FACS facilitates rapid data acquisition, specific multiparameter analysis, and functional separation with high accuracy (17, 18).

During the cell sorting process, a tunable transducer permits the fluid sheath to be broken into individual droplets such that each droplet encapsulates single cells. An electric circuitry places an electrical charge on the fluid stream and the individual droplets. The point at which a cell passes through the laser focus and enters into a droplet corresponds to a specific delay. Because the droplets carry a charge on their surface, a deflecting plate can redirect these charged droplets to collection tubes. Sorting criteria, region designation, multiparameter acquisition, and/or analysis are defined by a software system that includes display platforms.

An alternative to FACS system for cell sorting is magnetic cell separation. This technique is based on magnetic labeling of cells with very small microbeads that do not alter cell structure and function of cells. Separation of labeled cells takes place within a column that provides a magnetic field for cell sorting.

3.6.2. Sorting Cells with Endothelial Phenotype

As mentioned above, conditions that support endothelial differentiation of mESCs yield a heterogeneous population of cells with less than 2 % pure endothelial cells. Although hemangioblasts, which express Flk-1, can differentiate into both hematopoietic and endothelial cells, endothelial progenitor cells also express VE-Cadherin. Thus, to distinguish endothelial precursor cells from hemangioblasts, mesenchymal, epithelial, and other mixed cell types within the differentiating cultures, selecting for cells expressing both Flk-1 and VE-Cadherin is advisable (see Note 9). By labeling putative ECs with two fluorescent-tagged antibodies against Flk-1 and VE-Cadherin, ECs can be selected via FACS. We have used mESCs that were transfected with green fluorescent protein (GFP), which required initial compensation (18–20 %), to define negative populations before sorting by Flk-1 and VE-Cadherin surface markers.

Samples can then be analyzed on a Becton Dickinson FACS Vantage TM/DIVA, with a nozzle size of 80 μm, in which forward light scatter (FSC) is collected through a neutral density filter in the forward light scatter path, and side scatter (SSC) is collected through a neutral density filter at a 90° angle. The 488 nm lasers excite fluorescein isothiocyanate (FITC)/GFP and PE, while the 633 nm laser excites APC. Fluorescence emissions can be collected through the FITC (533/30 BP), PE (585/42 BP), and APC (660/20 BP) filters in fluorescence channels FL1, FL2, and FL4, respectively. Below are the steps for sample preparation for FACS sorting and expansion of sorted cells:

1. Remove the differentiation medium from 10 cm plates and wash the cells twice with D-PBS (5 plates × 10 cm should yield sufficient cell numbers for sorting and subsequent gene expression analysis of sorted cells).
2. Harvest the cells with 3–5 mL D-PBS–10 mM EDTA or cell dissociation medium and incubate for 5–10 min at 37 °C and 5 % CO_2.
3. Pipette the cells up and down with a 5 mL cell culture pipette and wash the plate to detach all the cells (use a cell scraper if needed).
4. Pass the cell suspension through a 22 g needle five times to dissociate the EBs into single cells, verified by an inverted microscope.
5. Centrifuge the cells at 4 °C, at 125 × *g* for 5 min.
6. Wash the cells with FACS washing buffer.
7. Centrifuge the cells at 4 °C and 125 × *g* for 5 min and remove the supernatant.
8. Resuspend the cells in 5 mL FACS blocking buffer and incubate them for 30 min on ice.
9. Centrifuge the cells at 4 °C, 125 × *g* for 5 min, and remove the supernatant.
10. Resuspend the cells in 500 μL blocking medium and count the cells using trypan blue stain and a hemocytometer.
11. Prepare six round-bottomed polystyrene 5-mL FACS tubes for unstained control, FLK-1-PE single color control, VE-Cadherin-APC single color control, double-stain FLK-1-PE + VE-Cadherin-APC, IgG-PE, and IgG-APC isotype controls to confirm specific antibody binding and blocking of nonspecific receptor binding (see Note 10).
12. Transfer 1–2 × 10^5 cells in the above-defined control tubes and transfer the rest of the cells to the tube for double-stained FLK-1-PE + VE-Cadherin-APC.
13. Add APC conjugated anti-mouse CD144 (VE-Cadherin) to the VE-Cadherin-APC single-control tube and add PE conjugated anti-mouse FLK-1(VEGF-R2) to the FLK-1-PE single-control tube. Add a mixture of both antibodies (1:1) to the double-stain FLK-1-PE + VE-Cadherin-APC tube. The final concentration of antibody should be 1 μg of antibody/10^6 cells in 100 μL block solution. Adjust the final concentration with blocking solution.
14. Use the unstained sample, which was not exposed to antibody, as a negative control. Use the EOMA cell line (ATTC # CRL-2586) that expresses both VE-Cadherin and Flk-1 as a positive control.

Label EOMA cells with both APC conjugated anti-mouse VE-Cadherin and PE conjugated anti-mouse Flk-1 under the same (see Note 11) labeling conditions.

15. Incubate all samples on ice for 30 min.
16. Centrifuge samples at 4 °C, 125 × *g* for 5 min, and remove the supernatant.
17. Wash the cells three times with washing solution and centrifuge to remove supernatant.
18. If cell clumps are observed, filter the cell suspension through a 40 μm cell strainer into a 5-mL round bottom cap tube to prevent blockages in the flow cytometer before sorting.
19. Keep all the samples on ice and in the dark until FACS analysis is carried out.
20. Vortex all FACS tubes for a short period of time before analysis.
21. Run unstained cells through the sorter as the first control.
22. Analyze 30,000 events for each sample.
23. Adjust the FSC and SSC settings until the cells appear in the middle of the FSC versus SSC dot plot, and exclude cell aggregate and cell debris. Then adjust the PMT voltage of the FITC, APC, and PE detectors until the cells appear within the lower quadrant of the different dot plots, thus setting the background or unstained fluorescence levels according to these parameters.
24. Run the single color sample and correct the spectral overlap using digital compensation by subtracting the spectral overlap signal of single colors from the overall signals detected in each channel.
25. Apply the correct amount of compensation to confirm that the positive population was directly horizontal or vertical to the negative population and not detectable in the other detectors. FITC compensation is not required for the APC detector, but should be checked.
26. Run the positive control and the isotype-matched antibody to verify antibody specificity.
27. After multiparameter acquisition setup for defining background/negative signal and compensation values, run double-stain FLK-1-PE + VE-Cadherin-APC sample.
28. Draw a PE versus APC two-color quadrant graph that displays the fluorescence profile of double-stained cells.
29. Select the cells residing in the double-positive FLK-1-PE/VE-Cadherin-APC quadrant (Q2) for sorting and collection (Fig. 3).
30. Using the previously defined setup, run the double-positive sample for sorting with low speed and collect the Q2 subpopulations in a FACS tube containing EC differentiation medium and 10 mg/mL gentamicin.

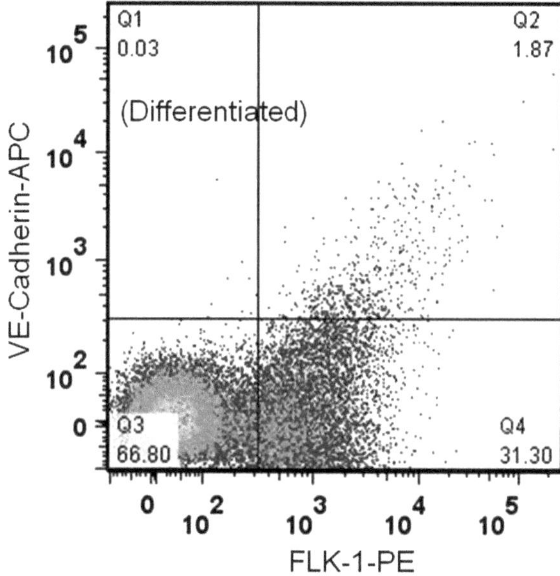

Fig. 3. The two-color fluorescence profile of differentiating ECs in ESC cultures stained for VE-Cadherin and FLK-1. The Q3 quadrant contained a negative population as determined by unstained and isotype controls. The Q1 and Q4 quadrants contained cells that only expressed VE-Cadherin and FLK-1 detected by APC conjugated anti-mouse CD144 (VE-Cadherin) and PE conjugated anti-mouse Flk-1 (VEGF-R2).

31. Centrifuge the collected subpopulations at 4 °C, 125 × *g* for 5 min, and replace the medium with fresh EC differentiation medium and 10 mg/mL gentamicin.
32. Plate the cells in a 0.1 % gelatinized plate and incubate the cells at 37 °C in 5 % CO_2. Change the medium every other day (see Note 12).

All the steps related to endothelial differentiation of mouse embryonic stem cells have been summarized in Fig. 4.

4. Notes

1. Mitomycin C solution is light sensitive. Store the mixture at 4 °C in a dark container for up to 6 weeks or at −20 °C for up to 4 months.
2. Freezing medium is light sensitive. Turn off the tissue culture hood lights during sample preparation. Aliquot the freezing medium into smaller batches (3–5 mL) and keep at 4 °C in a foil-covered container for up to 2 weeks, or store at −20 °C for up to 4 months.
3. Differentiation medium is light sensitive. Cover the medium container with aluminum foil. Also prepare fresh medium in

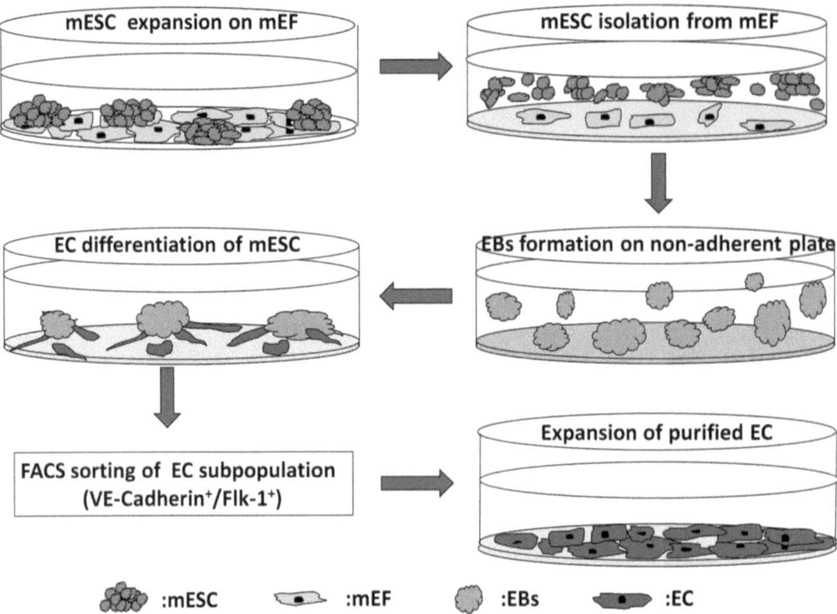

Fig. 4. Schematic diagram demonstrating all steps during EC differentiation of ESCs and expansion of purified cells. The mESCs expanded on the mEF feeder layer are first subcultured on 0.1 % gelatinized plates to separate mEFs from the mESCs before EB formation. The non-adherent mESCs are then transferred to non-adherent plates coated with Poly-HEMA; the loss of stromal contact and withdrawal of Leukemia inhibitory factor (LIF) induce formation of 3D EBs. After 5 days of culture, the optimally sized EBs are switched to EC differentiation medium and are transferred to gelatinized plates the next day. EBs are cultured on the gelatinized plates in EC differentiation medium for up to 7 additional days. FACS sorting is then used to purify the emerging endothelial subpopulation expressing the endothelial markers (Flk-1^+ and VE-Cadherin$^+$). Purified sorted ECs can then be further expanded by culturing on gelatinized coated plates.

smaller batches for each experiment and store at 4 °C for up to 4 weeks.

4. After dissolving the Collagenase IV in Knockout D-MEM medium, filter the solution using a 0.2 μm filter, then aliquot the solution into smaller batches (3–5 mL), and store at −20 °C. Samples can be kept up to 4 months.
5. To store poly-HEMA coated plates, wrap them in parafilm after exposing them to UV light and keep them at room temperature. Wash the plates three times with sterile D-PBS in a tissue culture hood before use.
6. Cultured, mitotically-inactivated mEF cells can be kept for up to 2 weeks in mEF medium while changing the medium every 2–3 days.
7. Primary mouse embryo fibroblasts, Hygro-resistant strain C57BL/6 (Millipore, Billerica, MA) also can be used as an alternative to mEF cells. These primary cells have been treated with Mitomycin-C and can be directly plated onto gelatin-coated culture vessels. $2–3 \times 10^6$ cells will produce the appropriate density feeder layer on a 10 cm plate.

8. It is important to monitor the cells and change their medium every day. mESCs can easily become contaminated.
9. One potential way of improving the percentage of ECs differentiated from ESCs is to first sort the Flk-1$^+$ cells and then continue differentiation of this subpopulation on collagen type IV coated plate prior to sorting for endothelial cells. The percentage of double-positive cells for VE-Cadherin and Flk-1 can be increased up to fivefold.
10. Perform the identical steps for staining and sample preparation on all control samples.
11. Control samples are used for defining the background/negative signal, compensation values, and multiparameter sort criteria. Sort region is defined on the double-positive quadrant of a double-stained sample.
12. The EC differentiation medium can be switched to EGM-2 medium after 1 week.

Acknowledgment

This work is supported by grants from NIH/NCI TMEN grant (U54CA126552.) to Nancy Boudreau and Mina J Bissell and U.S. Department of Energy, Office of Biological and Environmental Research (DE-AC02-05CH1123), a Distinguished Fellow Award and Low Dose Radiation Program (03-76SF00098) to Mina J. Bissell. Mandana Veiseh was supported by a postdoctoral fellowship from the NCI of the NIH (F32 CA132491A). We thank Pamela Derish in the Department of Surgery at UCSF for editorial review of the manuscript.

References

1. Doetschman TC, Eistetter H, Katz M, Schmidt W, Kemler RJ (1985) The in vitro development of blastocyst-derived embryonic stem cell lines: formation of visceral yolk sac, blood islands and myocardium. J Embryol Exp Morph 87:27–45
2. Thomson JA, Itskovitz-Eldor J, Shapiro SS, Waknitz MA, Swiergiel JJ, Marshall VS, Jones JM (1998) Embryonic stem cell lines derived from human blastocysts. Science 282:1145–1147
3. Vittet D, Prandini MH, Berthier R, Schweitzer A, Martin-Sisteron H, Uzan G, Dejana E (1996) Embryonic stem cells differentiate in vitro to endothelial cells through successive maturation steps. Blood 88:3424–3431
4. Feraud O, Cao Y, Vittet D (2001) Embryonic stem cell-derived embryoid bodies development in collagen gels recapitulates sprouting angiogenesis. Lab Invest 81:1669–1681
5. Blancas AA, Lauer NE, McCloskey KE (2008) Endothelial differentiation of embryonic stem cells. Curr Protoc Stem Cell Biol Chapter 1. Unit 1 F 5.
6. Marchetti S, Gimond C, Iljin K, Bourcier C, Alitalo K, Pouyssegur J, Pages G (2002) Endothelial cells genetically selected from differentiating mouse embryonic stem cells incorporate at sites of neovascularization in vivo. J Cell Sci 115:2075–2085
7. Nishikawa SI, Nishikawa S, Hirashima M, Matsuyoshi N, Kodama H (1998) Progressive lineage analysis by cell sorting and culture identifies FLK1 + VE-cadherin + cells at a diverging

point of endothelial and hemopoietic lineages. Development 125:1747–1757

8. Hirashima M, Kataoka H, Nishikawa S, Matsuyoshi N (1999) Maturation of embryonic stem cells into endothelial cells in an in vitro model of vasculogenesis. Blood 93: 1253–1263
9. Bahrami SB, Veiseh M, Dunn AA, Boundreau NJ (2011) Temporal changes in Hox gene expression accompany endothelial cell differentiation of embryonic stem cells. Cell Adhesion & Migration 5(2):133–141
10. Martinez-Ceballos E, Chambon P, Gudas LJ (2005) Differences in gene expression between wild type and Hoxa1 knockout embryonic stem cells after retinoic acid treatment or leukemia inhibitory factor (LIF) removal. J Biol Chem 280:16484–16498
11. Loring JF, Porter JG, Seilhammer J, Kaser MR, Wesselschmidt R (2001) A gene expression profile of embryonic stem cells and embryonic stem cell-derived neurons. Restor Neurol Neurosci 18:81–88
12. Lugus JJ, Park C, Choi K (2005) Developmental relationship between hematopoietic and endothelial cells. Immunol Res 32:57–74
13. Yamaguchi TP, Dumont DJ, Conlon RA, Breitman ML, Rossant J (1993) flk-1, an flt-related receptor tyrosine kinase is an early marker for endothelial cell precursors. Development 118:489–498
14. Williams RL, Hilton DJ, Pease S, Willson TA, Stewart CL, Gearing DP, Wagner EF, Metcalf D, Nicola NA, Gough NM (1988) Myeloid leukaemia inhibitory factor maintains the developmental potential of embryonic stem cells. Nature 336:684–687
15. Preffer F, Dombkowski D (2009) Advances in complex multiparameter flow cytometry technology: applications in stem cell research. Cytometry B Clin Cytom 76:295–314
16. Wang Y, Hammes F, De Roy K, Verstraete W, Boon N (2010) Past, present and future applications of flow cytometry in aquatic microbiology. Trends Biotechnol 28:416–424
17. Hulspas R, O'Gorman MR, Wood BL, Gratama JW, Sutherland DR (2009) Considerations for the control of background fluorescence in clinical flow cytometry. Cytometry B Clin Cytom 76:355–364
18. Lugli E, Roederer M, Cossarizza A (2010) Data analysis in flow cytometry: the future just started. Cytometry A 77:705–713

Part II

Nonmammalian Model Systems

Chapter 8

Transcriptome Analysis of *Drosophila* Neural Stem Cells

Katrina S. Gold and Andrea H. Brand

Abstract

In *Drosophila*, the central nervous system is populated by a set of asymmetrically dividing neural stem cells called neuroblasts. Neuroblasts are derived from epithelial or neuroepithelial precursors, and divide along their apico-basal axes to produce a large apical neuroblast and a smaller basal ganglion mother cell. The ganglion mother cell will divide once again to produce two post-mitotic neurons or glia. In this chapter we outline a method for labeling different types of neural precursors in the *Drosophila* central nervous system, followed by their extraction and processing for transcriptome analysis. This technique has allowed us to capture and compare the expression profiles of neuroblasts and neuroepithelial cells, resulting in the identification of key genes required for the regulation of self-renewal and differentiation.

Key words: *Drosophila*, Neural stem cell, GAL4/UAS, In situ cell picking, RT-PCR, Transcriptome analysis

1. Introduction

Understanding how a limited set of neural precursors can give rise to the complexity and diversity of the fully formed central nervous system is one of the major research topics in developmental neurobiology. A powerful tool for investigating this question is lineage analysis, in which the number and cell fate of the progeny arising from a single stem cell are determined. Lineage analysis can be combined with molecular genetic and bioinformatic techniques to elucidate the genetic networks that regulate stem cell behavior and decide cell fates.

There are two main approaches to labeling neural stem cell lineages in *Drosophila*: progenitors can either be physically labeled, or genetically marked. Methods for the former usually involve the injection of lineage tracers, such as horseradish peroxidase (1, 2), or the application of lipophilic vital dyes, such as DiI (3). More recently, a number of genetic labeling methods for lineage analysis have been developed and are still evolving (e.g., MARCM (4),

Kimberly A. Mace and Kristin M. Braun (eds.), *Progenitor Cells: Methods and Protocols*, Methods in Molecular Biology, vol. 916, DOI 10.1007/978-1-61779-980-8_8, © Springer Science+Business Media, LLC 2012

GTRACE (5), twin spot generator (6), twin spot MARCM (7); see also (8) for a brief review of some new techniques).

Lineage analysis often relies on antibody staining of fixed tissue samples. This approach has led to the identification of molecular markers for most embryonic neuroblasts, and detailed descriptions of many of their lineages (9–13). The advent of increasingly sophisticated time lapse microscopy techniques (14–18) means these kinds of data can now be complemented by live imaging to capture more dynamic aspects of stem cell behavior.

Research in our lab is focused on understanding the genetic regulation of neural stem cell self-renewal and differentiation (19–21). To this end, we have devised a method for isolating single or small groups of neural stem cells from living *Drosophila* brains, extracting total RNA, and synthesizing cDNA libraries for transcriptome analysis (22, 23). We developed this approach by combining a technique for cell extraction using a microcapillary needle (see Note 1), originally pioneered for embryonic neuroblast transplantation assays (24, 25), with a reverse transcription PCR protocol optimized for single cells (22, 26–29).

Recently, we used this method to investigate how the switch from symmetric to asymmetric neural stem cell division is regulated in the larval optic lobe (23). Symmetrically dividing neuroepithelial cells in the outer proliferation center of the optic lobe transform into asymmetrically dividing neuroblasts during larval development (Fig. 1) (30, 31). We reasoned that, as neuroepithelial cells and neuroblasts are clonally related, there might be a limited but crucial number of transcriptional changes governing the switch in division mode. In order to uncover these changes, we extracted around 50 GFP-labeled neuroepithelial cells or neuroblasts from individual late-third instar larval brains and isolated total RNA, which was reverse transcribed, amplified, and then hybridized to whole transcriptome oligonucleotide arrays (Fig. 2). Directly comparing the neuroepithelial cell and neuroblast transcriptomes enabled us to identify the Notch pathway as a key regulator of the transition from symmetric to asymmetric division (23).

The optic lobe is particularly suitable for investigating neural stem cell regulation in this manner because independent GAL4 driver lines exist for labeling the two populations of neural precursors, and the cells lie close to the brain surface, so they are easier to extract. However, although the method outlined below refers to the larval optic lobe, the technique is versatile and can easily be modified to isolate neural stem cells in different locations or at different developmental stages. For example, we have also isolated and carried out transcriptome analysis on neuroblasts from the embryonic ventral nerve cord (22) (see Note 2). In principal it could be further adapted for any progenitor, provided the cell type of interest is identifiable (either morphologically distinct or strongly and specifically labeled by a reporter gene), and accessible in situ (see Note 3).

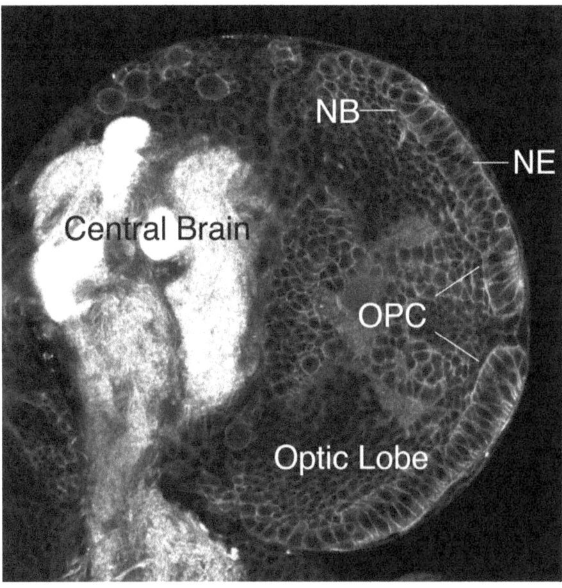

Fig. 1. Third larval instar brain lobe, consisting of the central brain and optic lobe. Cells are outlined by Dlg (Discs large) antibody staining. The outer proliferation center (OPC) of the optic lobe contains two populations of neural stem cells: symmetrically dividing neuroepithelial cells and asymmetrically dividing neuroblasts. Neuroepithelial cells (NE) transform into neuroblasts (NB) at the medial edge of the neuroepithelium. These neuroblasts give rise to the neurons of the medulla cortex, one of the visual integration centers of the adult brain. Adapted from (23).

2. Materials

2.1. Needle Preparation: Microcapillary Pulling and Beveling

1. Borosilicate glass microcapillaries, 1.0 mm outside diameter × 0.78 mm inside diameter (GC100TF-10, Harvard Apparatus, Edenbridge, UK).
2. Commercial micropipette puller (Flaming/Brown P-87 with 2.5 × 2.5 mm box filament, Sutter Instrument Company, Novato, USA).
3. Micropipette beveller (Bachhofer, Reutlingen, Germany) for sharpening microcapillaries.
4. Air-filled syringe and polyethylene tubing for controlling pressure in the microcapillary when extracting and expelling cells (25).
5. DEPC-treated ddH_2O.
6. 70% ethanol.

2.2. Drosophila Strains

Multiple GAL4 lines drive GAL4 expression in neural stem cells in the embryo and larva. There are also a variety of fluorescent UAS reporter lines for visualizing cells in living brain. Many of these can be ordered from the major fly stock centers such as the Bloomington

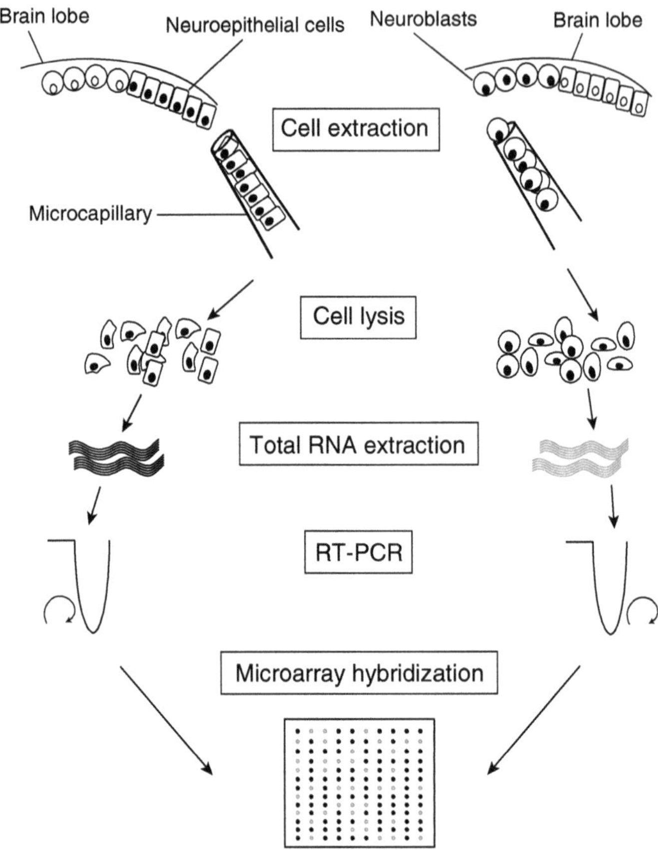

Fig. 2. Scheme showing major steps for optic lobe neural stem cell isolation and transcriptome analysis. Neuroepithelial cells or neuroblasts are genetically labeled, using different GAL4 lines. Cells are extracted from third instar larval brains using a beveled glass microcapillary. Cells are lysed in order to extract total RNA, which is reverse transcribed and amplified by PCR. The neuroepithelial and neuroblast cDNA libraries are fluorescently labeled and directly compared by hybridization to a full genome microarray. Adapted from (23).

Stock Centre (http://flystocks.bio.indiana.edu/) and Kyoto Stock Center (http://www.dgrc.kit.ac.jp/en/index.html). In order to visualize neural precursors in the optic lobe, we use two driver lines, which label the neuroepithelial and neuroblast populations distinctly: $GAL4^{c855a}$ and $GAL4^{1407}$ (*inscuteable-GAL4*) (see Note 4). We have used these to drive the expression of membrane-tethered mCD8-GFP and Histone-2B-mRFP1, so that both cell membranes and nuclei were labeled.

2.3. Staged Larval Collections

1. Fly cages.
2. Apple juice plates with wet yeast.
3. Fly food plates (35 mm Petri dishes filled with fly food), or fly food vials, with wet yeast.

2.4. Drosophila Larval Brain Dissection

1. Cold phosphate-buffered saline (PBS), pH 7.4.
2. Sharp, fine forceps (Dumont no. 5, Fine Science Tools).
3. Dissection needles (e.g., 0.4×13 mm syringe needles, BD Microlance, mounted on cotton buds).
4. Lids from plastic cell culture Petri dishes (e.g., Nunc).
5. 22×50 mm poly-L-lysine-coated coverslips (no. 1.5, VWR): prepare 20% (v/v) poly-L-lysine solution in ddH$_2$O from 0.1%w/v stock (Sigma-Aldrich). Pipette a 5 μl drop onto the center of each coverslip and leave to dry on a hot plate. Store in a coverslip rack in a dust-free container.

2.5. Cell Extraction and Lysis

1. Cell lysis buffer (945 μl): prepare stock solution in advance by mixing 100 μl 10× PCR buffer with MgCl$_2$ (Invitrogen), 10 μl NP-40 (American Bioanalytical), 50 μl 0.1 M DTT (Dithiothreitol, Invitrogen), 785 μl ddH$_2$O treated with DEPC (Diethylpyrocarbonate). Store at –20°C.
2. Cell lysis mix (50 μl): 1 μl RNase inhibitor mix (1:1 mixture of RNasin, Promega, and Stop, Flowgen Bioscience), 1 μl 10 ng/μl Anchor T amplification oligonucleotide primer (HPLC grade and resuspended in ddH$_2$O, sequence TAT AGA ATT CGC GGC CGC TCG CGA 24(T)), 1 μl 2.5 mM dNTPs (Takara), 47 μl cell lysis buffer (see Item 1 of Section 2.5). Keep on ice. This should be prepared on the day of cell extraction (see Step 3 of Section 3.3).
3. Mineral oil.
4. Inverted fluorescence microscope (Olympus 1X71, Olympus, Japan) with micromanipulator (MN-151 Joystick Micromanipulator, Narishige, Tokyo, Japan) and UV light source.
5. Glass microscope slides.

2.6. Reverse Transcription PCR

1. Reverse transcription (RT) working mix (2.5 μl): 0.3 μl Superscript II (Invitrogen), 0.1 μl RNase inhibitor mix (see Item 2 of Section 2.5), 2.1 μl lysis buffer (see Item 1 of Section 2.5). Keep on ice.
2. Poly(A) tailing reaction mix (5 μl): 0.15 μl 100 mM dATP (Promega), 0.5 μl 10× PCR buffer with MgCl$_2$ (Invitrogen), 3.85 μl DEPC-treated ddH$_2$O, 0.25 μl TdT (terminal deoxynucleotidyl transferase, Roche), and 0.25 μl RNaseH (Roche). Keep on ice.
3. PCR mix (50 μl): 5 μl 10× ExTaq buffer (Takara), 5 μl 2.5 mM dNTPs (Takara), 1 μl 1 μg/μl Anchor T primer (see Item 2 of Section 2.5), 38.5 μl DEPC-treated H$_2$O, 0.5 μl ExTaq polymerase (Takara). Keep on ice.
4. Commercial kit for PCR purification (e.g., Qiagen, Sigma).
5. Equipment for running standard DNA agarose gels.
6. Spectrophotometer.

3. Methods

3.1. Needle Preparation: Microcapillary Pulling and Beveling

One of the most important steps in this technique is the preparation of suitable needles for cell isolation. There is really no substitute for trial and error optimization, but we have laid out a few key principles below. Please note that gloves should be worn at all times when handling the microcapillaries.

1. Take borosilicate glass microcapillaries and pull them on a commercial micropipette puller (see Item 2 of Section 2.2).
2. The aim is to produce a needle with a long, narrow taper. It should be fine enough not to cause too much damage during insertion into the brain and wide enough to take up cells. The taper should not be so long that it bends rather than piercing and entering the brain (see Note 5).
3. Once the needle has been pulled, the tip should be cut back with fine forceps under a dissecting microscope. The needle diameter should be slightly larger than that of the cells you are isolating. For example, it should have an internal diameter of 12–15 μm to accommodate optic lobe neuroepithelial cells and neuroblasts.
4. Mount the needle on a commercial micropipette beveller and set the beveling angle to between 10 and 12° from the vertical. Mark the posterior upper surface of the needle with a pen, so that the needle has the correct angle when it is being fixed to the micromanipulator for cell extraction.
5. Clean the needles, first with DEPC-treated H_2O and then 70% ethanol, to remove glass shards. Store them on plasticine in a dust-free plastic container.

3.2. Staged Larval Collections

1. Set up fly cross in a cage at 25°C to obtain progeny with the appropriate genotype.
2. Collect embryos in 4-h time windows on apple juice plates with fresh yeast.
3. Remove any hatched larvae and yeast from apple juice plates at around ~22–23 h after egg laying.
4. Collect freshly hatched larvae 4 h later. Transfer from apple juice plates to fly food plates or vials.
5. Rear at 25°C to the desired stage.

3.3. Preparation of Cell Lysis Buffer and Microscope Setup

1. Fresh cell lysis mix should be prepared in an eppendorf tube before dissection begins (see Item 2 of Section 2.5), and should be kept on ice next to the microscope.
2. Take a clean, sharpened microcapillary and attach it to an air-filled syringe using polyethylene tubing.

3. Mount the microcapillary on the micromanipulator and fill with mineral oil. The easiest way to do this is to place a large drop of mineral oil onto a glass slide, lower the microcapillary into the oil, and slowly draw up the syringe plunger to take up the oil.

3.4. *Drosophila* Larval Brain Dissection

1. Pick third instar larvae from vials or food plates at the appropriate time point, using forceps.
2. Transfer the larvae to a Petri dish containing tissue paper soaked in water. Leave the larvae to crawl for a few minutes, so that they are clean of fly food and yeast.
3. Place a clean larva in a drop of PBS on a Petri dish lid.
4. Tear the larva in half using the forceps and discard the posterior half.
5. Grip the anterior end of the larva (at the mouth hooks) with one pair of forceps. Use the other pair of forceps to peel the body wall back over the mouth hooks, turning the larva inside out. Identify the brain and release it from the body wall by cutting the nerves and esophagus. Then slice off the imaginal discs using the dissecting needles, being especially careful not to damage the brain lobes when removing the eye discs.
6. Move the dissected brain to a clean drop of PBS and continue dissecting to obtain 2–3 brains in total.
7. Place a drop of PBS onto a poly-L-lysine-coated coverslip. Transfer dissected brains into the drop, using the forceps to hold them by the nerves. Orient the brains so that they are dorsal side up and push them down gently onto the coverslip. Check that they adhere securely to the poly-L-lysine.

3.5. Cell Extraction and Lysis

1. Place the coverslip on the inverted microscope stage.
2. Using Nomarski optics, focus on one of the dissected brains. We use either the 20× or 40× objectives in conjunction with 10× eyepieces.
3. Bring the needle into the same plane of focus. Take up a small volume of PBS by drawing up the syringe plunger, so that you can see a clear oil/PBS interface in the microcapillary.
4. Insert the needle into brain lobe. The easiest point of insertion is usually at the ventral-most level, since this is where the brain adheres most firmly to the coverslip (see Note 6). Once the outer glial sheath has been penetrated, slowly move the microcapillary tip to the region of interest. In the case of the optic lobe, this is the lateral portion of the brain, lying just underneath the surface.
5. Open the UV filter to visualize the exact location of the cells of interest (see Note 7). Position the needle tip next to the GFP-positive cells, and slowly draw up the syringe plunger to extract them (Fig. 3, see Note 8).

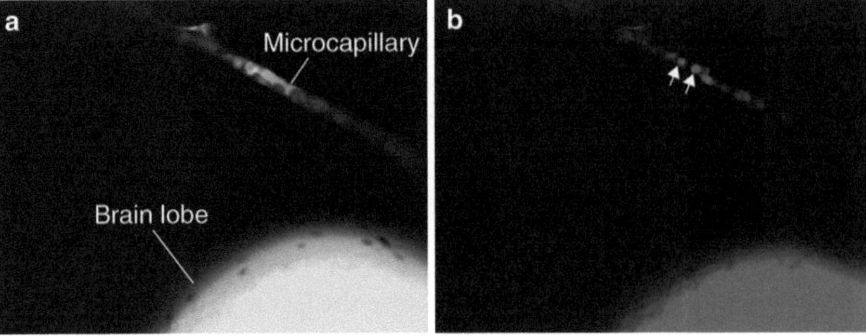

Fig. 3. Neural stem cell extraction in situ. (**a**) Neuroepithelial cells that have just been extracted from the optic lobe can be visualized inside the glass microcapillary. The cells are expressing membrane-tethered GFP and histone-bound RFP, driven by *GAL4*^{c855a}. (**b**) Individual cell nuclei, marked with Histone-2B-mRFP1, can be distinguished under UV illumination (see white *arrows*). Picture courtesy of Boris Egger.

6. Remove needle from the brain and remove coverslip from microscope.
7. Pipette 2.5 μl of ice-cold lysis mix from the eppendorf onto a clean glass slide and mount slide on microscope.
8. Expel cells from needle into the drop of cold lysis mix (see Note 9).
9. Take up 2.5 μl of the lysed material with a pipette and transfer to a PCR tube.
10. Incubate PCR tube with lysis mix at 65°C for 2 min to denature RNA.
11. After 2 min, snap cool on ice.

3.6. Reverse Transcription

1. Make up the reverse transcription mix on ice (see Item 1 of Section 2.6).
2. Carry out reverse transcription by adding 2.5 μl of RT mix to the lysed cells and incubating at 37°C for 90 min.
3. Terminate the RT reaction by heating to 65°C for 10 min, and then cool the samples to 4°C.
4. The first DNA strands synthesized by reverse transcription must be polyadenylated to allow second strand synthesis and PCR amplification. Make up TdT mix for poly(A) tailing on ice (see Item 2 of Section 2.6).
5. Add 5 μl of TdT mix to the RT reaction mixture. Place the tube in a PCR machine and incubate the reaction for 20 min at 37°C, followed by 10 min at 65°C.

3.7. PCR Amplification

1. Make up PCR mix on ice (see Item 3 of Section 2.6).
2. Add the 10 μl aliquot of polyA-tailed cDNA to the PCR mix.

3. Amplify the reverse transcribed and tailed samples using the following PCR program:
   (a) One cycle of 95°C for 1 min, 37°C for 5 min, and 72°C for 20 min.
   (b) 30–34 cycles of 95°C for 30 s, 67°C for 1 min, and 72°C for 6 min with a 6 s extension per cycle.
   (c) 72°C for 10 min.
   (d) Hold at 4°C, and store amplified samples at -20°C.
4. Run out 5 µl of the sample on an agarose gel to check quality. A homogenous smear of DNA should be visible.
5. Purify cDNA with a commercial kit, according to the manufacturer's instructions.
6. Measure cDNA concentration and check sample purity ($A_{260/280}$) with a spectrophotometer. The cDNA is now ready to be used for expression profiling (see Note 10).

4. Notes

1. We have experimented with other methods of neuroblast isolation in *Drosophila* embryos, in particular magnetic bead sorting and FACS. However, although enrichment of neuroblasts from the total cell fraction can be achieved with these methods, the mechanical shear and stress involved lead to widespread cell lysis (K. Edoff, personal communication). We find that extracting cells in situ using a microcapillary needle is faster and results in less cell death.
2. Embryonic neuroblasts can be identified by their size, shape, and position within the ventral nerve cord, in addition to reporter gene expression.
3. The proliferation of reporter lines (32, 33) and gene targeting systems in *Drosophila* (e.g., the Q system (34), LexA (35)) makes it increasingly likely that most cell types can be reproducibly identified and manipulated. Another possibility is to combine cell extraction with the MARCM system in order to isolate clones of wild-type or mutant cells.
4. Perdurance of GFP, as well as GAL4 protein, needs to be considered when choosing a driver line to label progenitors. In many cases, perdurance of these proteins can lead to the labeling of more differentiated cells further along the stem cell lineage, resulting in a less pure sample.
5. The heat, pulling strength, velocity, delay, and time are all variables which can be manipulated to produce needles with the

right kind of taper. A good guide to micropipette pulling can be found on the Sutter Instruments website (http://www.sutter.com/contact/faqs/pipette_cookbook.pdf). Prokop and Technau's chapter on cell transplantation also contains detailed information on needle preparation (25).

6. The glial sheath surrounding the larval brain can be tough and difficult to penetrate. It is worth experimenting with the taper dimensions and beveling angle to make sharper needles.
7. Try to reduce any UV-induced cell damage by extracting cells as quickly as possible to limit their exposure.
8. We extract around 50 cells per sample from the optic lobe, but our lab has also carried out single cell amplification and analysis using this protocol. However single cell data must be processed and analyzed with care, since there can be significant transcriptional variability between phenotypically identical cells (Subkhankulova et al. 2008).
9. It is possible to monitor the progress of cell lysis using the UV illumination to ensure that all the cells lyse.
10. Diagnostic PCRs can be carried out at this point to assess the likelihood of cDNA library contamination by other cell types. We recommend testing both positive and negative markers for the cell type of interest, using a low number of PCR cycles. For example, neuroblasts express genes such as *asense* and *deadpan*, but not the glial marker *repo*, while neuroepithelial cells should express epithelial markers such as *PatJ*.

Acknowledgments

The authors thank Karin Edoff, Boris Egger, Paul Wu, Adrian Carr, Hiroaki Matsunami, and Rick Livesey for sharing protocols and their considerable expertise. This work was supported by the Wellcome Trust.

References

1. Technau GM, Campos-Ortega JA (1985) Fate-mapping in wild-type Drosophila melanogaster. Dev Genes Evol 194:196–212
2. Technau GM (1987) A single cell approach to problems of cell lineage and commitment during embryogenesis of Drosophila melanogaster. Development 100:1–12
3. Bossing T, Technau GM (1994) The fate of the CNS midline progenitors in Drosophila as revealed by a new method for single cell labelling. Development 120:1895–1906
4. Lee T, Luo L (1999) Mosaic analysis with a repressible cell marker for studies of gene function in neuronal morphogenesis. Neuron 22:451–461
5. Evans CJ, Olson JM, Ngo KT, Kim E, Lee NE, Kuoy E, Patananan AN, Sitz D, Tran P, Do MT, Yackle K, Cespedes A, Hartenstein V, Call GB, Banerjee U (2009) G-TRACE: rapid Gal4-based cell lineage analysis in Drosophila. Nat Methods 6:603–605
6. Griffin R, Sustar A, Bonvin M, Binari R, del Valle Rodriguez A, Hohl AM, Bateman JR, Villalta C,

Heffern E, Grunwald D, Bakal C, Desplan C, Schubiger G, Wu CT, Perrimon N (2009) The twin spot generator for differential Drosophila lineage analysis. Nat Methods 6:600–602

7. Yu HH, Chen CH, Shi L, Huang Y, Lee T (2009) Twin-spot MARCM to reveal the developmental origin and identity of neurons. Nat Neurosci 12:947–953
8. Lee T (2009) New genetic tools for cell lineage analysis in Drosophila. Nat Methods 6:566–568
9. Doe CQ (1992) Molecular markers for identified neuroblasts and ganglion mother cells in the Drosophila central nervous system. Development 116:855–863
10. Broadus J, Skeath JB, Spana EP, Bossing T, Technau G, Doe CQ (1995) New neuroblast markers and the origin of the aCC/pCC neurons in the Drosophila central nervous system. Mech Dev 53:393–402
11. Bossing T, Udolph G, Doe CQ, Technau GM (1996) The embryonic central nervous system lineages of Drosophila melanogaster. I. Neuroblast lineages derived from the ventral half of the neuroectoderm. Dev Biol 179:41–64
12. Schmidt H, Rickert C, Bossing T, Vef O, Urban J, Technau GM (1997) The embryonic central nervous system lineages of Drosophila melanogaster. II. Neuroblast lineages derived from the dorsal part of the neuroectoderm. Dev Biol 189:186–204
13. Spindler SR, Hartenstein V (2010) The Drosophila neural lineages: a model system to study brain development and circuitry. Dev Genes Evol 220:1–10
14. Urbach R, Schnabel R, Technau GM (2003) The pattern of neuroblast formation, mitotic domains and proneural gene expression during early brain development in Drosophila. Development 130:3589–3606
15. Urbach R, Technau GM (2003) Segment polarity and DV patterning gene expression reveals segmental organization of the Drosophila brain. Development 130:3607–3620
16. Urbach R, Technau GM (2003) Molecular markers for identified neuroblasts in the developing brain of Drosophila. Development 130:3621–3637
17. Megason SG, Fraser SE (2003) Digitizing life at the level of the cell: high-performance laser-scanning microscopy and image analysis for in to imaging of development. Mech Dev 120: 1407–1420
18. Walter T, Shattuck DW, Baldock R, Bastin ME, Carpenter AE, Duce S, Ellenberg J, Fraser A, Hamilton N, Pieper S, Ragan MA, Schneider JE, Tomancak P, Heriche JK (2010) Visualization of image data from cells to organisms. Nat Methods 7:S26–41
19. Southall TD, Egger B, Gold KS, Brand AH (2008) Regulation of self-renewal and differentiation in the Drosophila nervous system. Cold Spring Harb Symp Quant Biol 73:523–528
20. Egger B, Chell JM, Brand AH (2008) Insights into neural stem cell biology from flies. Phil Trans R Soc Lond 363:39–56
21. Doe CQ (2008) Neural stem cells: balancing self-renewal with differentiation. Development 135:1575–1587
22. Choksi SP, Southall TD, Bossing T, Edoff K, de Wit E, Fischer BE, van Steensel B, Micklem G, Brand AH (2006) Prospero acts as a binary switch between self-renewal and differentiation in Drosophila neural stem cells. Dev Cell 11: 775–789
23. Egger B, Gold KS, Brand AH (2010) Notch regulates the switch from symmetric to asymmetric neural stem cell division in the Drosophila optic lobe. Development 137:2981–2987
24. Technau GM (1986) Lineage analysis of transplanted individual cells in embryos of Drosophila melanogaster. Dev Genes Evol 195: 389–398
25. Prokop A, Technau GM (1993) Cell transplantation. In: Hartley D (ed) Cellular interactions in development: a practical approach. Oxford University Press, p 33–57
26. Iscove NN, Barbara M, Gu M, Gibson M, Modi C, Winegarden N (2002) Representation is faithfully preserved in global cDNA amplified exponentially from sub-picogram quantities of mRNA. Nat Biotechnol 20:940–943
27. Osawa M, Egawa G, Mak SS, Moriyama M, Freter R, Yonetani S, Beermann F, Nishikawa S (2005) Molecular characterization of melanocyte stem cells in their niche. Development 132:5589–5599
28. Saito H, Kubota M, Roberts RW, Chi Q, Matsunami H (2004) RTP family members induce functional expression of mammalian odorant receptors. Cell 119:679–691
29. Subkhankulova T, Livesey FJ (2006) Comparative evaluation of linear and exponential amplification techniques for expression profiling at the single-cell level. Genome Biol 7:R18
30. Egger B, Boone JQ, Stevens NR, Brand AH, Doe CQ (2007) Regulation of spindle orientation and neural stem cell fate in the Drosophila optic lobe. Neural Dev 2:1
31. Hofbauer A, Campos-Ortega JA (1990) Proliferation pattern and early differentiation of the optic lobes in Drosophila melanogaster. Dev Genes Evol 198:264–274
32. Pfeiffer BD, Jenett A, Hammonds AS, Ngo TT, Misra S, Murphy C, Scully A, Carlson JW, Wan KH, Laverty TR, Mungall C, Svirskas R,

Kadonaga JT, Doe CQ, Eisen MB, Celniker SE, Rubin GM (2008) Tools for neuroanatomy and neurogenetics in Drosophila. Proc Natl Acad Sci USA 105:9715–9720

33. Pfeiffer BD, Ngo TT, Hibbard KL, Murphy C, Jenett A, Truman JW, Rubin GM (2010) Refinement of tools for targeted gene expression in Drosophila. Genetics 186:735–755

34. Potter CJ, Tasic B, Russler EV, Liang L, Luo L (2010) The Q system: a repressible binary system for transgene expression, lineage tracing, and mosaic analysis. Cell 141:536–548

35. Diegelmann S, Bate M, Landgraf M (2008) Gateway cloning vectors for the LexA-based binary expression system in Drosophila. Fly 2:236–239

Chapter 9

Live Imaging for Studying Asymmetric Cell Division in the *C. elegans* Embryo

Alexia Rabilotta, Rana Amini, and Jean-Claude Labbé

Abstract

Asymmetric cell division is essential during development to generate cell diversity and throughout adult life to maintain tissue homeostasis. For instance, many types of stem cells must divide asymmetrically to maintain their self-renewal capacities. Furthermore, recent studies suggest that the loss of asymmetric division could be used by cancer stem cells to trigger excessive proliferation of undifferentiated cells during tumorigenesis. The embryo of the nematode *Caenorhabditis elegans* is a simple and powerful model to study asymmetric cell division. After fertilization, the zygote undergoes a series of symmetric and asymmetric divisions regulated by highly reproducible events that can be followed and quantified by real-time microscopy. Deciphering the pathways involved in the control of asymmetric division in *C. elegans* embryos could lead to a better understanding of this process in stem cells and to more specific therapeutic approaches for certain human cancers.

Key words: Live imaging, *C. elegans*, DIC microscopy, Confocal microscopy, Asymmetric cell division, PAR proteins

1. Introduction

Asymmetric cell division is a process by which a single mother cell divides to generate two different daughter cells with distinct fates. It relies on the distribution of cell fate determinants along an axis of polarity that is established within the mother cell prior to its division. After division, each daughter cell inherits these determinants specifically. During development, this type of division is essential for generating cell diversity, allowing the formation of the different tissues in the organism (1). In adults, asymmetric division is essential to keep an adequate balance between self-renewal and differentiation of stem cells. In addition to its role in embryonic and adult development, asymmetric cell division could also be involved in tumorigenesis (2). Recent studies suggest that cancer

Kimberly A. Mace and Kristin M. Braun (eds.), *Progenitor Cells: Methods and Protocols*, Methods in Molecular Biology, vol. 916, DOI 10.1007/978-1-61779-980-8_9, © Springer Science+Business Media, LLC 2012

stem cells can lose their ability to divide asymmetrically and undergo excessive proliferation through symmetrical and self-renewing divisions (3, 4). This deregulation could drive the continuous expansion of malignant cells that form some tumors.

Caenorhabditis elegans is an outstanding model to study cell division since it allows visualization, by different imaging techniques, of the key features of embryonic symmetric and asymmetric divisions with high spatial and temporal resolution. The stereotypical cellular events taking place during early embryonic development in *C. elegans* have been well described (Fig. 1). After fertilization and completion of maternal meiosis, the sperm-associated centrosomes trigger cytoplasmic flows, an intracellular motility process involving actomyosin-dependent cortical rearrangements in the zygote. During these cytoplasmic flows, the male and female pronuclei travel towards each other and meet at the posterior end of the embryo. After meeting, the pronuclei move to the center of the zygote. The spindle forms at the center of the cell and undergoes displacement towards the posterior during metaphase and anaphase, resulting in two daughters of different sizes and fates after cytokinesis. The next division of the two blastomeres is both asynchronous and asymmetric: the larger, anterior cell divides ~2 min before the smaller, posterior cell, and the spindles are oriented perpendicular to each other. The anterior cells give rise to all of the ectoderm and some mesoderm, while the posterior cells are precursors of all of the endoderm and the germline.

The short life cycle of *C. elegans* (~48 h at 25°C) facilitates genetic studies and the function of one or several genes of interest can be assessed rapidly using double-stranded RNA-mediated interference (RNAi). The first asymmetric divisions of the embryo depend on asymmetric localization of the PAR proteins along an anterior–posterior axis of polarity (5). PAR proteins are conserved in other species including humans, where they are essential for maintaining epithelial cell polarity (6, 7). In the fertilized *C. elegans* embryo, they form two mutually exclusive cortical groups: the anterior complex composed of PAR-3, PAR-6, and PKC-3 (hereafter referred to as the PAR-3 complex) and the posterior group with PAR-2 and PAR-1. The loss of any PAR gene leads to a loss of polarity and to embryonic lethality (5, 8), with defects such as an abnormal first symmetric division, a central rather than a posterior mitotic spindle, mis-segregation of cell fate determinants, and deregulation of cell cycle timings during the second mitosis. The polarity phenotypes associated with asymmetric cell division defects can be monitored by filming embryonic divisions by Differential Interference Contrast (DIC) microscopy or by fluorescence microscopy, monitoring the localization of specific polarity markers fused to a fluorescent protein and transgenically expressed in embryos (Fig. 1). Here we describe methods for sample preparation and time-lapse analysis to monitor asymmetric divisions during *C. elegans* embryonic development and discuss the advantages/disadvantages of these different imaging methods.

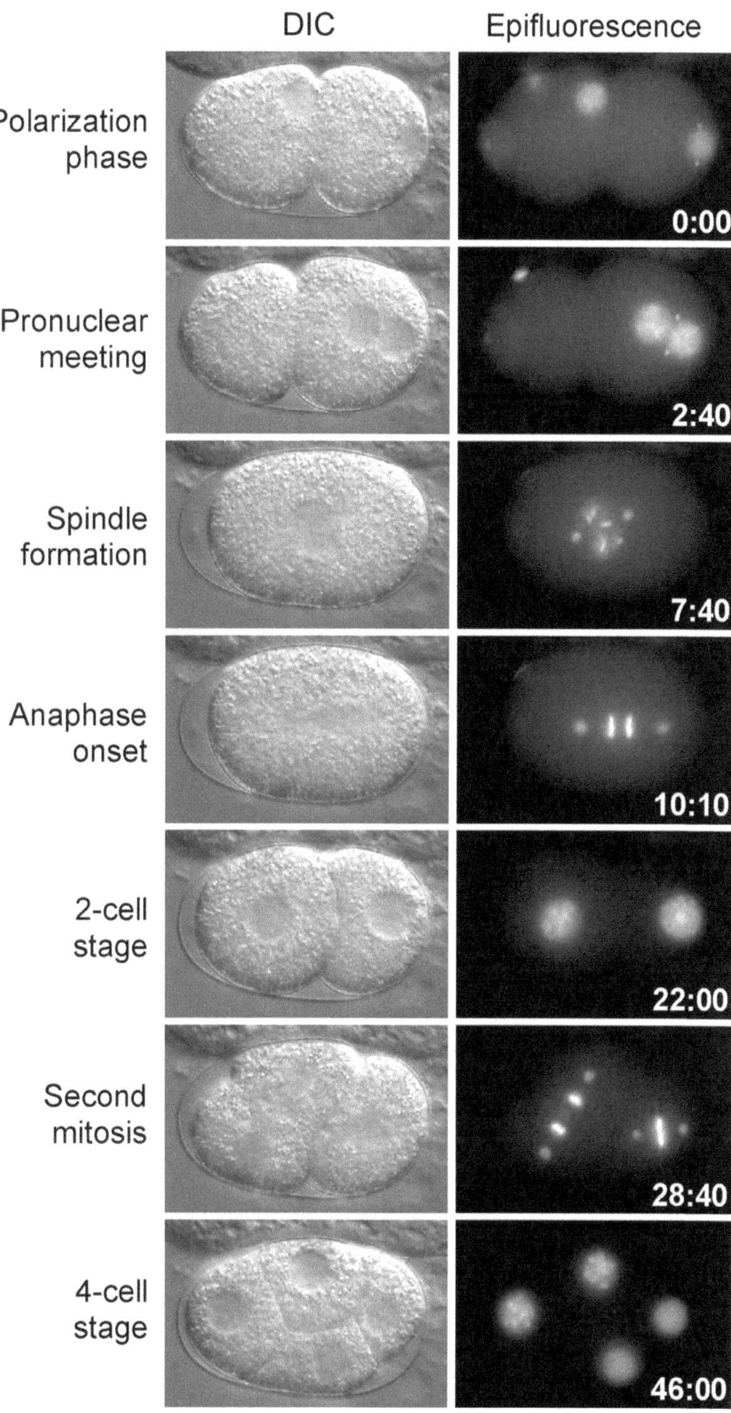

Fig. 1. Time-lapse analysis of *C. elegans* early development. The first two cell divisions of a transgenic embryo co-expressing γ-Tubulin::GFP and Histone H2B::GFP was imaged by DIC and epifluorescence microscopy. In this case, the embryo was mounted using the bulk manipulation method described in Subheading 3.2.1. Images were acquired by a Zeiss HRM camera mounted on a Zeiss Axio-Imager Z1 microscope, and the acquisition system was controlled by Axiovision software. DIC and epifluorescence images were sequentially acquired at 20 s intervals using a Plan Apochromat 63X/1.4 NA objective. Image manipulation and analysis were performed using ImageJ software. In all frames, anterior is to the left and the time indicated (min:sec) is from the start of the acquisition. The embryo is ~50 μm in length.

2. Materials

2.1. NGM Plates

1. Sodium chloride (NaCl).
2. Agar.
3. Bacto-Peptone.
4. 5 mg/ml cholesterol in ethanol (Do not autoclave!).
5. 1 M Phosphate buffer (KPO_4) pH 6.0.
6. 1 M calcium chloride ($CaCl_2$).
7. 1 M magnesium sulfate ($MgSO_4$).
8. Lysogeny broth (LB) culture medium.
9. Petri dishes (they vary in size).

2.2. RNAi Feeding Plates

1. Isopropyl b-D-thiogalactopyranoside (IPTG, 1 mM final concentration).
2. Carbenicillin (25 μg/ml final concentration).

2.3. Worm Dissection and Mounting

1. Egg Buffer (filter-sterilized): 48 mM KCl, 2 mM $MgCl_2$, 2 mM $CaCl_2$, 25 mM HEPES (pH 7.4).
2. Agarose 2%: Dissolve and boil 1 g of agarose in 50 ml of deionized water. Aliquot solution in small eppendorf tubes and melt in heating block as needed.
3. Poly-L-Lysine 0.1%: Dissolve 50 mg of Poly-L-Lysine in 50 ml of deionized water. The diluted Poly-L-Lysine solution can be stored at 4°C for several months.
4. Vaseline.
5. Heat block (70°C).
6. Glass slides (25 × 75 mm).
7. Optically flat coverslips (18 × 18 mm, no.1).
8. Glass capillaries (50 μl) and mouth pipette.
9. Depression slide (should be 25 × 75 mm and approximately 1.5 mm thick).
10. 25-gauge needles.
11. Dissection stereomicroscope: *C. elegans* can be visualized using a stereomicroscope equipped with a transmitted light source. Magnification should range from 6× to 50×.

3. Methods

3.1. Nematode Growth and Maintenance

C. elegans is typically grown on Petri dishes filled with Nematode Growth Medium (NGM) and inoculated with *E. coli* strain OP50 as food source. Worms can be maintained on these plates for several days.

1. Pouring NGM plates: Mix 3 g NaCl, 17 g agar, and 2.5 g peptone in a 2-L bottle. Add 975 ml of H_2O. Autoclave for 50 min. Cool medium in a 55°C water bath for 30 min. Add 1 ml of 1 M $CaCl_2$, 1 ml 5 mg/ml cholesterol in ethanol, 1 ml of 1 M $MgSO_4$, and 25 ml of 1 M phosphate buffer. Swirl to mix well. Using sterile procedures, dispense the NGM medium into Petri dishes (for instance using a peristaltic pump). Fill plates 2/3 full of agar. Leave plates at room temperature for 2–3 days before use to let them dry.
2. Seeding NGM plates: Inoculate sterile LB medium with *E. coli* strain OP50 and grow the culture overnight at 37°C. Dispense 1–4 drops (50–200 µl) of bacterial culture on each plate and put the plates at room temperature or 37°C for 24 h. Plates can be used immediately. Both the seeded plates and the *E. coli* culture can be stored at 4°C for up to 1 month.

3.2. RNAi

One effective technique to study gene function in *C. elegans* is RNAi. RNAi is a method that relies on introducing double-stranded RNA (dsRNA) into worms, thus activating an enzymatic cascade that results in specific depletion of an endogenous gene product with corresponding sequence to the dsRNA (9). RNAi can be used to provide useful information about the loss-of-function phenotype of an inactivated gene when mutant alleles do not exist. RNAi can be performed in *C. elegans* by injecting dsRNA into the gonad or gut of the animal, by soaking worms in a solution containing dsRNA or by feeding animals with *E. coli* expressing dsRNA. The latter method is particularly relevant since it is cost-effective and a bacterial collection containing ~85% of all predicted *C. elegans* open reading frames is available (10). Here we describe a short protocol to perform feeding RNAi.

3.2.1. RNAi Feeding Plates

Prepare NGM plates as described in Subheading 3.1 and add the following ingredients after autoclaving but before pouring: Isopropyl b-D-thiogalactopyranoside (IPTG, 1 mM final concentration) and carbenicillin (25 µg/ml final concentration). Plates can be used 2 days after pouring or can be stored at 4°C for up to 7 days.

3.2.2. Procedure

1. Day 0: Inoculate 1 ml of LB medium containing 100 µg/ml of ampicillin with *E. coli* strain HT115 transformed with a plasmid in which a sequence corresponding to a gene of interest has been inserted between two T7 promoters in reverse orientation (e.g., plasmid L4440; (10)). Incubate culture overnight with agitation at 37°C.
2. Day 1: Dilute the overnight, stationary bacterial culture 1:100 in 1 ml of fresh LB medium containing 100 µg/ml of ampicillin and incubate at 37°C with agitation until the culture is in early log phase (typically 2–3 h). Dispense 1–4 drops (50–200 µl) of early log phase bacterial culture on NGM plates containing IPTG

and carbenicillin and incubate the plates overnight at room temperature or at 37°C. This will induce production of dsRNA.

3. Day 2: Transfer L3–L4 stage hermaphrodites onto the RNAi-potent plates and incubate at the appropriate temperature (15–25°C). The choice of temperature will determine the time of embryo production, and thus the time of RNAi feeding. Plates are typically incubated 24–48 h at the appropriate temperature.
4. Day 3–4: When all animals have started laying eggs, the embryos can be extracted from their mothers and their phenotype can be analyzed by time-lapse microscopy.

3.3. Preparation for Dissection and Mounting

There are two main techniques that can be used to mount embryos for live imaging. The first method consists in bulk manipulation of the animals, whereas the second one relies on individual manipulation of the embryos. While they are described separately below, both methods require similar material preparation.

1. To prevent their displacement in the buffer while filming, the embryos are placed on a Poly-L-Lysine-coated coverslip to which they will attach. These coverslips are prepared before dissection. For this, place six coverslips side-to-side on a single glass slide by adding a small drop of water between the slide and each coverslip. Put the slide with the coverslips directly on a heating block at 70°C and add 5 μl of 0.1% Poly-L-Lysine on each coverslip. Rapidly spread the Poly-L-Lysine on each coverslip using a micropipette tip until it evaporates completely, leading to a homogenous thin layer of Poly-L-Lysine on their surface. The coated coverslips can be stored at room temperature for several months and used as needed.
2. A thin layer of agarose is needed to compensate for the compression forces that are applied on the mounted embryo. The agarose is kept melted in an eppendorf tube placed in a heating block at 70°C. To prepare the pad, take two glass slides and stick pieces of tape along their length. Put a fresh glass slide between the two taped ones so that the three slides are positioned parallel to each other along their length. Add one drop (50 μl) of melted 2% agarose in the center of the fresh glass slide and rapidly cover the agarose with another fresh glass slide positioned perpendicular to the three other slides, forming a cross between the two fresh slides. The pieces of tape on the exterior slides create a difference in height that will allow formation of a thin layer of agarose. The agarose pad is now ready to be used and can be recovered by gently gliding the two fresh glass slides along each other. Recovering of the agarose pad should be done only after the embryos have been mounted on the coverslip to prevent evaporation and drying. The agarose pad should be made fresh every time.

3. Heated Vaseline is used to seal the coverslip. This is important to prevent evaporation of the medium containing the embryos and displacement of the coverslip on the agarose pad while filming. Vaseline can be put in an eppendorf tube and kept heated at 70°C before and during dissection.

3.4. Mounting Embryos

3.4.1. Bulk Manipulation

This technique has the advantage of being easier for preparation of the embryos but does not allow to sort and film desired embryos in one single imaging field, as is the case for the individual manipulation technique described in Subheading 3.4.2.

1. Transferring the worms: Put a small drop (2–4 μl) of egg buffer on a Poly-L-Lysine-coated coverslip and transfer worms in the buffer. The number of worms to be transferred should be determined according to the number of animals available for dissection. Typically, 2–3 worms should be sufficient to find at least one early embryo that has yet to undergo its first asymmetric division.
2. Cutting the worms: Use two small needles to cut each worm, by placing the needles' pointed ends on each side of the worm and sectioning. Cutting a worm both at its neck and tail will allow younger embryos to come out of the uterus, increasing the possibility to get them before or during early polarization, whereas cutting them at their middle part will release older embryos first.
3. Assembly of the coverslip with the agarose pad and sealing with Vaseline: Recover the agarose pad by gently gliding the two glass slides along each other. Take the coverslip containing the embryos between thumb and index, flip it upside down, and gently put it on the agarose pad. Seal the coverslip with melted Vaseline, making sure that the slide is on a hard surface in order to diffuse the heat from the warm Vaseline and thus avoid damaging the embryos. The slide is now ready for live imaging of the embryos.

3.4.2. Individual Manipulation

This technique allows the individual manipulation and mounting of selected embryos. The major advantage is to increase the number of desired embryos in one field of view.

1. Glass capillary: Embryos can be individually manipulated with a capillary mounted on a plastic mouth pipette. Hold the tips of a 50 μl glass capillary between your hands, heat the middle part in a flame, and pull until it separates. Break the elongated tip of the capillary with your fingers to make an opening of a diameter large enough to let one embryo pass easily through the capillary. The size of the opening can be monitored under a dissection stereomicroscope. The capillary can be used with the plastic mouth pipette to manipulate embryos individually.

2. Transferring and cutting the worms: Add 50–70 μl of egg buffer in a depression slide. Transfer 4–6 worms to the buffer on the depression slide and cut them open to release the embryos, using 25-gauge needles (see Subheading 3.4.1).
3. Aspiration and disposition of cell stage specific embryos: Put a small drop (2–4 μl) of egg buffer on a Poly-L-Lysine-coated coverslip. Using the mouth pipette and capillary, carefully aspirate one embryo and move it to the coverslip. The selected embryos should be at an early stage of development, before the pronuclei have met. The mouth pipette can be used to control the deposition of the embryo. Repeat this step for as many embryos as needed and use the glass pipette to position them all in a compact area on the coverslip.
4. Assembly of the coverslip with the agarose pad and sealing with Vaseline: Recover the agarose pad, place the coverslip on the pad, and seal with Vaseline as described in Subheading 3.4.1.

3.5. Imaging *C. elegans* Embryos

The choice of specific acquisition parameters to image *C. elegans* early embryonic development depends on the question asked. Considerations must be given to the duration and the speed of the biological process that is followed in order to capture each aspect of it. For fluorescence imaging, further considerations should be given to avoid photoxicity and excessive photobleaching of the signal over time. To this end, light intensity, exposure time, acquisition speed, and detector gain should be optimized to take these considerations into account while maximizing image resolution. To quantify asymmetric division of the one-cell embryo, the embryos selected for imaging should be at an early stage of development, before the pronuclei have met. This will ensure that all of the major events of asymmetric cell division can be properly recorded.

The next sections will describe three common types of imaging for time-lapse analysis: DIC, epifluorescence, and confocal microscopy. The microscopes should be fitted with the proper filters for the chosen imaging method, a high-resolution camera, and software that controls the different microscope's components and that permits time-lapse imaging, capturing images at a precise interval time during a defined time period. While imaging can be carried out in a single plane of focus, motorized stages allow acquisitions in multiple planes (Z sections, see below), thus permitting the analysis of cellular events in three dimensions.

3.5.1. DIC Microscopy

The *C. elegans* embryo, which shows poor contrast under bright light, appears clear with enhanced contrast under DIC microscopy, and thus it is possible to visualize cellular structures such as yolk granules, nuclei, centrosomes, and mitotic spindles (Fig. 1). DIC microscopy is therefore a great option to follow many events during early *C. elegans* embryonic divisions, including cortical contractions, movements and positioning of nuclei, centrosomes, and spindle, cell cycle timings, and cell movements.

DIC microscopy relies on the principle of interferometry and detects the interference created by the superposition of two waves. The polarized light source, which comes from a tungsten–halogen lamp, is divided by a first prism into two orthogonally polarized rays that are then focused on the specimen by a condenser lens. In the specimen, the two rays are sheared and take different optical path lengths through areas that differ in refractive index and geometry. The change in optical path causes a change in phase that is translated by the filters into a change in amplitude, visible by eye, after recombination of the two rays. The image produced appears black to white on a gray background and the contrast is proportional to the path length gradient along the sheared direction. The pseudo three-dimensional effect of depth in a specimen when visualized under DIC optics corresponds to variations in optical density of the sample, and is thus not a true representation of the geometry of the specimen.

The mounted embryos can typically be visualized with a 63× or 100× objective, both of which provide enough magnification to observe subcellular structures. The light intensity and time of exposure can be set empirically to provide quality images, and the time interval between time-lapse acquisitions should be determined based on the required time resolution. For example, the first embryonic division occurs ~40 min after fertilization and the second division occurs 15 min after the first one (Fig. 1). A good time interval to monitor embryonic events can range from 5 to 30 s, while rapid, transient events such as cortical contractions require shorter time intervals. The use of a software-controlled shutter will allow illumination of the specimen only when the camera takes a picture, and will thus avoid overheating of the embryo (see Notes 1 and 2). Furthermore, the plane of focus can drift during the time-lapse capture, and therefore the focal plane can be readjusted on the specimen between acquisitions.

3.5.2. Fluorescence Live Imaging

Fluorescence microscopy is a useful technique to directly monitor the dynamic localization of fluorescently tagged proteins during asymmetric cell division in *C. elegans*. The tagged proteins can either be distributed asymmetrically, and thus report on polarization and asymmetries, or localize to discrete cellular structures, and thus allow monitoring of distinct cellular events.

Fluorescence microscopy relies on the light excitation of fluorophores in a sample. The light source can be a mercury lamp (for epifluorescence) or laser lines (for confocal). Fluorophores have distinct light excitation and emission properties, and the excitation and emission wavelengths must be accurately controlled by use of filters.

Illumination of live specimen should be minimized to decrease phototoxicity and photobleaching, and therefore light intensity should be kept as low as possible (see Note 1). As such, collection of the light emitted by the specimen should be maximized. A high numerical aperture on the microscope objective as well as high-quality fluorescence filters will increase light collection. The camera used for collection is also critical. In this respect, high-resolution charge-coupled device (CCD) cameras provide a high signal-to-noise ratio for fluorescence imaging.

3.5.3. Epifluorescence Microscopy

Epifluorescence microscopy is a widely used technique in live cell imaging. In this technique, the excitatory light is generated by a xenon arc (or a mercury-vapor) lamp and is focused onto the specimen by the microscope objective, which then collects the light emitted by specimen. Both excitation and emission lights are sorted out by dichroic filters, which select specific wavelengths. The mounted embryos can typically be visualized with a 63× objective with high numerical aperture. Light intensity can be controlled by addition of neutral-density filters between the light source and the objective, and the time of exposure and frequency of time-lapse acquisition should be optimized to maximize observations and minimize phototoxicity and photobleaching (Fig. 1). These parameters vary depending on the levels of the fluorescent protein and should be determined empirically.

The main advantage of epifluorescence microscopy is that the system allows for rapid and efficient collection of emitted light. However, one major disadvantage is that it does not exclude emitted light that comes from fluorophores positioned in other focal planes, leading to an increase in background fluorescence and decrease in signal-to-noise ratio. These parameters can be better controlled using confocal imaging systems.

3.5.4. Confocal Microscopy

In contrast to epifluorescence microscopy, in which the fluorescence emitted from the entire specimen is collected, confocal microscopy relies on the presence of a pinhole that excludes out-of-focus fluorescence and only allows collection of light emitted from the plane of focus (see Note 3). This results in an increase in image resolution but requires laser illumination of the specimen. Light intensity can be controlled by regulating laser power. Mounted embryos can be observed as described for epifluorescence microscopy, and the time of exposure and frequency of time-lapse acquisition should again be optimized to maximize observations and minimize phototoxicity and photobleaching (Fig. 2). Several systems can be used to perform time-lapse confocal imaging of *C. elegans* embryos. We will briefly introduce three of these systems.

1. Laser-Scanning Confocal (LSC) Microscopy: LSC microscopy relies on laser excitation of a single focal point on the specimen. Oscillating mirrors allow a positional control of the illuminating laser light, eventually permitting the sequential excitation of all regions of the specimen. The emitted fluorescence

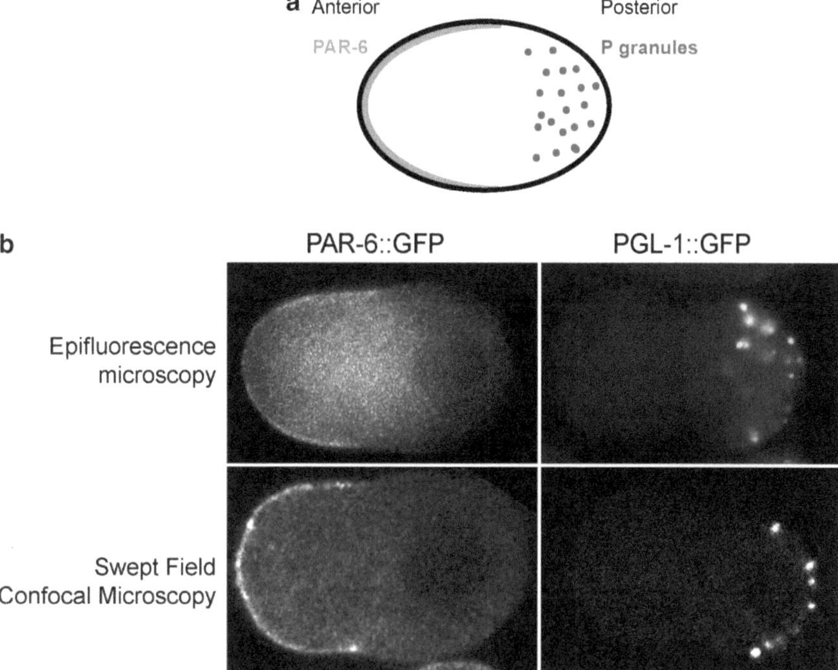

Fig. 2. Localization of polarity markers in the *C. elegans* embryo. (**a**) Schematic representation of a polarized one-cell *C. elegans* embryo. PAR-6 localizes to the anterior cortex while P granules are found in the posterior pole of the embryo. (**b**) Embryos expressing PAR-6::GFP or PGL-1::GFP (a P granule component) were imaged by epifluorescence (*top panels*) and swept field confocal microscopy (*bottom panels*).

is collected by a detector that allows digital reconstitution of the image. The advantage of this imaging system is that it allows a precise control of the region to be illuminated on the specimen. This, for instance, is useful when performing fluorescence recovery after photobleaching (FRAP) experiments. However, the major disadvantage of LSC microscopy is that scanning of the specimen can be slow, especially if the fluorescent signal is low. Given the rapid pace of embryonic development in *C. elegans*, this can result in important limitations in acquisition speed.

2. Spinning-disk confocal (SDC) microscopy: SDC microscopy is a much faster method to image the rapid events that occur during early *C. elegans* embryonic development. Whereas LSC microscopy is speed-limited by the time taken by the single laser beam to scan the specimen, SDC microscopy relies on a faster, multi-beam scanning approach. The excitation laser beam is initially diffracted into multiple beams that are refocused on a rotating disk containing 20,000 pinholes, with 1,000 of them illuminating the specimen at any time. Rapid rotation of the disk and the spatial arrangement of the pinholes allow the continuous illumination of the entire specimen, and emitted fluorescence is collected by a digital camera.

The main advantage of SFC microscopy is that it allows very fast acquisition of individual images, which is well adapted to the rapid pace of *C. elegans* embryonic development. One disadvantage of this method is that the size of the pinholes cannot be modified, which sets the focal depth at ~0.7 μm.

3. Swept Field confocal (SFC) microscopy: Like SDC microscopy, SFC microscopy allows fast acquisition of images and is also well adapted to imaging *C. elegans* early embryonic development. In SFC microscopy, the excitation laser is initially diffracted in multiple beams and these are refocused into pinholes that are arranged in a linear manner. Piezo-controlled mirrors provide a rapid and efficient system to continuously sweep the specimen with the laser beams, and emitted fluorescence is collected by a digital camera. One advantage of SFC microscopy is that the size of the pinholes can be modified, allowing good control of the optical resolution (Fig. 2). The pinholes can also be replaced with slits of varying sizes, thus increasing the speed of acquisition.

3.6. Image Analysis

Asymmetric division of the *C. elegans* embryo can be monitored visually, either by DIC and/or by following protein markers fused to fluorophores (Fig. 1). However, like most biological processes, quantification of various parameters should be sought to monitor if polarization and asymmetric division occurred normally. Quantification can be done using a variety of analysis softwares, including ImageJ, a freeware developed by the US National Institutes of Health.

Several parameters can be measured on embryos imaged by DIC. For instance, spindle position can be quantified by measuring the distance between the anterior cortex and both centrosomes at the first telophase (or at any other time, Fig. 3). Likewise, blastomere size can be easily quantified by measuring the size of each cell at the two-cell stage (Fig. 3). Other reporters of asymmetric cell division can be measured in the two blastomeres during second mitosis: the two spindle are oriented perpendicular to each other and the two blastomeres typically divide ~2 min apart.

Fluorescence time-lapse imaging provides a mean to quantify the position of markers that normally localize asymmetrically in the dividing embryo. These markers include cortically localized PAR proteins (such as anterior PAR-6 and posterior PAR-2), MEX-5 (which localizes to the anterior cytoplasm), and PIE-1 and P granules (which localize to the posterior cytoplasm) (5). Localization of these markers fused to fluorescent proteins can be done by quantifying their fluorescence intensity along the antero-posterior axis of the embryo and thus determining their asymmetric enrichment (Fig. 3). Other markers, such as tubulin or histones, can also be imaged when fused to fluorescent proteins (Fig. 1) and used to quantify various aspects of cell cycle progression during asymmetric cell division.

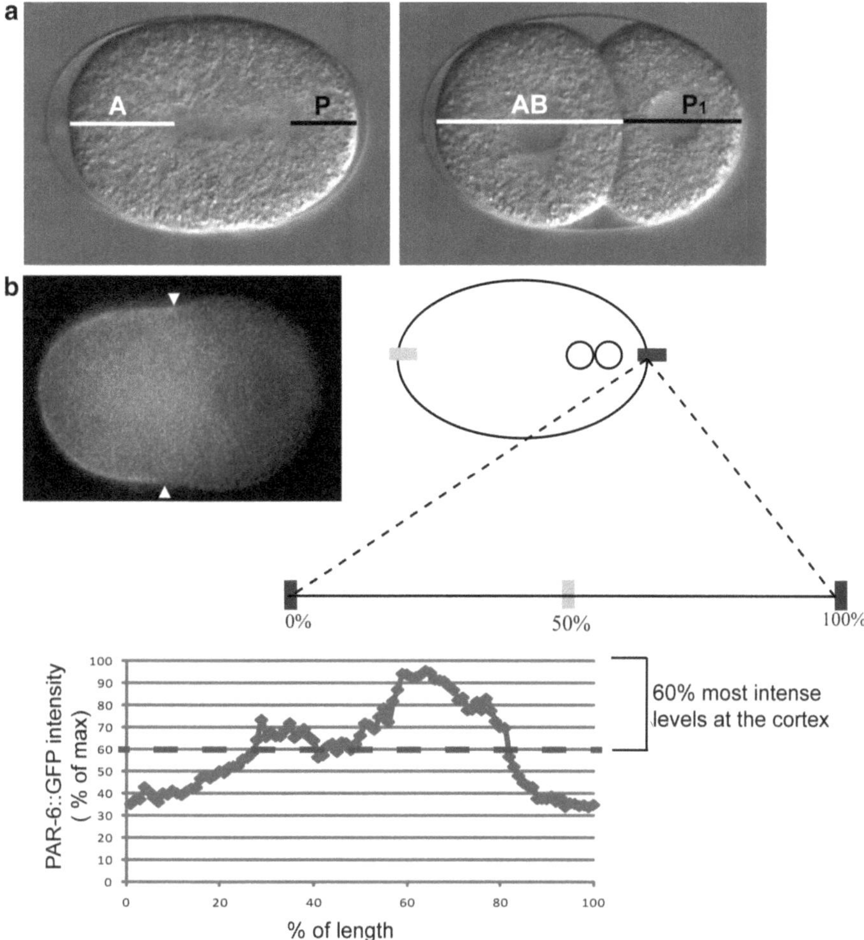

Fig. 3. Measuring asymmetric division of the *C. elegans* embryo. (**a**) DIC images of developing embryos allow quantification of posterior spindle positioning, by measuring the distance between the cell cortex and the anterior or posterior centrosome (A/P), and cell size, by measuring the distance between cell cortices at the two-cell stage (AB/P_1). (**b**) Fluorescence images of developing embryos allow quantification of the position of markers in the cell. For instance, for PAR-6::GFP, fluorescence intensity can be measured along a line drawn at the cortex of a polarized embryo at prophase, and represented as a percentage of maximal intensity along this line. The extent of the PAR-6::GFP domain can be determined using an arbitrary threshold, typically 50–70% of maximal intensity.

4. Notes

1. Keeping cells alive: As live cell imaging occurs over time, one major challenge is to maintain cells alive and dividing during the experiment. As such, the temperature of the developing embryos should be kept between 15 and 25°C, either by climatizing the room or using a temperature-controlled stage or environmental chamber. Exposure time must also be minimized to prevent phototoxicity and photobleaching. For most imaging systems, the use of shutters, neutral-density

filters, and low laser power conditions should be sought to minimize exposure. Digital cameras can also be set to increase their sensitivity, thus minimizing exposure time. For instance, 2×2 binning of the pixels on a digital camera increases sensitivity fourfold, but decreases resolution fourfold. Likewise, increasing the digital gain of the camera will increase sensitivity but will decrease the signal-to-noise ratio. Acquisition parameters depend on fluorophores and should be determined empirically.

2. Multichannel acquisition: All of these systems allow acquisition of multiple channels, whether in DIC or fluorescence mode, and this can be controlled by the acquisition software. In such case, the microscope takes some time to switch filters between each channel acquisition, and this should be taken into consideration when determining the time interval in time-lapse series.
3. Z scan: Imaging only one plane of focus can often limit the amount and quality of information that is obtained. This is especially true for the *C. elegans* embryo, which has a diameter of 25–30 μm. If the microscope is equipped with a motorized focus, all of the imaging systems described above allow multifocal acquisitions, or Z scans. The number of focal planes to be acquired and distance between each focal plane can be determined empirically, based on the method being used and the need for detailed image analysis. As Z scanning increases the illumination of the specimen, considerations should be given to exposure times and acquisition intervals, as mentioned above.

Acknowledgements

A.R. and R.A. contributed equally to this work. We would like to thank Christian Charbonneau from IRIC's bio-imaging platform for technical support. A.R. held scholarships from the Canadian Institutes of Health Research and the Cole Foundation. R.A. held a scholarship from Université de Montréal's Department of Pathology and Cell Biology. This work was made possible by grants from the Canadian Cancer Society (#19378) and the Canadian Institutes of Health Research (#158715) to J.-C.L., who holds the Canada Research Chair in Cell Division and Differentiation. IRIC is supported in part by the Canadian Center of Excellence in Commercialization and Research (CECR), the Canada Foundation for Innovation (CFI), and the Fonds de Recherche en Santé du Québec (FRSQ).

References

1. Betschinger J, Knoblich JA (2004) Dare to be different: asymmetric cell division in *Drosophila* C. elegans and vertebrates. Curr Biol 14: R674–685
2. Morrison SJ, Kimble J (2006) Asymmetric and symmetric stem-cell divisions in development and cancer. Nature 441:1068–1074
3. Wu M, Kwon HY, Rattis F, Blum J, Zhao C, Ashkenazi R, Jackson TL, Gaiano N, Oliver T, Reya T (2007) Imaging hematopoietic precursor division in real time. Cell Stem Cell 1: 541–554
4. Cicalese A, Bonizzi G, Pasi CE, Faretta M, Ronzoni S, Giulini B, Brisken C, Minucci S, Di Fiore PP, Pelicci PG (2009) The tumor suppressor p53 regulates polarity of self-renewing divisions in mammary stem cells. Cell 138:1083–1095
5. Cuenca AA, Schetter A, Aceto D, Kemphues K, Seydoux G (2003) Polarization of the *C. elegans* zygote proceeds via distinct establishment and maintenance phases. Development 130: 1255–1265
6. Goldstein B, Macara IG (2007) The PAR proteins: fundamental players in animal cell polarization. Dev Cell 13:609–622
7. Hyenne V, Chartier NT, Labbé JC (2010) Understanding the role of asymmetric cell division in cancer using *C elegans*. Dev Dyn 239:1378–1387
8. Gönczy P, Rose LS (2005) Asymmetric cell division and axis formation in the embryo. In: The *C. elegans* Research Community (ed) WormBook, doi/10.1895/wormbook.1.30.1, http://www.wormbook.org
9. Fire A, Xu S, Montgomery MK, Kostas SA, Driver SE, Mello CC (1998) Potent and specific genetic interference by double-stranded RNA in *Caenorhabditis elegans*. Nature 391: 806–811
10. Kamath RS, Fraser AG, Dong Y, Poulin G, Durbin R, Gotta M, Kanapin A, Le Bot N, Moreno S, Sohrmann M, Welchman DP, Zipperlen P, Ahringer J (2003) Systematic functional analysis of the *Caenorhabditis elegans* genome using RNAi. Nature 421:231–237

Chapter 10

Tol2-Mediated Gene Transfer and In Ovo Electroporation of the Otic Placode: A Powerful and Versatile Approach for Investigating Embryonic Development and Regeneration of the Chicken Inner Ear

Stephen Freeman, Elena Chrysostomou, Koichi Kawakami, Yoshiko Takahashi, and Nicolas Daudet

Abstract

The vertebrate inner ear is composed of several specialized epithelia containing mechanosensory "hair" cells, sensitive to sound and head movements. In mammals, the loss of hair cells for example during aging or after noise trauma is irreversible and results in permanent sensory deficits. By contrast, avian, fish, and amphibians can efficiently regenerate lost hair cells following trauma. The chicken inner ear is a classic model system to investigate the cellular and molecular mechanisms of inner ear development and regeneration, yet it suffered until recently from a relative lack of flexible tools for genetic studies. With the introduction of in ovo electroporation and of Tol2 transposon vectors for gene transfer in avian cells, the field of experimental possibilities has now expanded significantly in this model. Here we provide a general protocol for in ovo electroporation of the chicken otic placode and illustrate how this approach, combined with Tol2 vectors, can be used to drive long-term and inducible gene expression in the embryonic chicken inner ear. This method will be particularly useful to investigate the function of candidate genes regulating progenitor cell behavior and sensory cell differentiation in the inner ear.

Key words: Inner ear, Otic placode, Hair cell, Chicken embryo, Electroporation, Transfection, Tol2 transposon, Development, Regeneration

1. Introduction

The inner ear, with its complex three-dimensional architecture and high level of cell diversification and specialization, is a fascinating (yet challenging) model system for a developmental biologist. The vast majority of the cell types that compose the mature inner ear derive from the otic placode, an ectodermal structure located on both sides of the hindbrain and clearly visible at incubation day (E)

Kimberly A. Mace and Kristin M. Braun (eds.), *Progenitor Cells: Methods and Protocols*, Methods in Molecular Biology, vol. 916, DOI 10.1007/978-1-61779-980-8_10,

2 in the chicken embryo. The otic placode rapidly transforms into a "cup", which then invaginates into the underlying mesoderm to form a closed otocyst. The otocyst gradually transforms into the inner ear "labyrinth" formed of several fluid-filled cavities housing the vestibular and auditory sensory epithelia, containing mechanosensory "hair" cells and non-sensory supporting cells (reviewed in (1, 2)).

As opposed to mammals, birds can produce and regenerate lost hair cells throughout life. In the vestibular organs, hair cell production starts early during development and a continuous turnover of sensory cells occurs throughout life. In the auditory organ, named the basilar papilla, terminal mitoses are completed by E9 and the full complement of hair cells has differentiated by E12. However following hair cell death, supporting cells can regenerate new auditory hair cells through mitotic and non-mitotic processes (reviewed in (3, 4)). Owing to its easy access and amenability to surgical manipulation, the chicken embryo has been extensively used to investigate induction and early morphogenesis of the inner ear using tissue transplantation or rotation experiments (see for example (5, 6)). With in ovo electroporation of plasmid DNA (7, 8) it has also become possible to transiently manipulate gene expression in the early embryonic inner ear (9, 10). However the tools for genetic manipulation at later stages of development of the chicken inner ear were until recently fairly limited. This represented an important obstacle to the investigation of the late differentiation and the post-traumatic regeneration of hair cells in the avian inner ear (reviewed in (3, 4)).

Until now, the most popular vectors used for long-term transgene expression in avian cells were RCAN and RCAS retrovirus, derived from the Rous sarcoma virus and originally developed by Hughes and colleagues (11). Infection with retrovirus (12, 13) or electroporation with RCAS proviral DNA (14) of the chicken otic cup/otocyst can lead to sustained gene expression in the embryonic inner ear up to pre-hatching stages, yet this approach has significant limitations in terms of transgene size, which cannot exceed 2.5 kb. In addition, virus-infected cells become refractory to secondary infection. Co-expression of several transgenes can only be achieved through infection by retroviruses harboring different envelope proteins. With the recent demonstration that Tol2 transposon vectors can be used successfully to achieve gene transfer in avian cells (15, 16), these limitations have been lifted.

Transposons are mobile genetic elements that can stably integrate into the cell genome when transposase activity is present. Vectors derived from the Tol2 transposon, originally discovered in Medaka fish (17), have been successfully used to insert foreign genes into a variety of animal cells (18). Recently, a series of Tol2 vectors has been developed for efficient gene transfer and inducible expression in chicken embryos (15, 16). Here, we show that following in ovo electroporation at placodal stages, these Tol2 vectors can drive long-term and inducible expression of transgenes in the developing chicken inner ear up to pre-hatching stages.

The great flexibility and efficiency of Tol2-mediated gene transfer will undoubtedly lead to renewed interest in avian models to investigate the molecular mechanisms of inner ear development and hair cell regeneration.

The first protocol is intended to provide a set of general guidelines and advices for in ovo electroporation of the chicken otic placode with plasmid DNA. In the second protocol, we provide some instructions relating to the use of the Tol2 transposon system for tet-on inducible gene expression in the chicken inner ear in ovo and in vitro. Transfected inner ear tissue can be used in a variety of experimental applications (immunohistochemistry, tissue culture, in vitro regeneration studies, live-imaging microscopy) whose detailed description is beyond the scope of the present chapter.

2. Material

2.1. In Ovo Electroporation of the Otic Cup

1. Fertilized chicken eggs obtained from a commercial source. These can be stored at 12–14°C for up to 1 week before starting the incubation.
2. One egg incubator (37.5°C).
3. Two pairs of forceps; one pair of small scissors.
4. 20 ml syringe with 18 G needle; 1 ml syringe with 25 G needle.
5. Sterile phosphate-buffered saline (PBS) pH 7.4.
6. Plasmid DNA, purified using anion exchange columns and ethanol-precipitated before resuspension in distilled nuclease-free water at a final concentration of 1–3 μg/μl.
7. Injection solution 10×: distilled water containing 20% sucrose and 1–2% Fast Green (Sigma). Prepare 10 ml of solution, filter using a Nalgene 0.4 μm syringe filter, and aliquot in 0.5 ml tubes. These can be stored at –20°C for extended period of time.
8. Eppendorf 0.5–20 μl GELoader tips.
9. Cellulose tape (e.g., Sellotape, Scotch tape, 15 and 50 mm wide).
10. Electroporator able to deliver square wave pulses of current (e.g., BTX ECM830, Sonidel CUY-21, Intracel TSS20).
11. One "chopstick" or L-shaped platinum electrode (e.g., Sonidel CUY611 and CUY613 series) as an anode. One tungsten electrode with a sharp tip as a cathode. Electrodes can be purchased from electroporation device suppliers or custom-made (see Note 1).

12. Borosilicate glass needles (e.g., Harvard Apparatus ref 30-0050 with OD-1.2 mm and ID-0.94 mm) for injecting the plasmid DNA solution.
13. A one-port holder (e.g., Harvard Apparatus MP series Microinjection Electrode Holder) for the glass needle, connected to a standard three-dimensional manipulator mounted on a vertical stand. The port of the holder is connected to a ~50 cm long piece of tubing ended by a P200 pipette tip through which air is mouth-blown to force the DNA solution out of the glass needle.
14. One small 3D micromanipulator (You-2, Narishige) mounted onto the (glass needle) one-port holder handle and holding the tungsten electrode.
15. One standard three-dimensional micromanipulator on a vertical stand to hold the positive electrode.
16. One stereomicroscope (at least 40× magnification) and a cold-light source with a flexible light guide.
17. One small handheld lens attached to a suitable holder to focus the incident light on the surface of the embryo (see Fig. 1).

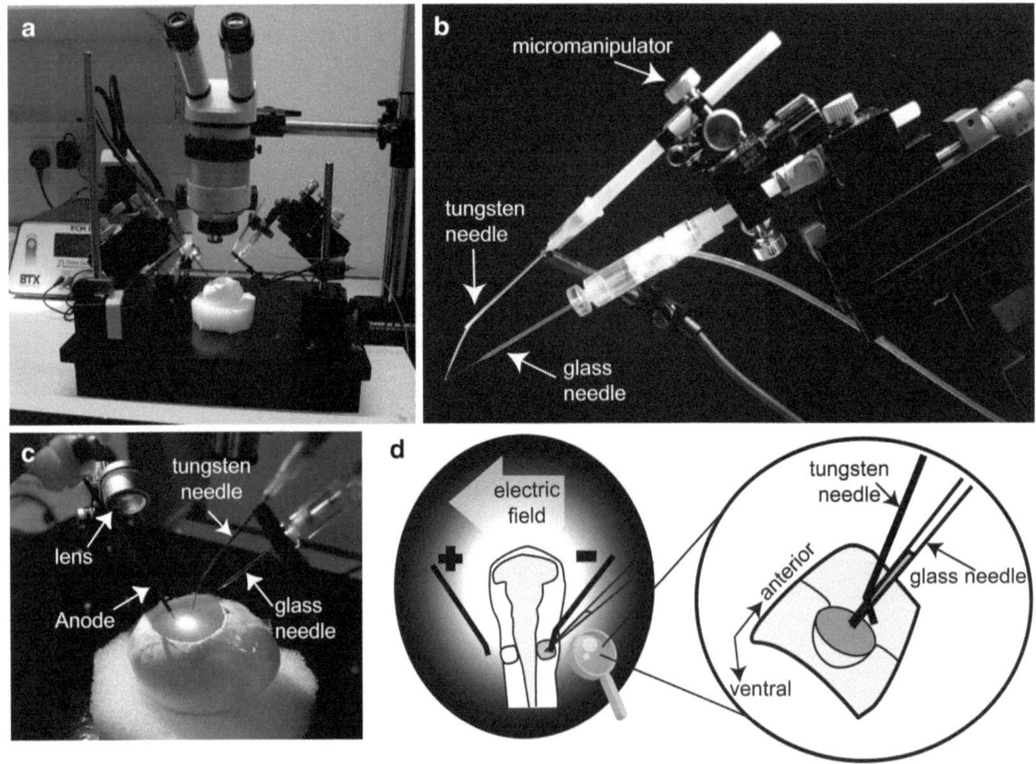

Fig. 1. (**a**) General view of the setup used for in ovo electroporation of the otic cup. (**b**) Higher magnification view of the manipulator holding the glass injection needle and the tungsten needle (cathode). (**c**) The electrodes in position. Note the small magnifying lens used to focus the incident light on the surface of the embryo. (**d**) Drawing illustrating the positioning of the electrodes and injection needle with respect to the otic placode or cup.

2.2. Inducible Gene Expression Using Tol2 Transposon Vectors and the Tet-ON System

1. Purified Tol2 plasmids: pT2K-CAGGS-rtTA-M2; pT2K-BI-TRE-EGFP; pCAGGS-T2TP (see Figs. 2–3 and Sato et al. 2007 (15) for further information).
2. Doxycycline hyclate (Sigma). Prepare a fresh stock solution at 1 mg/ml in PBS and store at 4°C for up to a week.
3. One sterile 1 ml syringe with a 25 G needle.
4. 6-well tissue culture dishes.
5. 35 mm and 70 mm sterile petri dishes.
6. A set of sterile dissecting tools (scissors, scalpel, forceps, fine tweezers, micro-spoon).
7. One fine tungsten needle mounted onto a holder.
8. Dubbelco's Modified Eagle's Medium (DMEM) and DMEM/F-12 (1:1)+L-glutamine; +15 mM HEPES (Gibco) culture medium.
9. Penicillin and ciprofloxacin (Sigma).

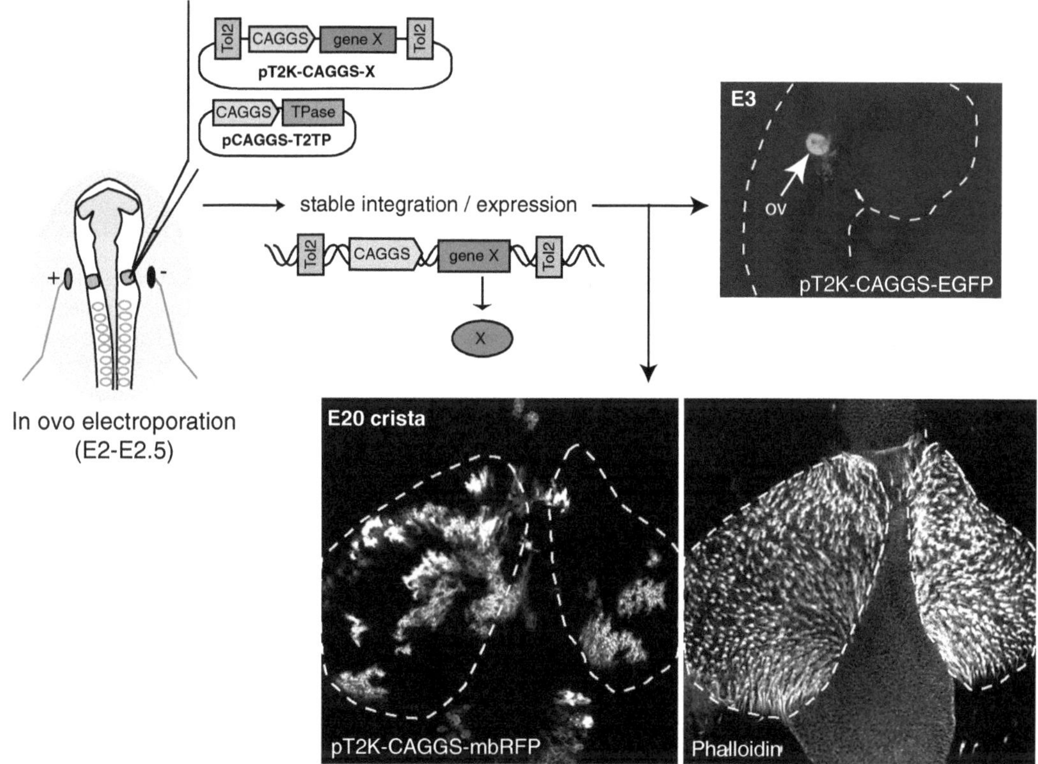

Fig. 2. Tol2 transposon vectors drive long-term gene expression in the developing inner ear. In the first example, strong EGFP fluorescence is seen in the otocyst 24 h after electroporation with pT2K-CAGGS-EGFP. In the second example, the embryo was transfected with pT2K-CAGGS-mbRFP (membrane-localized, red fluorescent protein) and let to develop until pre-hatch stages (E20). The inner ear was stained with fluorescent phalloidin to label the hair cell actin-rich bundles, and one sensory crista was visualized with confocal microscopy. Note that the pattern of transfection is very mosaic and groups of transfected cells are found within both sensory (delineated with dashed lines) and non-sensory domains.

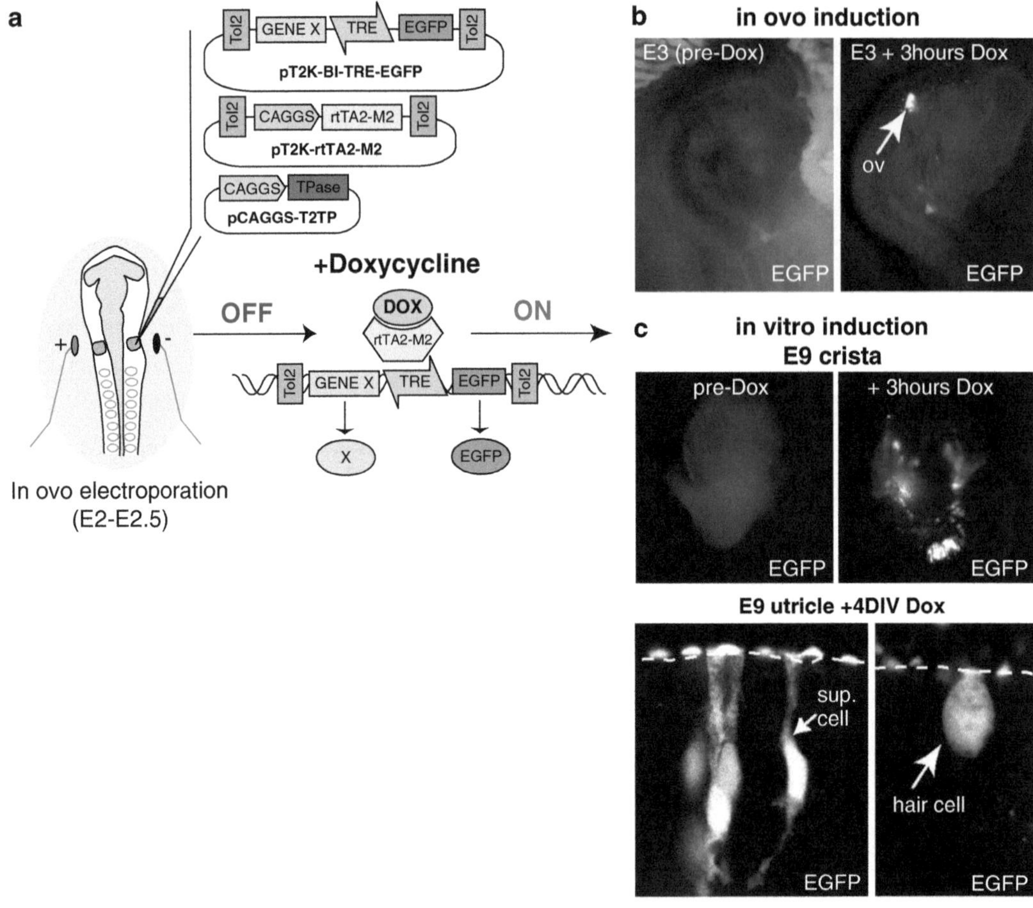

Fig. 3. Tol2 transposon vectors and tet-on inducible gene expression. (**a**) The Tet-on system can be used to trigger expression of a transgene X and EGFP at selected times of development. Induction of gene expression can be triggered in ovo (**b**) or in vitro (**c**) by doxycycline treatment. In (**c**) the *bottom* panels show side views of EGFP-positive hair cells and supporting cells in an E9 utricle that has been treated for 4 days in vitro with doxycycline. The apical surface of the epithelium is delineated with a *dashed line* and the stereocilia bundles of hair cells have been stained with a specific marker.

10. One stereomicroscope (at least 40× magnification) and a cold-light source with a flexible light guide, placed under a horizontal flow hood.

3. Methods

3.1. In Ovo Electroporation of the Otic Cup

The following protocol is intended to give general instructions for in ovo electroporation of the otic cup with plasmid DNA or any negatively charged molecule. When using a standard plasmid unable to integrate into the cell genome, transgene expression will be transient and should diminish after 2–3 days. However with a

Tol2 vector, long-lasting expression of the transgene should be seen up to pre-hatching stages (see Fig. 2).

1. Incubate fertilized chicken eggs in a humidified incubator at 38°C for approximately 48 h, or until stage Hamburger–Hamilton (HH) 12–15 (see Note 2).
2. Spray the eggs with 70% ethanol and let them dry for approximately 5 min horizontally. Apply a band of 15 mm wide Sellotape tape on the surface of the eggs along its long axis and using a pair of forceps make two holes within the tape: one at the top and one at the broad end of the egg. Remove from the broad end of the egg 3–4 ml of albumen with a 20 ml syringe and an 18 G needle. Let the egg sit for another 15 min; then open a 15 × 30 mm window on the top of the egg with a pair of fine scissors (see Note 3).
3. Preparation of the DNA solution: thaw at room temperature a frozen aliquot of 10× injection solution; it can be refrozen and reused at least ten times. Fast Green is included for easy visualization of the DNA solution during in ovo electroporation. For 10 μl of DNA solution, pipette 1 μl of the 10× injection solution into a 0.5 ml Eppendorf tube and add the required amount of purified plasmid DNA (see Note 4) and water up to 10 μl, and then pipette up and down several times to mix well.
4. Using a P10 pipette fitted with a GELoader microloader tip, backfill the glass needle with 5–10 μl of DNA solution (avoid forming bubbles) and mount it onto the pipette holder. This volume should be sufficient to electroporate at least 30 embryos.
5. Mount the glass needle onto the one-port holder. Cut the tip of the glass needle with a fine pair of tweezers and adjust the position of the tungsten needle so that its tip is located slightly posterior to that of the glass needle and almost contacting it (see Fig. 1). When pulling the glass needles, use heating/pulling parameters that produce needles with a long and fine tip, which can then be easily cut down in case the needle would get clogged during the injection.
6. Place the egg within a suitable holder (such as a ring of blue-tack inside a petri dish) and under the stereomicroscope. Adjust lighting by changing the position of the light guide and the magnifying lens so that a small spot of light is focused on the surface of the embryo and the right otic placode/cup is clearly visible (see Note 5).
7. Using a fine tungsten (or 25 G) needle, poke a small hole through the vitelline membrane lateral to the embryo and close to the hindbrain level; then apply one or two drops of sterile PBS. The PBS will infiltrate through the hole and the vitelline membrane should be lifted from the surface of the embryo.

It is then possible to enlarge the hole and to access easily the hindbrain/otic region with the electrodes and injection glass needle.

8. Using the micromanipulators, bring the injection glass needle and the tungsten cathode on top of the right otic placode/cup and place the anode on the opposite (left) side of the embryo, making sure that the needle and electrodes do not touch any embryonic tissue. Readjust the position of the tungsten needle relative to that of the injection glass needle if needed (see Note 6). Using the You-2 micromanipulator to hold the tungsten needle greatly facilitates this operation.
9. Inject by air-blowing the DNA solution into the otic cup that should become blue and clearly visible now. If the otic cup is not clearly filled or the solution fails to remain in the cup, check that the membranes that may be covering the otic region have been removed properly.
10. Apply the train of electric pulses (see Note 7) and keep gently injecting fresh DNA solution into the otic cup during the pulses. Bubbles should form at the tip of the tungsten needle. If not, check that electric connections are correct.
11. Remove carefully the electrodes and apply 2–3 drops of PBS on top of the embryo; then reseal immediately the egg with 50 mm wide Sellotape tape (see Note 8).
12. Return the eggs to the incubator and let to develop for 24 h or until required for analysis or further experimental work (see Note 9).

3.2. Inducible Gene Expression Using Tol2 Transposon Vectors and the Tet-ON System

Comprehensive information about the design and validation of the Tol2 transposon Dox-inducible system in chicken embryos can be found in Sato et al. (15). Briefly, one Tol2 vector drives constitutive expression of the rtTA2^{S}M2 tet-on activator (pT2K-CAGGS-rtTA-M2) and another one contains a bidirectional tet-responsive promoter (TRE) controlling the expression of EGFP and, if required, an additional transgene X (pT2K-B1-TRE-EGFP). When these two vectors are stably integrated into the same cell, doxycycline treatment transiently induces expression of EGFP and the transgene X (Fig. 3).

1. Electroporate embryos with the appropriate Tol2 plasmids (see Note 10).
2. Incubate the electroporated embryos until the required embryonic stage for induction of gene expression.
3. Prepare the working solution of doxycycline by diluting the stock solution in sterile DMEM. Do not store and reuse this working solution. For in ovo treatment, we find that the total volume of solution administered is important to subsequent

embryo survival. As a general rule of thumb, the earlier the embryonic stage, the smaller the volume administered (do not exceed 100 μl at E3-E4; 200 μl at E5-E6; 300 μl at E7–E10). Regardless of the chosen volume of injection, the amount of doxycycline administered in ovo is the same—we administer 30 μg per embryo. For stages past E10, we use in vitro treatments of transfected inner ear tissue with 500 ng/ml doxycycline (see below).

The rest of this protocol will be split into two parts. The first will deal with the procedures to administer doxycycline to the chick embryo in ovo. The second will deal with the procedures necessary to dissect out the inner ear from a late stage (E9–E10) embryo, set up an organotypic culture, and induce gene expression via doxycycline treatment in vitro.

3.2.1. In Ovo Procedure

1. Prepare a working solution of 500 ng/ml doxycycline in DMEM and keep at 37°C until needed.
2. Take the electroporated eggs out of the incubator and clean them with 70% ethanol.
3. Under a horizontal flow hood, open the egg by cutting away the Sellotape window with a clean pair of small scissors. Place the egg in a suitable holder and under the stereomicroscope.
4. Attach a 25 G needle to a sterile 1 ml syringe. Fill the syringe with the appropriate volume of the doxycyclin working solution (see Step 3). Ensure that there are no bubbles in the needle.
5. Working under the microscope, carefully pierce the extraembryonic membranes surrounding the embryo with the point of the 25 G needle (see Note 11). Work the needle point into the space underneath the chorion and slowly inject the doxycycline solution.
6. Reseal the window of the egg with Sellotape and return the egg to the incubator. Fluorescence is clearly visible 3 h after injection in E3–E5 embryos (Fig. 2). From E6, in ovo visualization of fluorescence becomes more difficult, as the embryo increases its movements and the inner ear moves away from the embryo's outer surface.
7. In order to maintain a reasonably constant concentration of doxycyclin within the extraembryonic environment, re-administer the doxycycline at 12-h intervals if long-term induction of gene expression is needed.

3.2.2. In Vitro Procedure

The following procedure is performed under a horizontal flow hood in sterile conditions.

1. Prepare some tissue culture medium (DMEM/F-12 with 10 units/ml penicillin and 10 units/ml ciprofloxacin) with and without 500 ng/ml doxycycline.

2. Dispense media into 6-well culture dishes (3 ml/well) and maintain the dishes within a tissue culture incubator (37°C, 5% CO_2) until needed.
3. Remove the embryo from the egg using a pair of curved forceps and place it in a 70 mm petri dish.
4. Section at the base of the neck with a pair of tweezer, and carefully cut the head along its longitudinal axis in half with a scalpel. Place the right half of the head (containing the electroporated inner ear) into a 35 mm petri dish filled with room temperature DMEM.
5. Dissect the membranous part of the inner ear out of the surrounding cartilage using fine tweezers and tungsten needles.
6. Using a micro-spoon or a pair of fine tweezers, transfer the dissected inner ear samples into the 6-well culture dish prefilled with culture medium prepared in Step 12. Ensure that the tissue has sunk to the bottom of the wells; if the tissue is floating, gently poke it down to the bottom using dissecting tools or by dispensing a few drops of culture medium on its surface.
7. Place the 6-well culture dish into a tissue culture incubator (37°C, 5% CO_2) and maintain until required for analysis or further experimentation. Fluorescence should become visible in doxycycline-treated samples after 3–6 h (Fig. 3).
8. Replace the culture media containing doxycycline every 24 h.

4. Notes

1. We use a "chopstick" CUY611P3-1 platinum electrode with a 1 mm tip as an anode and a custom-made tungsten electrode as a cathode. To make a tungsten electrode, cut a 3–4 cm long segment of 0.5 mm diameter tungsten; then sharpen its tip by electrolysis in a bath of 1 M sodium hydroxide. Connect the tungsten rod to the positive cable and a metal paper clip to the negative cable using banana clips, and apply a 12 V current between them until a sharp tip is obtained. The tungsten needle can be fitted inside a syringe needle (18-22G) to which the negative electric wire is connected via a banana clip.
2. Stage HH13 is optimal—the otic cup is easy to locate and to access with minimal dissection and damage to the embryo. Past this stage, the embryo turns on its left side, the otic cup starts to close, and the amniotic folds that cover the head region need to be removed to access the otic cup. It remains possible to electroporate the closed otocyst; however higher voltage conditions are required and the risks of damaging the tissue and killing the embryo are increased.

3. Windowing of the eggs is a step that should not be overlooked, in particular for long-term survival of electroporated embryos. Sellotape is useful to prevent shell debris from falling inside the egg during the opening. Try to minimize the time the eggs are left open to the air before proceeding to electroporation or the embryos will dry and die.
4. The final concentration of plasmid DNA used for electroporation depends on the type of construct and promoter. For most CMV-based plasmids we have used, 0.2–0.5 μg/μl is sufficient for strong expression, but we recommend 1 μg/μl as a starting point. When co-electroporating several plasmids, try to maintain a final DNA concentration <3 μg/μl or the solution may become very viscous and difficult to inject through the glass needle. When using Tol2 vectors, the vector encoding the transposase (pCAGGS-T2TP) needs to be co-electroporated (at a final concentration >0.3 μg/μl) together with the Tol2 plasmid(s) (each one at a final concentration >0.3 μg/μl) for stable integration of the Tol2 transgene(s). If the total volume of your combined DNA solution exceeds 10 μl, you may use a centrifugal vacuum concentrator to evaporate excess water. Briefly heating the solution at 80°C can also help to resuspend precipitated DNA.
5. Good visualization of the otic placode/cup and optimal placement of the electrodes are essential for successful electroporation. We use a small magnifying lens (e.g., 10 × 18 mm geology handheld pocket lens) to focus the incident light on the surface of the embryo, which drastically improves contrast and visualization of the embryonic structures.
6. The electrical field will form in between the two electrodes. Depending on the developmental stage, the otic cup will be more or less closed and the position of the tungsten needle relative to that of the injection needle should be modified accordingly to maintain the two electrodes and injection site along a roughly linear axis. In all cases, do not allow the tungsten needle to touch any tissue.
7. We use a BTX ECM830 electroporator and the following conditions: 7 V, 3 × 50 ms or 3 × 100 ms pulses with 100 ms time interval between each pulse. These values are provided as guidelines only and should be adapted to each individual setup to reach a good compromise between efficiency of transfection and embryo survival.
8. Provided that the concentration of Fast Green was high enough, you should still be able to see some traces of the dye in the otic cup region. This can be very helpful when initially practicing the method. Before resealing the egg, in particular when using an automatic egg-turning system, ensure that there is sufficient

space between the embryo and the window to avoid contact during the incubation. If not, remove some albumen.

9. In our hands, the survival rate of electroporated embryos decreases approximately from 80 to 100% at 1–4 days post-electroporation to 10–25% at 16 days post-electroporation, with maximum embryonic death occurring at about 5–6 days post-electroporation. If performed incorrectly (poor placement of the electrodes, too high voltage, damage during dissection/DNA injection) electroporation will result in very poor embryo survival or abnormal development of the inner ear. It is essential to practice the technique sufficiently and determine the optimal voltage conditions for your setup with a control plasmid construct (e.g., CMV-driven EGFP expression vector) before embarking on further experiments. For long-term survival, we recommend to (a) use fresh eggs, and be careful during the windowing procedure; (b) clean all instruments, electrodes, and working space with 70% ethanol; (c) if possible, place the electroporation setup under a horizontal flow hood to minimize the risks of infection; (d) ensure that the embryo does not come into contact with the tape by removing enough albumen after electroporation; and (e) maintain egg incubators and trays clean.
10. For inducible Tet-on expression, the inducible plasmid vector (pT2K-BI-TRE-EGFP or derivative) must be electroporated with both pCAGGS-T2TP plasmid vector and the pT2K-CAGGS-rtTA-M2 plasmid vector. As mentioned before we try to keep the final concentration of the DNA solution at no more than 3 μg/μl. For the inducible system, in our hands, the highest transfection efficiency is achieved using a ratio of 2:1 between pT2K-BI-TRE-EGFP and pT2K-CAGGS-rtTA-M2. This ratio appears to be extremely important, and alterations to it result in drastically reduced transfection efficiency.
11. As the embryo turns onto its left side during development, the heart will be located on the right-hand side if looking at the embryo with the anterior facing upwards. We pierce the chorion to the left of the embryo, adjacent to its dorsal side. This avoids potentially damaging the heart or any major veins/arteries during the injection process.

Acknowledgements

N.D. thanks Dr. Alexander Davies for teaching him the basics of in ovo electroporation. This work was supported by Deafness Research UK and the Biotechnology and Biological Sciences Research Council (BBSRC).

References

1. Fekete DM, Wu DK (2002) Revisiting cell fate specification in the inner ear. Curr Opin Neurobiol 12:35–42
2. Bok J, Chang W, Wu DK (2007) Patterning and morphogenesis of the vertebrate inner ear. Int J Dev Biol 51:521–533
3. Stone J, Rubel E (2000) Cellular studies of auditory hair cell regeneration in birds. Proc Natl Acad Sci USA 97:11714–11721
4. Stone JS, Cotanche DA (2007) Hair cell regeneration in the avian auditory epithelium. Int J Dev Biol 51:633–647
5. Bok J, Bronner-Fraser M, Wu DK (2005) Role of the hindbrain in dorsoventral but not anteroposterior axial specification of the inner ear. Development 132:2115–2124
6. Groves A, Bronner-Fraser M (2000) Competence, specification and commitment in otic placode induction. Development 127: 3489–3499
7. Nakamura H, Funahashi J (2001) Introduction of DNA into chick embryos by in ovo electroporation. Methods 24:43–48
8. Momose T, Tonegawa A, Takeuchi J, Ogawa H, Umesono K, Yasuda K (1999) Efficient targeting of gene expression in chick embryos by microelectroporation. Dev Growth Differ 41:335–344
9. Alsina B, Abelló G, Ulloa E, Henrique D, Pujades C, Giraldez F (2004) FGF signaling is required for determination of otic neuroblasts in the chick embryo. Dev Biol 267:119–134
10. Daudet N, Lewis J (2005) Two contrasting roles for Notch activity in chick inner ear development: specification of prosensory patches and lateral inhibition of hair-cell differentiation. Development 132:541–551
11. Hughes S, Greenhouse J, Petropoulos C, Sutrave P (1987) Adaptor plasmids simplify the insertion of foreign DNA into helper-independent retroviral vectors. J Virol 61:3004–3012
12. Kiernan A, Fekete D (1997) In vivo gene transfer into the embryonic inner ear using retroviral vectors. Audiol Neurootol 2:12–24
13. Eddison M, Le Roux I, Lewis J (2000) Notch signaling in the development of the inner ear: lessons from Drosophila. Proc Natl Acad Sci USA 97:11692–11699
14. Bird JE, Daudet N, Warchol ME, Gale JE (2010) Supporting cells eliminate dying sensory hair cells to maintain epithelial integrity in the avian inner ear. J Neurosci 30:12545–12556
15. Sato Y, Kasai T, Nakagawa S, Tanabe K, Watanabe T, Kawakami K, Takahashi Y (2007) Stable integration and conditional expression of electroporated transgenes in chicken embryos. Dev Biol 305:616–624
16. Watanabe T, Saito D, Tanabe K, Suetsugu R, Nakaya Y, Nakagawa S, Takahashi Y (2007) Tet-on inducible system combined with in ovo electroporation dissects multiple roles of genes in somitogenesis of chicken embryos. Dev Biol 305:625–636
17. Koga A, Iida A, Hori H, Shimada A, Shima A (2006) Vertebrate DNA transposon as a natural mutator: the medaka fish Tol2 element contributes to genetic variation without recognizable traces. Mol Biol Evol 23:1414–1419
18. Kawakami K (2005) Transposon tools and methods in zebrafish. Dev Dyn 234:244–254

Chapter 11

Labeling Primitive Myeloid Progenitor Cells in *Xenopus*

Ricardo Costa, Yaoyao Chen, Roberto Paredes, and Enrique Amaya

Abstract

In *Xenopus* the first blood cells to differentiate in the embryo are the primitive myeloid lineages, which arise from the anterior ventral blood islands during the neurula stages. Primitive myeloid cells (PMCs) will give rise to the embryonic pool of neutrophils and macrophages, a highly migratory population of cells with various functions during development and tissue repair. Understanding the development and behavior of PMCs depends on our ability to label, manipulate, and image these cells. *Xenopus* embryos have several advantages in the study of PMCs, including a well-established fate map and the possibility of performing transplants in order to label these cells. In addition, *Xenopus* embryos are easy to manipulate and their external development and transparency at the tadpole stages make them amenable to imaging techniques. Here we describe two methods for labeling primitive myeloid progenitor cells during early *Xenopus* development.

Key words: Embryonic myelopoiesis, *Cepba*, *Spib*, Primitive hematopoiesis, Myeloid progenitors, Cell migration, *Xenopus* laevis, *Xenopus* tropicalis

1. Introduction

Hematopoiesis is a continuous process, which occurs during two temporally and spatially distinct waves in vertebrates (1). In *Xenopus*, the first wave of hematopoiesis is initiated in the anterior ventral blood islands (aVBI), adjacent to the cardiac field, by the end of gastrulation (2–7). About a day later, primitive erythroid cells are specified and differentiate in the posterior ventral blood islands (pVBI). In *Xenopus*, the rostral and caudal domains of primitive hematopoiesis arise from separate regions of the gastrula, but they eventually end up juxtaposed in the ventral blood islands by the neurula stages (5, 8).

Cells in the aVBI differentiate into macrophages and neutrophils about a day before the primitive erythroid cells differentiate in the pVBI. Indeed, PMCs initiate their migratory behavior, including their ability to respond to wounds, well before the

Kimberly A. Mace and Kristin M. Braun (eds.), *Progenitor Cells: Methods and Protocols*, Methods in Molecular Biology, vol. 916, DOI 10.1007/978-1-61779-980-8_11, © Springer Science+Business Media, LLC 2012

formation of the vasculature, a functional heart, or the onset of globin expression (7). The development of primitive myeloid progenitors is similar in frogs and fish. For example, the embryonic macrophages and neutrophils develop from the zebrafish ALM (*a*nterior *l*ateral *m*esoderm) (9–11), the corresponding anatomical position to the *Xenopus* aVBI. The homology between primitive myeloid cells in frogs and fishes also includes the shared expression of a number of orthologous genes, such as myeloperoxidase (*mpo*), neutrophil cytosolic factors (*ncf1, ncf2*), L-plastin (*lcp1)*, metalloproteases (*mmp7, mmp13*), coronin (*coro*), lysozyme (*lyz)*, several transcription factors (*spib, spi1, cebpa*), as well as growth factor receptors such as the macrophage colony-stimulating factor receptor (*csf1r*) (6, 12–19). It is very likely that this early population of embryonic myeloid cells are homologous to those born in the mammalian yolk sac (20, 21).

Most of our current understanding of PMC development has focused on the transcriptional network and signaling pathways that lead to their specification, and to a lesser extent their behavior (6, 7, 9, 16, 18, 22, 23). In contrast, less is known about their various roles during development and following injury. Recently these cells have been eliminated in *Xenopus* embryos, resulting in high death rates and severe developmental abnormalities (24). In addition, tissue macrophages have been implicated in tissue homeostasis and vascularization of the hindbrain, potentially acting as cellular chaperones in both zebrafish and mice (25). Data gathered so far suggest that primitive myeloid progenitors give rise to multiple lineages important in development and homeostasis. The protocols presented allow the labeling and manipulation of the fate of primitive myeloid progenitors through transplantation and the microinjection of mRNA and morpholinos into blastomeres fated to give rise to PMC progenitors. These important tools will facilitate the characterization of the roles and behaviors of these cells during normal development and following tissue injury.

2. Materials

2.1. Labware and Equipment

1. Microinjection system (e.g., Picospritzer II, Parker Precision Fluidics—http://www.parker.com).
2. Needle puller (e.g., Sutter Instruments Co. Model P-97—http://www.sutter.com/).
3. Stereomicroscope for all the manipulations (e.g., Leica MZ series).
4. UV stereomicroscope for selection of injected donor embryos and imaging (e.g., Leica MZ FLIII—http://www.leica-microsystems.com).

5. Traditional microsurgery equipment (hair loop, eyebrow knife, glass needle, tungsten knife).
6. Borosilicate glass capillaries (e.g., Harvard Apparatus part no. 30-0038, 1.0 mm O.D. × 0.78 mm I.D.).
7. Gel sequencing tip to backload injection glass needle (Eppendorf no. 5242 956 003).
8. Watchmaker forceps (Dumont no. 5 and no. 55, Fine Science Tools).
9. Material to sharpen forceps (Dan's Black Arkansas stone—http://www.danswhetstone.com/ and Thomas Scientific abrasive film "sharpening paper" Cat. No. 6775—http://www.thomassci.com/).
10. *Xenopus laevis* and *Xenopus tropicalis* animals can be obtained commercially or through the community—http://www.xenbase.org/other/obtain.do.
11. Syringes and needles [26 ½ gauge for *Xenopus laevis* and 27 gauge or finer for *Xenopus tropicalis* (Terumo)].
12. Eyepiece micrometer (Leica).
13. Plasticware and disposable pasteur pipettes (3 ml), although we prefer to work with fire polished glass pipettes.
14. Modeling clay (Fimo Effect, Eberhard Faber).
15. Plastic petri dishes (10 cm diameter) coated with 3–5 mm of 1% standard agarose in 0.1× MMR (*Xenopus laevis*) or 0.01× MMR (*Xenopus tropicalis*), or nylon mesh if you prefer.
16. Dissection tools for the removal of *Xenopus* testis [surgical scissors, e.g., Fine Science Tools Cat. No. 14568-09 (*Xenopus tropicalis*) and 14568-15 (*Xenopus laevis*); standard curve forceps, e.g., Fine Science Tools Cat. No 11008-13].
17. Incubators (14–18°C for *Xenopus laevis* and 22–26°C for *Xenopus tropicalis*).
18. Glass-bottomed imaging dishes (MatTek Corporation).
19. 1–2 l plastic containers with lids for *Xenopus tropicalis* natural matings.

2.2. Reagents

1. Micro-ruby (Invitrogen Cat. No. D-7162, 2% aliquots stored at −20°C).
2. Tricaine/MS222 (Ethyl 3-aminobenzoate methanesulfonate, Sigma-Aldrich Cat. No. E10521).
3. Pigment Inhibitor (*N*-benzyl-*N*′-phenylthiourea, ChemBridge Europe Cat. No. 5101231, dissolved in DMSO (Sigma-Aldrich, Cat. No. D2650) at 5 mg/ml store in aliquots at −20°C and use at 2 μg/ml).
4. PMSG (Pregnant Mare Serum Gonadotrophin, Intervet, UK).

5. hCG (Human Chorionic Gonadotrophin, Intervet, UK).
6. 10× MMR stock solution (1 M NaCl, 20 mM KCl, 20 mM $CaCl_2$, 10 mM $MgCl_2$, 50 mM HEPES, pH 7.4, sterile, store at room temperature).
7. L-cysteine hydrochloride (Sigma Cat. No. C7880) for *Xenopus laevis* and L-cysteine anhydrous (Aldrich Cat. No 168149) for *Xenopus tropicalis.*
8. mMessage Machine Kit (Ambion, Applied Biosystems).
9. Ficoll PM 400 (Sigma, Cat. No F4365).
10. BSA (Bovine Serum Albumin, Sigma-Aldrich, Cat. No. A7906).
11. L-15 medium Leibovitz (Sigma L5520); keep sterile at 4°C, or aliquot and store at -20°C.
12. Methylcellulose (Sigma Cat. No. 274429); stir a 2% solution in the cold for about a week, and aliquot and store at -20°C.
13. Agarose (e.g., SeaKem LE agarose, Lonza).

3. Methods

For a general description of husbandry and methods for *Xenopus* embryology, see the Cold Spring Harbor Book on Early Development of Xenopus laevis—a laboratory manual (26) or see Xenbase (http://www.xenbase.org/other/methods.do). General descriptions and methods for *Xenopus* hematopoiesis can be found in a book edited by Margaret Baron (27–30).

This chapter is divided into five sections, (1) how to obtain embryos; (2) labeling embryos by microinjection; (3) transplantation of the anterior ventral blood islands; (4) labeling primitive myeloid progenitors by microinjection at the 32-cell stage; and (5) imaging of labeled primitive myeloid cells in live embryos.

3.1. How to Obtain Xenopus Embryos

1. *Xenopus laevis* females are pre-primed by injection of 100–150 units of PMSG into the dorsal lymph sac 3–10 days before embryos are needed. *Xenopus tropicalis* are pre-primed with 20–30 units of PMSG, 48–24 h before embryos are needed. Table 1 provides an overview of different methods that can be used to obtain *Xenopus* embryos.
2. The day before embryos are required, prime two adult *Xenopus laevis* females with 500–800 units of hCG, depending on the size of the aninal. Ovulation normally starts 12–14 h after injection and can be delayed by a few hours if the frogs are kept at 16°C overnight. *Xenopus tropicalis* are induced to ovulate by injection of 100–150 units of hCG, about 3–4 h

Table 1
Overview of methods to obtain *Xenopus* embryos

	Xenopus laevis (artificial fertilization)	*Xenopus tropicalis* (artificial fertilization)	*Xenopus tropicalis* (natural mating)
Eggs laid in	1× MMR or "squeeze"	"Squeeze"	System water
Time after priming	12–16 h	3–5 h	3–5 h
Testis kept in	L15	L15 + serum (cull)	Advantage—No need!
Dejellying solution	2% cysteine, 0.1×MMR, pH = 8.0	2% L-cysteine hydrochloride, 0.01×MMR, pH = 8.0	
Embryos kept at	14–23°C	22–32°C	
Optimal for transplants	16–18°C	22–23°C	

before eggs are required. Maintain frogs above 20°C, as lower temperatures lead to poor egg quality. Also note that egg quality falls sharply when females take longer than 4 h to start laying (see Note 1).

3. For artificial fertilizations, euthanize a male *Xenopus laevis* frog with an overdose of tricaine/MS222. This can be done either by immersion in 0.1% tricaine/MS222 (neutralized with 0.1% $NaCO_3$) or by injection of 1 ml of 40% tricaine/MS222. After the male is euthanized, the testes are removed and stored at 4°C in L-15 medium for up to a week. For in vitro fertilization, a tiny piece of testis is macerated and rubbed over freshly ovulated eggs, such that the eggs are dispersed as a monolayer in a petri dish, using watchmaker forceps. Leave the testis/egg mixture standing for 5 min. Flood with 0.1× MMR and wait for 15–20 min before dejellying. (For *Xenopus tropicalis* artificial fertilization, please see Notes).
4. Dejelly *Xenopus laevis* artificial fertilizations by flooding the petri dish containing the embryos with 0.1× MMR 2% cysteine (pH = 8.0 with NaOH). For *Xenopus tropicalis* use anhydrous L-cysteine (for which less NaOH is required to bring the pH to 8.0) and rinse the embryos twice with 0.01× MMR. Always make the cysteine solution fresh. For embryos from *Xenopus tropicalis* natural matings and after removing the mating pair, empty the contents of the 1–2 L container into a 2 L beaker. Floating embryos will settle at the bottom of the beaker. While this happens the embryos, which are stuck against the walls of recently emptied container can be dejellyed with a large volume of 0.01× MMR 2% L-cysteine (pH 8.0) and the help of a pasteur pipette. The 2 L beaker can now be emptied slowly and the embryos that settled on the bottom of the beaker can be dejellyed with the cysteine solution of the container, and moved

to a smaller beaker (reuse the cysteine solution if necessary). Rinse resulting dejellyed embryos before pouring them into an agarose dish and embryos are ready for selection. Place the embryos in 0.1× MMR 2% Ficoll for injection (*Xenopus laevis* and *Xenopus tropicalis*).

3.2. Labeling Xenopus Embryos by Microinjection

Embryos can be fluorescently labeled by microinjection of a fluorescently labeled dextran, such as micro-ruby, which is stable and can be detected in single cells up to stage 50. Micro-ruby stocks are made at 2% (m/v) in distilled water and injected as a 1:10 dilution in water. This concentration does not result in detrimental developmental defects when up to 10 nl are injected into one-cell stage *Xenopus laevis* embryos or up to 2 nl are injected into one-cell stage *Xenopus tropicalis* embryos.

1. Prepare injection setup (e.g., Picospritzer II). Load a borosilicate glass needle with 0.2% solution of micro-ruby dextran dye using an Eppendorf gel-loading tip. Attach needle to the injection setup, check the system pressure, and clip the tip of the needle with fine watchmaker forceps (a good starting point is immediately below the point where the pulled glass capillary becomes rigid). Calibrate this injection volume by clipping and adjusting the time of injection (50–200 ms), such that one injection delivers up to 10 nl for *Xenopus laevis* single cell stage embryos and up to 2 nl for *Xenopus tropicalis* single cell staged embryos. For this purpose, it is useful to install a calibration ruler in the eyepiece and use a table of volume correlations based on the diameter of the injection bubbles to measure injection volumes.
2. Transfer *Xenopus laevis* embryos into 0.1× MMR 2% Ficoll and inject 10 nl once at one-cell or twice with 5 nl in each blastomere at the two-cell stage. For *Xenopus tropicalis* embryos, adjust these volumes to 2 nl for single cell staged embryos or 1 nl twice for two-cell staged embryos. Move embryos with forceps on one hand while controlling the micromanipulator holding the needle with the other. Aim to inject 30–40 embryos per 30 min slot. After injection, pair stage-matched labeled and unlabeled control embryos, and keep them close within the incubator in case of temperature fluctuations between different regions of the incubator. Spread the developmental stage of labeled and control embryos over several hours by modulating the temperature of the embryos (14–23°C for *Xenopus laevis* and 22–28°C for *Xenopus tropicalis*). Transplants are done between stage 14 and 16, and *Xenopus laevis* are usually at this stage 20–22 h after fertilization at 18°C and *Xenopus tropicalis* are usually at this stage 16–18 h after fertilization at 22°C. For experimental timing constrains (see Note 2) and all the staging according with Nieuwkoop and Faber, 1994.

3.3. Transplantation of the Anterior Ventral Blood Islands

1. Prepare transplantation dishes by flattening a ball of modeling clay over a petri dish to a thickness of 2–3 mm thickness. Pour sterile 0.1× MMR supplemented with 1% BSA in the modeling clay coated dish, about 2 h before it is needed, rinse shortly, and equilibrate to room temperature before use.
2. Transplantation should be performed at the early neurula stages (stage 14–16). Before the labeled embryos reach this stage, screen the embryos for intense and uniform fluorescence under a fluorescent dissecting microscope, using a rhodamine filter.
3. Remove the vitelline membranes from the host and donor embryos, using sharp watchmaker forceps. For minimal injury remove the vitelline membrane around the dorsal neural plate. It is common for this procedure to produce small wounds, but the embryos will heal from these small injuries within 30 min.
4. Carve embryo-sized holes side by side in the modeling clay. A convenient way to do this is to melt the end of a Pasteur pipette and fire-polish it closed, such that a small embryo-sized ball of glass forms at the end of the pipette. This glass ball can be used to carve embryo-sized holes in the modeling clay. Note that the glass balls need to be around 1.3 mm in diameter for *Xenopus laevis* and 0.8 mm for *Xenopus tropicalis.*
5. Fill each transplantation petri dish with 0.4× MMR. Place embryos without vitelline membranes dorsal side downward and ventral side up, such that each embryo is more than halfway into the modeling clay hole. It is important that embryos do not have open wounds facing the modeling clay, as the wounds stick to the modeling clay. Place both host and donor embryos side-by-side in modelling clay, and immobilize embryos by gently pressing the modeling clay around them.
6. With classical tools, hold the already immobilized embryo, puncture, and cut around the dotted area shown in Fig. 1a, b with an eyebrow knife or a tungsten needle. At this stage, the aVBI is not under the bulky endoderm and the tissue to transplant is only of a few cell layers. Hence, using the blastocoel space, cut around the aVBI with fast but steady upward movements, holding the embryo with forceps. Be systematic and have host and donor embryos side by side. It is relatively straightforward to exchange the explanted tissue between donor and host embryos with the help of forceps.
7. Move onto the next transplant pair, but be careful moving the petri dish as fluid movement can lead to displacement of the transplants. One can also use fire bent coverslips to hold the transplants in place. Allow the transplants to heal for at least 5 min. This will occur faster when the strength of the solution is lowered by topping up with 0.1× (*Xenopus laevis*) or 0.01× MMR (*Xenopus tropicalis*). 30 min later, clean debris and

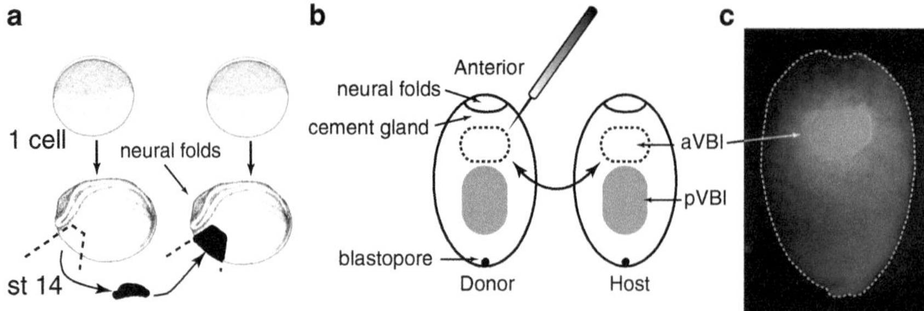

Fig. 1. Experimental design: Anterior ventral blood islands are transplanted from stage 14 labeled donor embryos into unlabeled host embryos. (**a**) Schematic ventral view of a transplant pair. (**b**) Reference to the localization points in order to dissect the appropriate region of the aVBI. (**c**) Image of a transplanted embryo, with white light here shown as fluorescent signal over a brightfield image of the embryo. White *dashed line* shows embryo contour.

"unlock" transplanted embryos from modeling clay, change solutions to lower salt strength, and avoid any unnecessary movement before transporting transplanted embryos to the incubator. The migration of isolated cells from the aVBI (anterior ventral blood islands) starts between stage 18 and 20, which is around 2–3 h after stage 16. Always use sterile solutions, and work under aseptic conditions, to avoid the need to use antibiotics.

3.4. Labeling Primitive Myeloid Progenitors by Microinjection at the 32-Cell Stage

An alternative way to label primitive myeloid progenitors is by injecting a tracer (*GFP* mRNA, *lacZ* mRNA or micro-ruby) into the C1 blastomere of regularly cleaving embryos (Fig. 2). The C1 blastomere does not only give rise to primitive myeloid progenitors, but given that PMCs are highly migratory, this approach can lead to a very good and easy method to label PMCs in the embryo. This technique also allows for the manipulation of the genetic program that controls myeloid development in chimeric embryos, by the injection of mRNA or morpholinos (18).

1. If using mRNA, prepare linear templates for mRNA synthesis by restriction digest, purify templates with phenol:chloroform and ethanol precipitation, and store at 0.5 μg/μl in RNase-free water. Produce 5′ capped mRNA using the mMessage Machine Kit (Ambion, SP6, T7, T3). Store in aliquots at –80°C and defrost just before injecting; keep on ice.
2. Obtain embryos as in the above sections. Since this technique relies on the availability of regularly cleaving embryos, it is essential that embryos be prescreened at the four-cell stage for regular cleavage patterns and distinct pigmentation between their dorsal versus ventral side (Fig. 2). Obtaining regularly cleaving embryos depends on the batch of eggs; so several females may be needed to identify those that give regularly cleaving embryos.

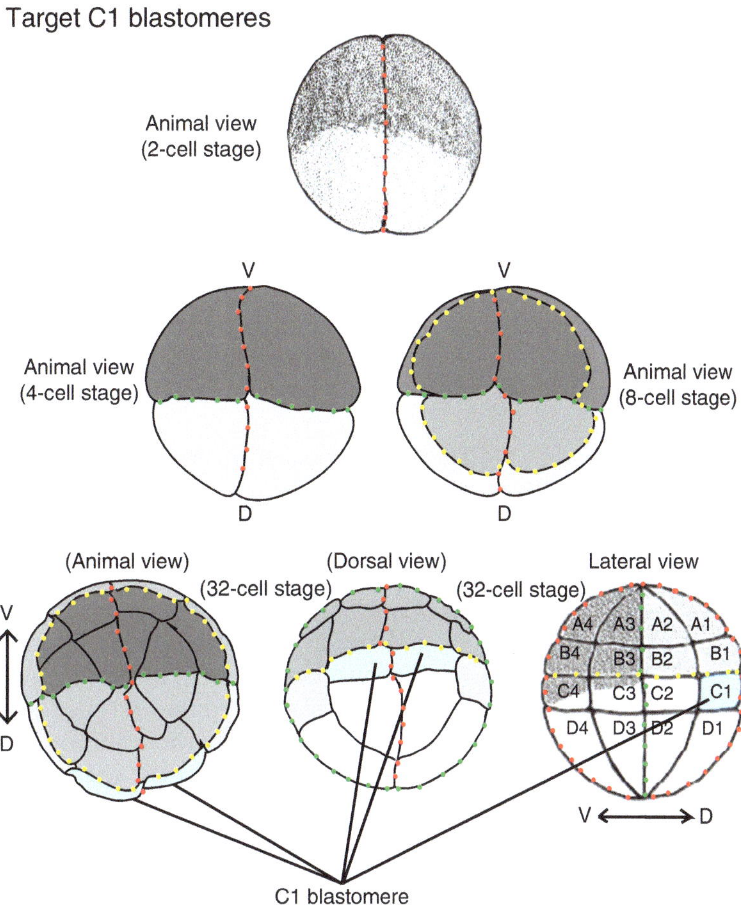

Fig. 2. Targeting C1 blastomere. The C1 blastomeres locate in the dorsal side of the vegetal hemisphere. Pigment, size, and shape can help you identify these two blastomeres, provided that the embryo cleaves regularly. Therefore, one should only select embryos that are symmetrically dividing from the two-cell stage. When reaching four-cell stage, orient the embryos so that the dorsal side, where blastomeres are smaller and less pigmented, is facing down from the animal view. Keep the embryos in the same position until they reach 32-cell stage. From animal view, the C1 blastomeres (*light blue*) are two cells located at the very bottom of the embryos. Rotate the embryos so that you can see their dorsal side. From there one should observe two slim C1 blastomeres with regular rectangle shape and no pigmentation located on both sides of the dorsoanterior midline (*red dot*). *Red*, *green*, and *yellow dots* outline the first, second, and third cleavages, respectively. *V* ventral, *D* dorsal.

3. Move regularly cleaving embryos to nylon mesh dish with injection buffer (2% ficoll in 0.1× MMR). If available, use a nylon mesh, as this allows embryos to remain immobile and it is easier to keep them oriented for longer. Gently orient embryos as shown in Fig. 2 and wait for the embryos to reach the 32-cell stage. When using *Xenopus tropicalis* natural matings, one can set one dish aside and start screening and

orienting another dish every 30 min. That will provide a constant supply of embryos at the 32-cell stage for injections over a period of several hours.

4. Prepare injection setup as described above (e.g., Picospritzer II) and inject dye, mRNA or morpholinos into the C1 blastomere, remove injection buffer, and incubate at the appropriate temperature, 16–18°C for *Xenopus laevis* and 23–26°C for *Xenopus tropicalis.* If a fluorescent dye or a fast folding fluorescent protein is used, embryos can be screened at neurula stages for good labeling within the aVBI. Single cell migration away from the aVBI can be visualized and scored under a UV stereomicroscope.

3.5. Imaging the Primitive Myeloid Cells in Live Embryos

3.5.1. General Considerations

The best approach for imaging fluorescently labeled PMC migration in living embryos depends largely on the stage of development. Cursory imaging and screening of embryos can be done using a fluorescence stereomicroscope.

1. *Xenopus* embryos are opaque at the early developmental stages and the higher yolk content of early embryos leads to more autofluorescence than older embryos. At stages up to the mid-tail bud stages, epifluorescence microscopy is the preferred imaging method, as the embryos are largely opaque. However, after the tail bud stages, the embryos become progressively more translucent; therefore at these stages, transmitted light (fluorescent and confocal microscopy) is preferred. To analyze single cell behaviors, time-lapse fluorescence video microscopy at a high resolution and magnification is required. This can be done using standard or confocal inverted fluorescence microscopes combined with the use of water or oil lenses with high numerical aperture (NA). However, when inverted microscopes are used, the embryo will need to be mounted in glass-bottomed petri dishes.
2. An additional consideration is the working distance of the objective lenses, especially when the embryos get larger, as higher magnification generally means shorter working distances. An additional advantage of inverted microscopes is that one can fit a temperature-controlled device, which is particularly important if *Xenopus tropicalis* embryos will be used for a long-term imaging. Upright microscopes with water dipping lenses are a good choice to visualize myeloid migration as they offer a very good balance between NA and working distance, but they limit the kind of dishes, plates, or chambers that can be used as devices that are too tall may damage the objective lenses. Therefore special care must be taken when using upright microscopes.
3. A powerful comprehensive analysis of myeloid cell behavior comes from the four-dimensional (4D) imaging. 4D reconstructed images taken with confocal systems offer great detail

in shape and volume. High sensitivity and speed scanners make possible Z-stack acquisition in time lapses. An unavoidable problem in long-term imaging is drifting. Expensive image analysis software packages such as Imaris or Volocity have drift correction algorithms. A good budget option is the use of the TurboReg PlugIn for ImageJ (Thevenaz P., Biomedical Imaging Group, Ecole Polytechnique Federale de Lausanne, http://bigwww.epfl.ch/thevenaz/turboreg/ and http://rsbweb.nih.gov/ij/index.html).

3.5.2. Anesthesia and Immobilization of Embryos

A key aspect of long-term imaging (e.g., time-lapse microscopy) is keeping the embryos from moving. This can be achieved by anesthesia, mounting the embryos in a viscous solution, using a solid grid, housing them in modeling clay or agarose wells, or any combination of the above. Which mounting procedure is best largely depends on the stage of development and the time and type of the imaging protocol (Please see Notes 3 to 7). There are two main tips to keep the embryos static over long periods of imaging: (1) tailor the size of the hole or groove so no excessive space is allowed for embryos to move or drift; (2) fine tune the anesthesia protocol as this is essential for good quality imaging (details will be explained below).

1. For imaging the very first steps of primitive myeloid progenitor migration, which occurs before muscle contractions initiate, there is no need to anesthetize the embryos. Mount the embryos in modeling clay or in agarose wells. For embryos younger than stage 23, make round holes, but for older embryos, make grooves so that the elongating embryo can grow. For tail bud stage embryos, since their length expands quickly during development, it is necessary to leave extra space along the anterior–posterior axis of the grooves. Holes in the agarose can be generated by melting the agarose locally on a pre-casted agarose-coated petri dish with the heated tip of a glass Pasteur pipette made into a glass rod. A similar approach to the one used to make holes in modeling clay. To make grooves or textured grip surfaces, use forceps or blades on the agarose or modelling clay.
2. For embryos that are older than the mid-tail bud stage (i.e., stage 25), one needs to anesthetize them. Embryos up to the early tadpole stages (i.e., stage 35) can be maintained in an anesthetic solution of 0.005–0.01% tricaine/MS222 in MMR with good results.
3. With swimming tadpoles, induce anesthesia with a high concentration of anesthetic for a very short period of time, rinse, and then maintain anesthesia at a lower level. Place the tadpoles in 0.1% MS222/tricaine solution only until the tadpoles stop swimming or stop responding to tapping stimulus. Wash them well in 0.1 or 0.01× MMR (*Xenopus laevis* or *Xenopus*

tropicalis, respectively). If the evaluation requires longer than 10 min, transfer the tadpoles into an anesthetic solution containing between 0.005 and 0.01% tricaine/MS222.

4. For sensitive imaging tasks (e.g., long-term time lapses and confocal microscopy) embed embryos in a viscous solution of 2% methylcellulose and 0.005% tricaine/MS222. This can be made in advance and stored at −20°C. Prior to use, defrost the 2% methylcellulose solution well in advance to allow time for bubbles to float to the top. At this concentration, the embryos are immobilized very well. However it may be hard to orient the embryos properly. If this is the case, try lowering the concentration locally with droplets of MMR solution and orient the embryo with the help of forceps.

4. Notes

1. *Xenopus tropicalis* eggs can also be "squeezed" and fertilized in vitro in an identical manner to that described above for *Xenopus laevis*. However, in vitro fertilization of *Xenopus tropicalis* eggs requires a fair amount of experience, as the fertilization rates can be lower, due to improper or careless isolation of the testes (which are smaller and more fragile than the testes of *Xenopus laevis*). In addition, fertilization rates are highest with freshly isolated testes, rather than stored testes. Therefore it is often preferable to set up natural matings for *Xenopus tropicalis* rather than to perform in vitro fertilizations. Natural matings also give a steady supply of embryos over a time window of 2–3 h. For natural matings, prime both male and female *Xenopus tropicalis* as above, and leave them to mate in a container with fresh system water, placing two females and three males per session. Once the males have amplexed the females, move each mating pair to a fresh container of system water. Once the female starts laying eggs, move the pair into fresh system water every 30 min. In this way, fertilized eggs can be collected every 30 min. It is important to be quick while transferring the mating pairs from one container to another. This is facilitated by the fact that the mating pair tends to float when not stressed by the handler. The transfer can be done with a transparent plastic cup or using both hands below and above the pair covering the eyes of both male and female.
2. Timing of *Xenopus tropicalis* transplantation is important; if animals are primed in the morning, embryos will be ready for transplants during the middle of the night due to the speed of their development. Since imaging and following the fate of your transplants is time consuming we prefer to prime later in

the first day and label embryos late in the evening. If the embryos are placed at low temperature (22°C), transplants can be performed very early the following morning.

3. We strongly recommend testing the ideal anesthetic conditions on control tadpoles and embryos before using them on experimental embryos, as anesthesia can be terminal if not titrated properly.
4. Embryos need to be prepared slightly differently if using inverted microscopes. We use MatTek glass covered bottom petri dishes to achieve the best optical transparency. For the same reason, the mounting material between embryos and the glass bottom of the petri dish need to be reduced to a minimum. Since the inverted microscopes have a limited working distance, especially at high magnification, embryos need to be positioned as close as possible to the glass bottom. When using higher magnification lenses the embryo may need to be pushed nearer the bottom of the glass by placing a coverslip over the top of the embryo and pressing down gently. Young embryos and tadpoles may be too fragile to take the entire weight of the coverslip; therefore modeling clay can be used below the coverslip to give sufficient support and avoid damage to the embryos.
5. We find that a 2% methylcellulose solution is ideal for long imaging runs as it limits big movements and twitching of the tadpoles. An alternative technique, if more complete immobilization is required, is to embed the embryos in low melting temperature agarose. Prepare 1.2% agarose stock solution and store frozen. Before use, melt 1 ml aliquots at 70°C. Then, allow for the temperature to stabilize to 37°C in a water bath close to the scope. Transfer the anesthetized embryos with a fire polished glass pipette into the agarose solution and allow a quick exchange of the aqueous solution for the agarose solution for a couple of seconds. This short time minimizes heat shock. Immediately place the tadpoles flat on a conventional dish (for upright microscopy) or a glass-bottomed one (for inverted microscopy), orienting tadpoles with the help of forceps. Wait a few minutes for the agarose to cool and gel, and top up with anesthetic solution. Now the embryos are ready for imaging. A range from 0.8 to 1.2% of agarose can be used. At 0.8% it is possible to lower the temperature of the melted agarose to 30°C, thus decreasing the level of thermal shock experienced by the embryos, especially when mounting *Xenopus laevis* embryos.
6. *Xenopus* tadpoles are mostly translucent, once the yolk reserves have been depleted. Nevertheless, the tadpoles contain pigmented melanophores all along the skin, internal organs, and blood vessels. The melanophores mask the fluorescent signal within and below the tissues that contain them. To overcome

this one can place the embryos from stage 32 in a solution containing the tyrosinase inhibitor *N*-benzyl-*N*′-phenylthiourea (PTU, ChemBridge Europe) at a concentration of 2 μg/ml.

7. *Xenopus* embryos are very sensitive to light exposure, especially UV light. For long-term time-lapse imaging, it is essential to use a computer-controlled shutter system, as well as software and camera systems with a signal integration function (e.g., Northern Eclipse, Empix). Low light and shorter exposure times will allow embryos to develop normally, with minimal phototoxicity and photobleaching.

Acknowledgements

We would like to thank Stuart Smith (NIMR, London) for tips on the use of the tyrosinase inhibitor, PTU.

References

1. Dzierzak E, Speck NA (2008) Of lineage and legacy: the development of mammalian hematopoietic stem cells. Nat Immunol 9: 129–136
2. Ciau-Uitz A, Walmsley M, Patient R (2000) Distinct origins of adult and embryonic blood in Xenopus. Cell 102:787–796
3. Lane MC, Sheets MD (2002) Rethinking axial patterning in amphibians. Dev Dyn 225: 434–447
4. Lane MC, Sheets MD (2006) Heading in a new direction: implications of the revised fate map for understanding Xenopus laevis development. Dev Biol 296:12–28
5. Lane MC, Sheets MD (2002) Primitive and definitive blood share a common origin in Xenopus: a comparison of lineage techniques used to construct fate maps. Dev Biol 248: 52–67
6. Smith SJ, Kotecha S, Towers N, Latinkic BV, Mohun TJ (2002) XPOX2-peroxidase expression and the XLURP-1 promoter reveal the site of embryonic myeloid cell development in Xenopus. Mech Dev 117:173–186
7. Costa RM, Soto X, Chen Y, Zorn AM, Amaya E (2008) spib is required for primitive myeloid development in Xenopus. Blood 112: 2287–2296
8. Ciau-Uitz A, Liu F, Patient R (2010) Genetic control of hematopoietic development in Xenopus and zebrafish. Int J Dev Biol 54: 1139–1149
9. Le Guyader D, Redd MJ, Colucci-Guyon E, Murayama E, Kissa K, Briolat V, Mordelet E, Zapata A, Shinomiya H, Herbomel P (2008) Origins and unconventional behavior of neutrophils in developing zebrafish. Blood 111:132–141
10. Warga RM, Kane DA, Ho RK (2009) Fate mapping embryonic blood in zebrafish: multi- and unipotential lineages are segregated at gastrulation. Dev Cell 16:744–755
11. Herbomel P, Thisse B, Thisse C (1999) Ontogeny and behaviour of early macrophages in the zebrafish embryo. Development 126: 3735–3745
12. Lieschke GJ, Oates AC, Crowhurst MO, Ward AC, Layton JE (2001) Morphologic and functional characterization of granulocytes and macrophages in embryonic and adult zebrafish. Blood 98:3087–3096
13. Carradice D, Lieschke GJ (2008) Zebrafish in hematology: sushi or science? Blood 111: 3331–3342
14. Crowhurst MO, Layton JE, Lieschke GJ (2002) Developmental biology of zebrafish myeloid cells. Int J Dev Biol 46:483–492
15. Bennett CM, Kanki JP, Rhodes J, Liu TX, Paw BH, Kieran MW, Langenau DM, Delahaye-Brown A, Zon LI, Fleming MD, Look AT (2001) Myelopoiesis in the zebrafish. Danio rerio. Blood 98:643–651
16. Rhodes J, Hagen A, Hsu K, Deng M, Liu TX, Look AT, Kanki JP (2005) Interplay of pu.1

and gata1 determines myelo-erythroid progenitor cell fate in zebrafish. Dev Cell 8:97–108

17. Lyons SE, Shue BC, Oates AC, Zon LI, Liu PP (2001) A novel myeloid-restricted zebrafish CCAAT/enhancer-binding protein with a potent transcriptional activation domain. Blood 97:2611–2617
18. Chen Y, Costa RM, Love NR, Soto X, Roth M, Paredes R, Amaya E (2009) C/EBPalpha initiates primitive myelopoiesis in pluripotent embryonic cells. Blood 114:40–48
19. Bukrinsky A, Griffin KJ, Zhao Y, Lin S, Banerjee U (2009) Essential role of spi-1-like (spi-1 l) in zebrafish myeloid cell differentiation. Blood 113:2038–2046
20. Palis J, Yoder MC (2001) Yolk-sac hematopoiesis: the first blood cells of mouse and man. Exp Hematol 29:927–936
21. Bertrand JY, Jalil A, Klaine M, Jung S, Cumano A, Godin IE (2005) Three pathways to mature macrophages in the early mouse yolk sac. Blood 106:3004–3011
22. Tashiro S, Sedohara A, Asashima M, Izutsu Y, Maéno M (2006) Characterization of myeloid cells derived from the anterior ventral mesoderm in the Xenopus laevis embryo. Dev Growth Differ 48:499–512
23. Hogan BM, Layton JE, Pyati UJ, Nutt SL, Hayman JW, Varma S, Heath JK, Kimelman D, Lieschke GJ (2006) Specification of the primitive myeloid precursor pool requires signaling through Alk8 in zebrafish. Curr Biol 16: 506–511
24. Smith SJ, Kotecha S, Towers N, Mohun TJ (2007) Targeted cell-ablation in Xenopus embryos using the conditional, toxic viral protein M2(H37A). Dev Dyn 236:2159–2171
25. Fantin A, Vieira JM, Gestri G, Denti L, Schwarz Q, Prykhozhij S, Peri F, Wilson SW, Ruhrberg C (2010) Tissue macrophages act as cellular chaperones for vascular anastomosis downstream of VEGF-mediated endothelial tip cell induction. Blood 116:829–840
26. Sive A, Grainger RM, Harland RM (2000) Early development of Xenopus laevis – a laboratory manual. Cold Spring Harbor Laboratory Press, Cold Spring Harbor, New York
27. Lane MC, Sheets MD (2005) Fate mapping hematopoietic lineages in the Xenopus embryo. Methods Mol Med 105:137–148
28. Walmsley M, Ciau-Uitz A, Patient R (2005) Tracking and programming early hematopoietic cells in Xenopus embryos. Methods Mol Med 105:123–136
29. Izutsu Y, Maeno M (2005) Analyses of immune responses to ontogeny-specific antigens using an inbred strain of Xenopus laevis (J strain). Methods Mol Med 105:149–158
30. Turpen JB (2005) Use of flow cytometry and combined DNA surface staining for analysis of hematopoietic development in the Xenopus embryo. Methods Mol Med 105:159–170

Chapter 12

Identification of Oocyte Progenitor Cells in the Zebrafish Ovary

Bruce W. Draper

Abstract

Zebrafish breed year round and females are capable of producing thousands of eggs during their lifetime. This amazing fecundity is due to the fact that the adult ovary, contains premeiotic oocyte progenitor cells, called oogonia, which produce a continuous supply of new oocytes throughout adult life. Oocyte progenitor cells can be easily identified based on their expression of Vasa, and their characteristic nuclear morphology. Thus, the zebrafish ovary provides a unique and powerful system to study the genetic regulation of oocyte production in a vertebrate animal. A method is presented here for identifying oocyte progenitor cells in the zebrafish ovary using whole-mount confocal immunofluorescence that is simple and accurate.

Key words: Oocyte progenitors, Oogonia, Germ cell, Zebrafish, Ovary, Vasa, Confocal immunofluorescence

1. Introduction

The zebrafish ovary is becoming an increasingly popular system for the study of ovarian development and function. Zebrafish females are capable of producing many thousands of mature eggs during their life span, laying several hundred eggs each mating. The reason for this fecundity is that the zebrafish ovary is able to produce new oocytes as adults, and therefore contains germ cells at all stages of development, from premeiotic oogonia to meiotic oocytes (1). Thus, oocyte progenitor cells are present and can be studied in the zebrafish ovary.

To date, zebrafish mutants have been identified in both forward and reverse genetic screens that affect germ cell development at many different stages. For example, mutations in several genes have been identified that affect oocyte progenitor production and/ or maintenance. Examples include the cytokine receptor CXCR4,

Kimberly A. Mace and Kristin M. Braun (eds.), *Progenitor Cells: Methods and Protocols*, Methods in Molecular Biology, vol. 916, DOI 10.1007/978-1-61779-980-8_12, © Springer Science+Business Media, LLC 2012

which is required for primordial germ cell migration to the gonad (2): *ziwi*, which is required for gonocyte survival and encodes a Piwi-like protein that is involved in piRNA function (3); and *nanos1*, which encodes the zebrafish ortholog of the Drosophila RNA-binding protein Nanos and is required for the maintenance of oocyte progenitor cells and therefore continuous oocyte production in adults (4). In addition, mutations that affect oocyte development and function have also been isolated, such as *bucky ball*, a novel, but highly conserved gene that is involved in Balbiani body assembly and the specification of the animal–vegetal axis (5, 6), and *brac1*, which is required for meiosis (7). Finally, recent advances in transgenesis technologies (8) and the identification of germ cell-specific promoters (e.g., (9)) allow for the expression of heterologous genes in germ cells to test specific hypotheses related to oocyte development and function.

The zebrafish ovary is a bilobed structure of the cystovarian type (10). The two halves join together at their posterior end to form a common oviduct that connects to the cloaca. Ovaries in zebrafish are not formed until after ~18–20 days of development (11). Prior to this age, the zebrafish gonad is bipotential and contains only symmetrically diving gonocytes (11). Sex in zebrafish is determined by a nonchromosomal mechanism that is not well understood, but all animals appear to initially activate a female developmental program as gonads of all larval fish that are between 18 and 25 days begin producing early-stage oocytes (11). It is known that germ cells are required for female development as agametic animals invariably develop as phenotypic males (12). There is also growing support for a model which proposes that there are a threshold number of early-stage oocytes that need to be produced during the 18–25 days time period in order to stabilize female development (7). In the absence of attaining this threshold, female development is arrested, the oocytes present die by apoptosis, and male development initiates. Thus, in normal development, fluctuations in the numbers of gonocytes produced by 18 days and/or the numbers of gonocytes that enter meiosis may be the main sex determinant in zebrafish. By 30 days of development, however, sex has been established and ovaries and testes are clearly distinguishable based on morphology and gene expression patterns (7, 13).

While zebrafish are not considered adults until they are capable of spawning, their ovaries have begun continuous production of new oocytes by 25 days of development. Oocyte development in zebrafish has been subdivided into five main stages based on morphological, physiological, and biochemical criteria (1). The earliest stage oocytes, called Stage IA, are 7–20 μm in diameter, have entered meiosis, and are arrested at the diplotene stage, but have not yet recruited follicle cells. By contrast, the fully grown and mature eggs that are ready to spawn, called Stage V, are 730–750 μm in diameter, which equates to a greater than 10^6-fold increase in volume relative to early-stage IA oocytes (1). Oocyte progenitor cells are premeiotic germ cells that are capable of mitotic proliferation. In zebrafish, oocyte progenitor cells, also called oogonia, can be identified based on their size (7–10 μm in diameter), their expression of germ cell-specific markers, such as Vasa and *ziwi*, and their characteristic nuclear morphology (1, 3, 4, 9, 14). In the adult ovary, oocyte progenitor cells also localize to a discrete zone on the lateral surface of the ovary that has been termed the germinal zone (4). While it is clear that oocyte progenitor cells are present in the adult ovary, for practical reasons they are easier to image in younger fish whose ovaries do not yet contain large, late-stage oocytes. In addition, most mutations that affect oocyte progenitor development and function have defects that manifest in young ovaries. For example, *nanos1* mutant ovaries contain few or no oocyte progenitors by 40 dpf. Therefore this chapter will outline a procedure for identifying oocyte progenitors in 30–40 day-old ovaries.

2. Materials

2.1. Tissue Fixation and Permeabilization

1. Phosphate-Buffered Saline (PBS): Prepare a 10× stock using 1.37 M NaCl; 27 mM KCl; 100 mM Na_2HPO_4; 17.6 mM KH_2PO_4. Autoclave and store at room temperature. Prepare a working solution by diluting tenfold distilled and autoclaved water.
2. Paraformaldehyde (PFA) Stock: Prepare a 8% paraformaldehyde stock solution as follows: 40 g paraformaldehyde in 400 ml water. While stirring heat to 60°C. Add 1 N NaOH drop-wise until paraformaldehyde is dissolved. Adjust to 500 ml, pour through Whatman filter, aliquot into 50 ml conical tubes, and freeze at -20°C. Once thawed, use within 1 week.
3. 1× Fixative: 4% PFA in 1× PBS.
4. Tricaine stock (3-amino benzoic acid ethyl; a.k.a MS222 and Finquel): Prepare tricaine stock solution by mixing 4 g tricaine powder with 900 ml water. Adjust pH to 7.0 with sodium bicarbonate. Adjust final volume to 1 L, aliquot into 50 ml conical tubes, and store at -20°C.
5. Sylgard-coated Petri dishes: Mix Sylgard 184 Silicone Elastomer parts A and B in 10:1 ratio (w/w). Fill Petri dishes (10 cm and 3.5 cm) one-third full and incubate at 50°C overnight, or until hardened.
6. Dumont No. 5 forceps.
7. Angled scissors: e.g., Supercut 10 cm angled scissors.
8. Permeabilization solution: PBS + Triton X-100 (PBT): 0.5% (v/v) Triton X-100 in 1× PBS.
9. Proteinase K: Make a 10 mg/ml proteinase K stock solution in PBT. Store 500 μl aliquots at -20°C.

2.2. Confocal Immunofluorescence

1. Antibody block solution: 10% normal goat serum (NGS) in PBT.
2. Antibody incubation solution: 2% NGS in PBT.
3. Primary antibody: rabbit anti-Vasa polyclonal sera (14).
4. Secondary antibody: Alexa Fluor 594 goat anti-rabbit IgG.
5. Nuclear stain: 4′,6-diamidino-2-phenylindole (DAPI): Prepare a 10 mg/ml stock in water and store at 4°C.
6. Vacuum grease.
7. 25 × 75 × 1 mm Microscope slides.
8. 22 × 22 mm coverslips.

3. Methods

Oocyte progenitor cells localize to relatively discreet regions in the zebrafish ovary. In a 40–50 day post-fertilization ovary, most of the oogonia (i.e., premeiotic cells) localize to the lateral and medial margins of the relatively planar ovary. By contrast, cells that have entered meiosis are most often found on the dorsal surface of ovary and away from the margin. In later stage ovaries (e.g., 3 months old) both oogonia and early meiotic oocytes (Stage IA) localize to a lateral zone on the surface of the ovary called the germinal zone (4, 9). It is likely the margins of the early ovary are the precursors to the germinal zone of the adult ovary. Because of the overall organization of oocyte precursor cells within the ovary, it is important to maintain the ultrastructure during the immunostaining procedure. The ovaries of zebrafish are very fragile and will readily fall apart if isolated from the fish prior to fixation. For this reason, it is important that the ovaries be fixed in situ, and then isolated after fixation.

The method outlined below details how to identify oocyte progenitors in the zebrafish ovary using a combination of immunostaining for the germ cell-specific marker Vasa, and staining nuclei with the DNA dye DAPI, which allows accurate germ cell staging based on nuclear morphology. This procedure takes 3 days to complete.

3.1. Preparation of Ovaries for Immunostaining

1. Fix ovaries in situ overnight in 1× Phosphate-buffered saline/4% paraformaldehyde as follows: First, euthanize fish with an overdose of tricaine for 5 min (10 ml tricaine stock in 100 ml fish water) followed by 5 min in ice water. Next, use angled scissors to decapitate fish, cutting just posterior to the gills, and then open the body cavity by cutting along the ventral midline. Fix no more than 40–50 day-old fish/25 ml fixative.
2. Following fixation, wash fish 3 × 10 min in PBT.

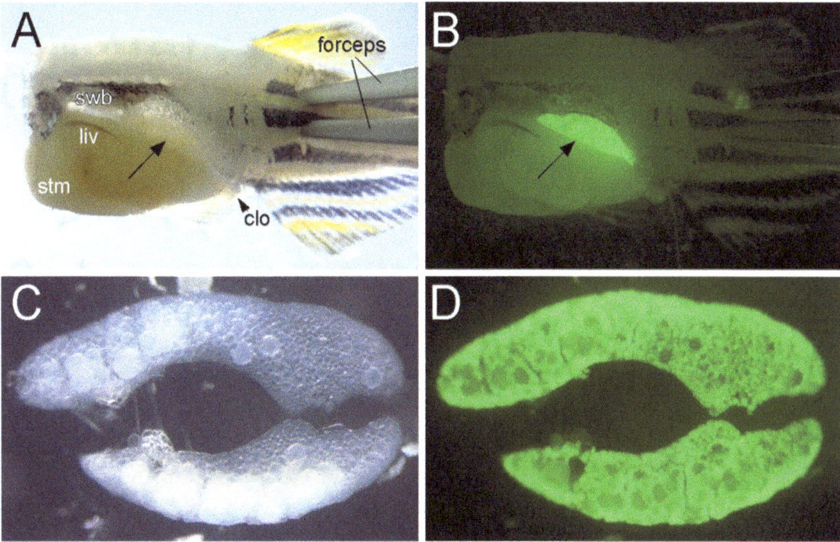

Fig. 1. Isolation of ovaries from 45 day-old Zebrafish. (**a** and **b**) After removal of the skin and body wall musculature that covers the coelomic cavity, the internal organs can be identified. The *arrow* points to the left ovary. (**b**) For ease of visualizing the location of the ovaries, the fish shown here carries a transgene that expresses EGFP from a germ cell-specific promoter (9). Dark-field (**c**) and fluorescent (**d**) images of the ovaries removed from the fish in (**a**). In all images, anterior is to the *left*. *clo* cloaca, *liv* liver, *stm* stomach, *swm* swim bladder.

3. Dissect ovaries from the body cavity (Fig. 1a, b). Place fish in a Sylgard-coated Petri dish in PBT. While viewing fish with the aid of a dissecting stereomicroscope, use one set of No. 5 Dumont forceps to stabilize the fish by impaling the forceps in the tail just posterior to the cloaca (Fig. 1a). With another set of forceps, carefully grasp skin and body wall muscle at the ventral midline cut and gently peel skin and muscle off of torso to expose the internal organs. The ovaries are located on each side of the fish, dorsal to the stomach and liver and ventrolateral to the swim bladder (Fig. 1a, b). After removing skin and muscle, remove stomach with liver attached. Next, remove swim bladder. Finally, both ovaries can be removed together by grabbing the cloaca and gently peeling them out of the body cavity. The ovaries may be covered by a lipid layer on their dorsal surfaces, which will cause them to float in the PBT solution. This layer can be easily removed using forceps. Fig. 1c, d shows a pair of isolated ovaries.
4. Wash dissected ovaries in PBT 3×15 min.
5. Methanol permeabilization: Rinse ovaries 2×5 min in 100% MeOH; then incubate for 1 h at –20°C. Next, rehydrate ovaries in a graded MeOH series in PBS (70%, 50%, 30%) (see Note 1).

6. Proteinase K permeabilization: Place ovaries in a solution of 10 μg proteinase K/ml PBT for 10 min at room temperature (see Note 2).
7. Inactivate the proteinase K by refixing ovaries in 4% PFA/1× PBS for 20 min.
8. Wash ovaries 3 × 10 min in PBT.

3.2. Whole-Mount Antibody Staining of Ovaries

1. Block 1 h in PBT + 10% Normal Goat Serum (NGS) at room temperature with gentle rocking.
2. Remove block and apply primary antibody diluted in PBT + 2% NGS. Optimal concentration of primary antibody should be determined empirically. For rabbit polyclonal serum, between 1:1,000 and 1:5,000 are good starting points. For anti-Vasa antibody, 1:2,000 gives excellent staining (Fig. 3). Incubate overnight at 4°C with gentle rocking.
3. Remove primary antibody (see Note 3) and wash two times briefly with PBT, then wash 4 × 15 min, followed by 2 × 30 min in PBT at room temperature with gentle rocking.
4. Apply secondary antibody (e.g., goat anti-rabbit Alexa Fluor 594) at appropriate dilution in PBT + 2% NGS (1:500–1:1,000 for Alexa Fluor-conjugated secondary antibodies), for a minimum of 4 h at room temperature, or overnight at 4°C with gentle rocking.
5. Wash 2 × quickly, then 4 × 15 min, and then 2 × 30 min in PBT + 2 μg/ml DAPI at room temperature with gentle rocking.

3.3. Mounting Ovaries for Confocal Analysis

1. To mount ovaries for microscopy, first clear gonads by dehydrating in a glycerol series, as follows:
   - 30% glycerol in PBT + 2 μg/ml DAPI 1 h at room temperature (minimum).
   - 50% glycerol in PBT + 2 μg/ml DAPI 1 h at room temperature (minimum).
   - 75% glycerol in PBT 1 h at room temperature RT (minimum).

     Ovaries will initially float in the 30% and 50% glycerol solutions, but will sink as they equilibrate. By contrast, after equilibration in the 75% glycerol solution ovaries may not always sink.
2. Mount ovaries on microscope slides in 75% glycerol/PBT as follows: place four spots ("post") of vacuum grease on a microscope slide in a square pattern that is smaller than the coverslip (Fig. 2). A 3 ml syringe filled with vacuum grease and fitted with an 18-gauge needle that has been cut short with wire cutters makes a good dispenser. Next place the ovaries in the

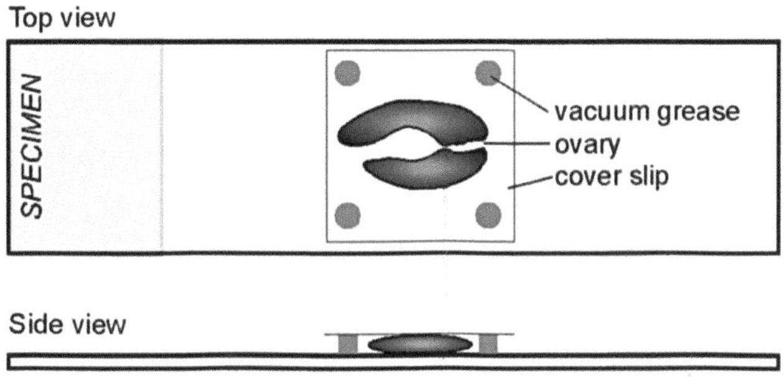

Top view

SPECIMEN

vacuum grease
ovary
cover slip

Side view

Fig. 2. Mounting ovaries for confocal microscopy. A *top* and *side* view of a microscope slide showing the positions of the vacuum grease "post" that help support the coverslip over the ovary.

middle of the square in a minimum volume of 75% glycerol/PBT. Arrange the ovaries so that dorsal is up (see Note 4). Finally, place a coverslip on the vacuum grease post and gently depress until the coverslip is in contact with, and slightly depresses, the ovary. Do not over-depress coverslip as this will distort the sample. Add enough 75% glycerol/PBT to fill underneath the coverslip.

3. Collect images of the stained sample using a confocal microscope. Figure 3 shows representative images of anti-Vasa stained 45 day post-fertilization ovaries that were counterstained with DAPI. For staging purposes, it is recommended to present the nuclear staining (DAPI) as a black and white image for maximum contrast (Fig. 3c, f).

4. Notes

1. Ovaries can be stored for several months in 100% MeOH at –20°C without significant loss of antigen.
2. Each batch of proteinase K must be tested for optimal concentration to use in this step.
3. The primary antibody can be reused several times without significant reduction in signal. If reusing, add sodium azide to 0.02% final concentration.
4. 40–50 day-old ovaries, while mostly planar, have a slight curvature such that the convex and concave surfaces represent the dorsal and ventral surfaces, respectively. Once ovaries are on the slide, they can be flipped and positioned using fine eyelashes glued to toothpicks (hair from some short-haired dog breeds work well also).

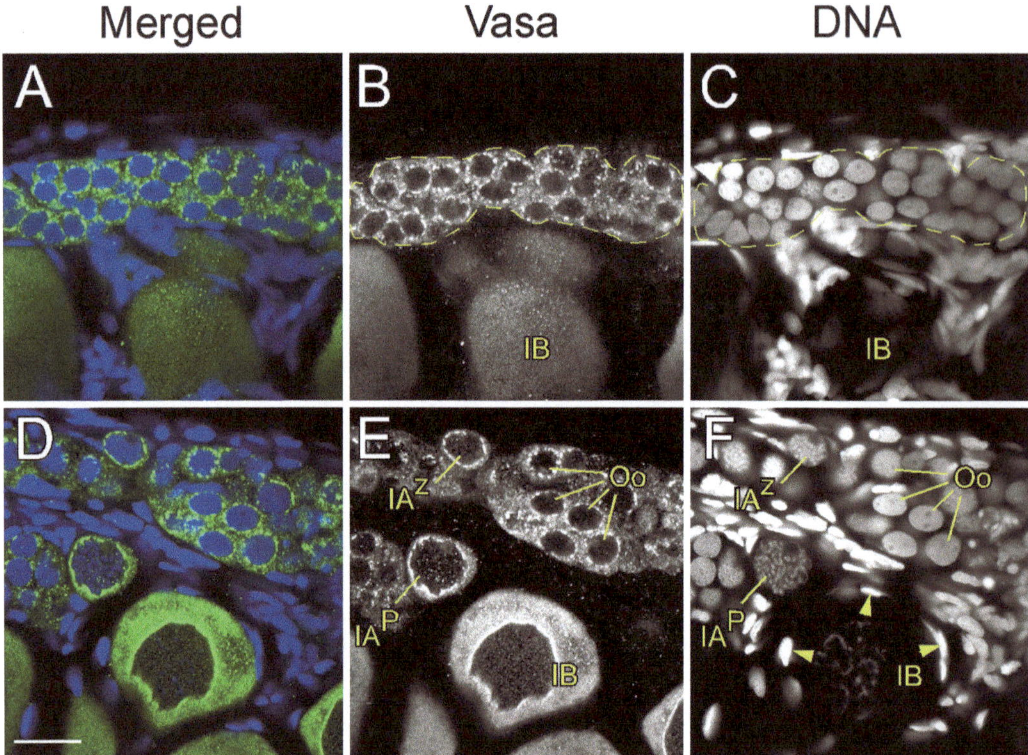

Fig. 3. Identification of oocyte progenitor cells using Vasa immunolocalization and nuclear morphology. Merged panels (**a** and **d**) show Vasa (*green*) and DNA (*blue*). Panels **b** and **e** show Vasa only. Panels **c** and **f** show DNA only. In **a–c** a cluster of premeiotic oogonia (*outlined*) are found at the lateral edge of a 45 day-old ovary. The nuclei of these cells stain uniformly with DAPI and have a single prominent nucleolus with several smaller nucleoli, which appear as holes in the staining. In **d–f** oogonia are found in close proximity to early meiotic oocytes. Chromosomes in zygotene-stage oocytes organize into a characteristic "bouquet" arrangement, while those in pachytene-stage oocytes are more loosely arranged in a larger nucleus. Stage 1B oocytes are arrested at the diplotene stage of meiosis I, are larger than Stage IA oocytes, and are surrounded by follicle cells (*arrowheads* in **f**). *Oo* oogonia, *IA^z and IA^p* zygotene- and pachytene-stage IA oocytes, respectively, *IB* stage IB oocyte. *Scale bar*: 20 μm in **c** for **a–f**.

Acknowledgments

The author would like to thank Holger Knaut for the anti-Vasa antibodies. This work was supported by a National Science Foundation grant IOS-0920637.

References

1. Selman K, Wallace RA, Sarka A, Qi X (1993) Stages of oocyte development in the zebrafish, *Brachydanio rerio*. J Morphol 218:203–224
2. Knaut H, Werz C, Geisler R, The Tubingen Screen Consortium, Nusslein-Volhard C (2003) A zebrafish homologue of the chemokine receptor Cxcr4 is a germ-cell guidance receptor. Nature 421:279–282
3. Houwing S, Kamminga LM, Berezikov E, Cronembold D, Girard A, van den Elst H, Filippov DV, Blaser H, Raz E, Moens CB, Plasterk RHA, Hannon GJ, Draper BW, Ketting

RF (2007) A role for Piwi and piRNAs in germ cell maintenance and transposon silencing in Zebrafish. Cell 129:69–82

4. Draper BW, McCallum CM, Moens CB (2007) nanos1 is required to maintain oocyte production in adult zebrafish. Dev Biol 305:589–598
5. Marlow FL, Mullins MC (2008) Bucky ball functions in Balbiani body assembly and animal-vegetal polarity in the oocyte and follicle cell layer in zebrafish. Dev Biol 321:40–50
6. Bontems F, Stein A, Marlow F, Lyautey J, Gupta T, Mullins MC, Dosch R (2009) Bucky ball organizes germ plasm assembly in Zebrafish. Curr Biol 19:414–422
7. Rodriguez-Mari A, Cañestro C, BreMiller RA, Nguyen-Johnson A, Asakawa K, Kawakami K, Postlethwait JH (2010) Sex reversal in Zebrafish *fancl* mutants is caused by Tp53-mediated germ cell apoptosis. PLoS Genet 6:e1001034
8. Kawakami K (2007) Tol2: a versatile gene transfer vector in vertebrates. Genome Biol 8:S7
9. Leu DH, Draper BW (2010) The ziwi promoter drives germline-specific gene expression in zebrafish. Dev Dyn 239:2714–2721
10. Dodd JM (1977) The ovary of nonmammalian vertebrates. In: Zukkerman S, Weir BJ (eds) The ovary, 2nd edn. Academic, New York, pp 219–263
11. Uchida D, Yamashita M, Kitano T, Iguchi T (2002) Oocyte apoptosis during the transition from ovary-like tissue to testes during sex differentiation of juvenile zebrafish. J Exp Biol 205:711–718
12. Slanchev K, Stebler J, de la Cueva-Mendez G, Raz E (2005) Development without germ cells: The role of the germ line in zebrafish sex differentiation. Proc Natl Acad Sci USA 102:4074–4079
13. Siegfried KR, Nüsslein-Volhard C (2008) Germ line control of female sex determination in zebrafish. Dev Biol 324:277–287
14. Knaut H, Pelegri F, Bohmann K, Schwarz H, Nusslein-Volhard C (2000) Zebrafish vasa RNA but not its protein is a component of the germ plasm and segregates asymmetrically before germline specification. J Cell Biol 149:875–888

Chapter 13

FACS Analysis of the Planarian Stem Cell Compartment as a Tool to Understand Regenerative Mechanisms

Belen Tejada Romero, Deborah J. Evans, and A. Aziz Aboobaker

Abstract

Planarians provide a relatively simple model system in which to study stem cell dynamics and regenerative phenomena. As with other systems understanding the dynamics of stem cell and stem cell progeny is crucial in order to get at the molecular mechanisms orchestrating stem cell biology. Planarians have an abundant adult stem cell population that can be observed using Fluorescence-Activated Cell Sorting (FACS). This approach allows different subpopulations of stem cells and their progeny to be monitored and sorted for downstream studies in response to different regenerative scenarios, drug treatments, or RNAi knockdown of genes required for regenerative events.

Key words: Stem cells, Progenitor cells, Neoblasts, Planarians, Flow cytometry cell sorting, FACS, Radiation sensitive cells, Regeneration

1. Introduction

Planarians have become a powerful system with which to investigate basic questions about regeneration of new structures from preexisting adult tissue (1). Like other biological processes the mechanisms of regeneration may be conserved between different phyla, so beginning to understand the molecular basis of regenerative phenomena in a simple metazoan is likely to have broad implications for understanding stem cell maintenance, differentiation, and functional integration within the context of adult regeneration (2).

Planarians have undergone significant maturation as a molecular genetic model system over the last decade (3). The application of RNAi to planarians has allowed gene function to be studied directly (4, 5), making it possible to develop a mechanistic description of the regenerative process. Allied to the existence of a draft sequence of *Schmidtea mediterranea* and EST resources for other species,

Kimberly A. Mace and Kristin M. Braun (eds.), *Progenitor Cells: Methods and Protocols*, Methods in Molecular Biology, vol. 916, DOI 10.1007/978-1-61779-980-8_13, © Springer Science+Business Media, LLC 2012

robust sensitive in situ hybridization and immunohistochemistry approaches to query mRNA and protein expression, a growing number of useful markers, and the transfer of other relevant technologies have led to a rapid expansion in planarian research.

Planarian regenerative capacity is facilitated by a population of planarian adult stem cells (pASCs), classically called neoblasts (1, 3). These cells are collectively totipotent proliferating cells that can replace any planarian tissue, including the brain, reproductive system, and germ line. Now famous classical experiments repeated in schools the world over show that whole animals can be reconstituted by the regeneration process from small starting pieces from almost anywhere along the body plan.

Amputation in planarians triggers a swift wound healing response to minimize tissue loss. A strong muscular contraction occurs near the wound site, which reduces the surface area of the wound. Within 30–45 min of amputation, a thin layer of dorsal and ventral epithelial cells spread and come together to cover the wound. These cells lose their morphological characteristics as they cover the site of injury (6).

After wound closure pASCs increase proliferative rates in two characteristic peaks over the next 3 days (7). Post-mitotic progeny of proliferating pASCs form an unpigmented mass of initially undifferentiated cells at the wound site called a blastema. Lost structures are replaced within the blastema, such that after decapitation for example, anterior structures are morphologically and functionally restored after 7 days. Over the next week the remaining tissues remodel and rescale to the new size of the animal (8).

To further investigate pASCs and their progeny, and to address the possible heterogeneity of stem cells, a collection of genes expressed in pASCs and their descendants have been identified for use as specific cell markers (9–11). Identification of these genes has been performed by microarray, and more recently, RNA sequencing analysis (9–11) of genes expressed in normal animals and animals subjected to a lethal dose of γ-irradiation to remove proliferating cells. This approach identifies genes potentially expressed in pASCs and their progeny. By combining in situ hybridization with irradiation and BrdU labeling of pASC progeny, genes that specifically label early and late pASC progeny were identified (10). The results of this work by Eisenhoffer et al. (summarized in Fig. 1) are currently the only proven source of pASC progeny markers. This work broadly defined three categories of pASC and progeny markers. Markers that specifically labeled pASC were called Category 1, those that labeled early progeny (that were quick to disappear after irradiation and quick to become labeled with BrdU) were called Category 2, and those that labeled later progeny were called Category 3. These three groups of cells also have different spatial locations with the planarian mesenchyme, with pASC progeny (Category 2 and 3) positioned more peripherally than pASCs (Category 1).

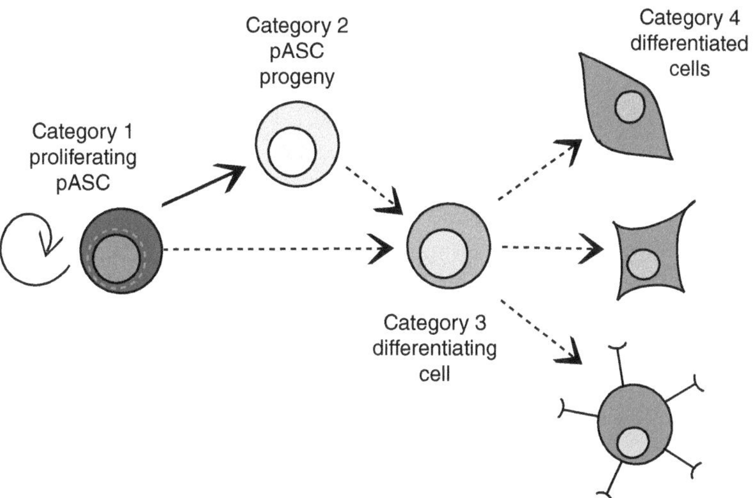

Fig. 1. Schematic representation of the simple cell lineage model proposed by Eisenhoffer et al. (9). Proliferating pASCs express Category 1 genes and are able to both self-renew and differentiate. Category 2 genes expressing cells are thought to be the pASC progeny, which are in the process of differentiating into Category 3 gene expressing cells. Category 3 cells will then differentiate into Category 4 gene expressing cells, which are differentiated. It has not yet been shown that all Category 1 cells have to pass through a Category 2 stage before expressing Category 3 genes.

Another method to look at the dynamics of pASC and their progeny is to use fluorescence-activated cell sorting (FACS) analysis. Dissociation of planarian cells and sorting by FACS were originally developed to isolate planarian neurons for in vitro culturing (12). Further development revealed that FACS analysis on planarians can also be used to robustly identify two populations; radiation-sensitive cells representing pASCs and their undifferentiated progeny, and radiation-insensitive cells representing differentiated cells (13). The radiation-sensitive population can be further subdivided into X1 cells, pASCs in G2/M of the cell cycle, and *X2* cells, which are a mixture of pASCs and their progeny (10). The subpopulations of cells separated by FACS are somewhat related to the three categories described by Eisenhoffer et al. (10). X1 cells expressing some of the Category 1 markers further confirm their stem cell identity. Likewise, *X2* cells show a mixture of Categories 1, 2, and 3 markers, as would be expected from a mixed population. Finally, the radiation-insensitive cells express markers of differentiated cells. Therefore, FACS analysis is a powerful tool for measuring changes in different planarian stem cell populations. These studies can be performed during regeneration and during homeostasis, and in different experimental backgrounds. Importantly, FACS analysis can be used to compare and understand the effects of irradiation, drug treatment, and RNAi gene knockdown (14) on each of the different cell populations. Sorting of the cell populations also allows for further profiling of gene expression by in situ hybridization or by quantitative RT-PCR (15).

The success of the FACS method mainly relies on the combination of dyes used for the analysis. Firstly, Hoechst 33342 is used to stain DNA. Hoechst is excited by ultraviolet light at 350 nm, and emits in the blue channel at a maximum of 461 nm. Hoechst is more suitable for these analyses than DAPI, since it is more lipophilic and thus enters the cells through the membrane more readily, where it then binds the minor groove of DNA. Additionally, Hoechst 33342 can be analyzed in the red channel (617 nm) to determine cell size and complexity. It is also possible to use calcein acetoxymethyl ester for this purpose, but we find that this reduces cell viability and requires a more sophisticated FACS machine than some researchers might have available. Finally, propidium iodide (PI) is used. PI intercalates into DNA, and when excited with a wavelength of 488 nm it emits at 562–588 nm. PI does not readily cross the cell membrane. A short incubation with PI therefore will only stain the DNA of cells with a broken cell membrane, i.e., dead cells. This excludes any dead cells from the analyses without difficulty.

2. Materials

2.1. Planarian Culture and Preparation of Samples

1. Planarian animals of the species *S. mediterranea* (from the laboratory of Professor E. Saló, University of Barcelona, Spain) or *Dugesia japonica* (from the laboratory of Professor K. Agata, University of Kobé, Japan) are used for FACS analysis.
2. Planarian water: Autoclaved tap water filtered through activated charcoal and buffered with 0.5 ml/L 1 M $NaHCO_3$.
3. Fresh organic calf liver.
4. Irradiation: Gamma irradiation source.
5. HU treatment: HU stock: 20 mM HU in planarian water. Store at 4°C (see Note 1).

2.2. Cell Dissociation

1. 2% L-cysteine–HCl in deionized water, brought to pH 7.2 with NaOH (see Note 2).
2. 10× CMF (Calcium magnesium free buffer): 25.6 mM $NaH_2PO_4{\cdot}2H_2O$; 142.8 mM NaCl; 102.1 mM KCl; 94.2 mM $NaHCO_3$ in deionized water, pH 7.2. Store at 4°C.
3. $CMFHE^{2+}$: 0.1% BSA, 0.5% Glucose, 30 µg/ml Trypsin inhibitor ovomucoid (Sigma); 0.5 µg/ml DNase I; 15 mM Hepes pH 7.2; 3 mM EDTA pH 8 in 1× CMF. Make fresh for each use (see Note 3).

4. 5/8 Holtfreter: 37.4 mM NaCl; 0.416 mM KCl; 0.568 mM $CaCl_2$; 0.25 mM $NaHCO_3$ in autoclaved deionized water.
5. Digestion solution (Papain): 15 U/ml Papain (Worthington Biochemical Corp.); 0.5 mM L-cysteine in $CMFHE^{2+}$. Store in single-use aliquots (300 μl) at −20°C (see Notes 4 and 5).
6. Digestion solution (Trypsin): 0.25%w/v Trypsin; 1 mM EDTA in PBS solution OR 0.25%w/v Trypsin in 5/8 Holtfreter solution. 30 μg/ml of Type II ovomucoid trypsin inhibitor may be added.
7. Stop solution: 1 mg/ml Trypsin inhibitor; 40 μg/ml DNase I; 50 μg/ml BSA in $CMFHE^{2+}$. Make fresh for each use.
8. 10× BSA and HEPES stock solutions: (1% and 150 mM, pH 7.5, respectively). Store at 4°C for a maximum of 1 month.
9. EDTA stock solution: 0.5 mM pH 8 (Invitrogen). Store at room temperature.
10. DNase I stock solution: 1 mg/ml in H_2O (2,000× for $CMFHE^{2+}$ and 25× for stop solution). Store at −20°C.
11. Trypsin inhibitor stock solution: 30 mg/ml in H_2O (1,000× for $CMFHE^{2+}$ and 30× for stop solution). Store at −20°C.
12. Hoechst 33342 stock solution: 10 mg/ml in PBS. Store at 4°C in opaque container (see Note 6).
13. Propidium iodide stock solution: 1 mg/ml in H_2O. Store at 4°C in opaque container (Sigma) (see Note 7).
14. Sharp disposable scalpels.
15. 1.5 ml low binding tubes.
16. 100 μm nylon net filter (see Note 8).

3. Methods

3.1. Planarian Culture

1. Keep planarians in an incubator at 20°C in autoclaved tap water, filtered through activated charcoal and buffered with 0.5 ml/L 1 M $NaHCO_3$.
2. Feed fresh organic calf liver once a week and change water once a week after feeding.
3. Starve for 1 week prior to experiments and throughout the course of any analysis.

3.2. Preparation of Samples for FACS

Previous to FACS analysis, the planarians must be treated in the desired way depending on the experiment performed. Some important treatments include irradiation, drug treatment, and RNAi injections. Gamma irradiation severely damages DNA, which kills proliferating cells. The pASCs begin to be lost immediately and the planarians lose almost all their proliferating stem cells after just 1 day. This is a useful control in FACS analysis to show that cells gated as pASCs are in fact proliferating cells. FACS analysis can be used to determine the effects of different drugs on the animals' different cell populations with respect to a control. This is particularly useful in the analysis of the effects of certain drugs on regeneration and on differentiation. Here, we show FACS analysis on hydroxyurea (HU)-treated worms. HU is a well-characterized drug that inhibits ribonuclease reductase and thereby the production of dNTPs. After treatment with HU, proliferating cells cannot pass through S phase of the cell cycle and eventually die.

1. Irradiate planarians with a γ-radiation source at 100 Gy to remove all stem cells. Worms are placed in a plastic dish and exposed on mass for the required time, depending on the activity of the γ-radiation source (16).
2. Treat planarians with 20 mM of HU in planarian water for 15 h before cutting; this can be done by simply diluting and HU stock solution into a final volume of 3.5 ml of planarian water in six well plates with between 3 and 4 planarians per well (17).

3.3. Cell Dissociation

The treated planarians must be dissociated into single cells and stained before they can be analyzed by flow cytometry. A schematic representation of the dissociation and the staining protocol is shown in Fig. 2 (see Note 9).

1. Prepare solutions and appropriate number of aliquots of digestion solution. Defrost on ice. Label one 1.5 ml low binding tube per sample.
2. Incubate a minimum of seven planarians of 7–8 mm in size per sample in a small Petri dish in 2% L-cysteine hydrochloride (Sigma) pH 7.0 for 2 min at room temperature to remove their external mucus.
3. Pour away the L-cysteine solution and replace with CMFHE^{2+}, taking care to remove all of the L-cysteine solution. If the digestion will be performed with trypsin, 5/8 Holtfreter or PBS containing 1 mM EDTA should be used instead of CMFHE^{2+}.
4. Cut planarians into small visible pieces with a scalpel. The scalpel should be frequently wiped to prevent the accumulation of mucus.
5. Carefully transfer the pieces to 1.5 ml low binding tubes. This is most easily performed with a plastic Pasteur pipette, transferring the small pieces with a large amount of liquid. Remove excess CMFHE^{2+} so that only 300 μl remain in the tube. Add 300 μl of digestion solution and digest the samples for 1 h in

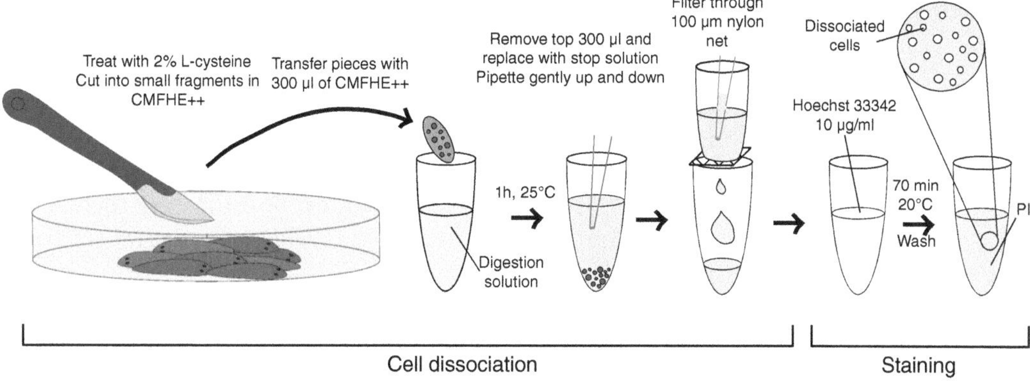

Fig. 2. Schematic representation of the FACS protocol. Planarians are treated with L-cysteine and cut into small visible pieces. These are transferred into an Eppendorf tube and digested with papain for 1 h. The digestion is then stopped and the cells are mechanically dissociated by pipetting up and down. The mixture is subsequently filtered and stained with Hoechst 33342 for 70 min and with PI for 5 min. The cells are then ready for FACS analysis.

at 25°C. The solution should not be mixed and the tubes should not be moved. In the case of digestion with trypsin, remove all of the liquid and replace with the digestion solution (trypsin).

6. Stop the digestion by removing half of the digestion solution and replacing with the same amount of stop solution at 25°C. Care must be taken to remove the top 300 µl, so that the pieces are not displaced and the cells are not lost. This step is not performed if digesting with trypsin.
7. Mechanically dissociate the digested pieces by pipetting up and down and the suspension filtered by passing though a 100 µm nylon net (see Note 10).
8. If digesting with trypsin, wash the cells twice at this point by centrifugation at 500×*g* for 5 min and resuspend in 5/8 Holtfreter.

3.4. Staining

1. Adjust the cell concentration to 5×10^5 cells/ml and incubate cells in 10 µg/ml of Hoechst 33342 for 70 min at room temperature, under slow agitation (10 rpm), in the dark.
2. Wash the dye off by centrifuging at 500×*g* for 5 min at 4°C and resuspend in $CMFHE^{2+}$.
3. Add 1 µg/ml propidium iodide.
4. The cells are now ready to undergo FACS analysis on a machine capable of measuring the fluorescence of Hoechst 33342 and PI. Keep cells at 4°C until processed to maximize cell viability (see Notes 11 and 12). Processing should be done within 45 min of PI addition.

3.5. Analysis of FACS Output

1. The analysis of cell counts is performed restrictively so as to exclude rather that include counts within the pASC compartment. Firstly, examine the Side Scatter versus the Forward Scatter plots (Fig. 3a). This analysis distinguishes cells by their internal granularity and complexity and cell shape and size, permitting the exclusion of dead cells and debris. Fig. 3a shows this plot. As a guide Gate A is drawn by estimating the size and complexity of the planarian stem cells to be similar to that of mouse or human lymphocytes.
2. Plot gated cells according to the Hoechst blue peak versus the Hoechst linear parameter or area plot (Fig. 3b). This is useful to exclude doublets or triplets of cells that did not dissociate fully. This is important, since a doublet of two G1 cells has the same DNA content as a single G2 cell. In this detection, the G2 peak has the same area (total fluorescence) as 2× G1 cells, but higher peak fluorescence. On the peak versus area plot single cells will be on the diagonal, but doublets will be below and can be gated out from further analysis.
3. Finally, plot the gated cells according to the Hoechst fluorescence on the blue channel versus fluorescence on the red channel (Fig. 3c). In the dot plot four distinct populations can be distinguished. Firstly, the proliferating radiation-sensitive cells in the cell cycle (Xsens) can be distinguished in the sample as a whole (Fig 3c). These cells are pASCs because they are cells with a small size but high DNA content (and they are irradiation sensitive, see later). There are two subpopulations within this Xsens group, X1 and *X2*. X1 cells are G2 pASCs, and *X2* cells are a combination of G1 pASCs and post-mitotic pASC progeny. The next group of cells are the Xins or radiation-insensitive cells. These post-mitotic cells are thought to represent some of the differentiated cells of the worm. The final group of cells are the PI-positive ones or dead cells. These are better observed in the log version of this plot, where a large population of them can be found in the region of low blue fluorescence and high red fluorescence (not shown). The last plot shown here is a histogram of the DNA content (Hoechst blue) of the pASCs (Xsens) (Fig. 3d). This is very similar to a typical histogram of proliferating cells, confirming that the selected cells (Xsens) are in the cell cycle. The G1 population will be a mix between pASCs in G1 and post-mitotic pASC progeny yet to develop differentiated morphology. Schematic representations of the different cells that form the cell populations are shown (Fig. 3e), also indicating where in the cell cycle these cells are (Fig. 3f).
4. After gating results can then be replotted with appropriate off machine software (we use the Weasel software, Walter and

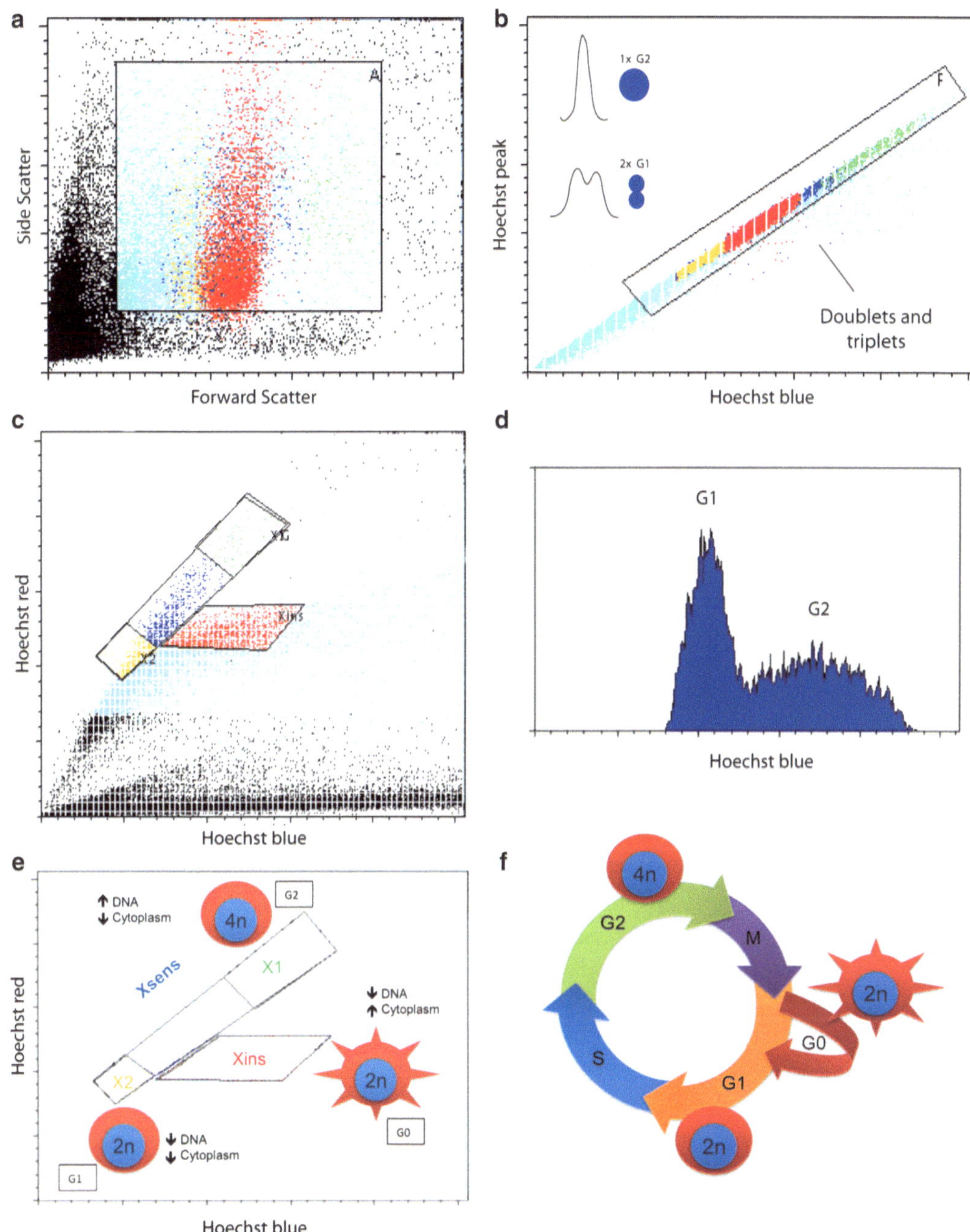

Fig. 3. Analysis of the FACS output. (**a**) Side Scatter versus Forward Scatter dot plot. Box A gates live cells, based on cell size and cellular complexity. (**b**) Hoechst peak versus Hoechst *blue dot plot*. This plot is used to exclude doublets and triplets, which fall out of gate F. This is based on the knowledge that even if two G1 cells as a doublet give the same DNA content as a single G2 cell on the blue channel (*x*-axis), the peak height will be half. This is represented on the cartoon in the *top left* corner. (**c**) Hoechst red versus Hoechst blue channel dot plots. This plot is used to identify the different cell populations: radiation-sensitive cells (Xsens), which contain X1 and *X*2 cell populations, and radiation-insensitive cells (Xins). (**d**) Histogram of the Xsens cell population. The G1 and G2 peaks are marked on the graph. (**e**) Schematic representation of the FACS output, including the significance of each of the cell populations and where they are on the plot. (**f**) Cell cycle diagram including what stage the cell populations from **e** are representative of.

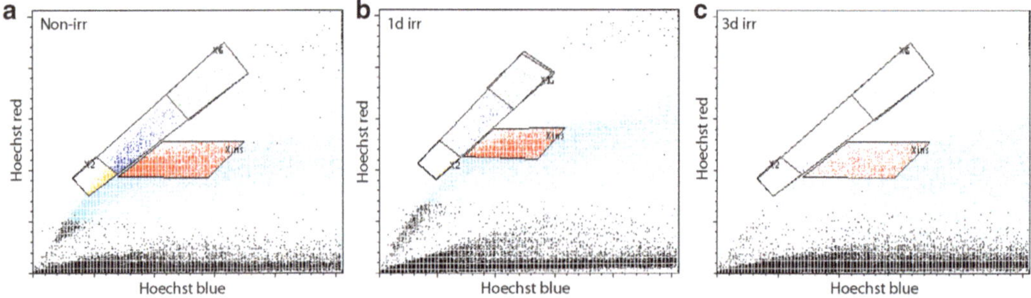

Fig. 4. Hoechst red versus Hoechst blue fluorescence dot plots showing the depletion of proliferating cells after irradiation. (**a**) Non-irradiated wild-type planarians; (**b**) Wild-type planarians 1 day after irradiation. (**c**) Wild-type planarians 3 days after irradiation.

Eliza Hall Institute of Medical Research). For an easier interpretation, we have excluded all non-proliferating and dead cells from the dot plots. Thereby, the cells shown are only the pASCs and their progeny. One of the purposes of FACS analysis is to compare the number of pASCs in different experimental backgrounds, such as irradiation and drug treatment. This technique highlights differences in the number of proliferating cells versus the number of differentiating cells by allowing direct comparison of X1 and *X2* cell populations. For ease of analysis for cell populations in RNAi backgrounds, the proportion of cells in the X1 and *X2* populations of RNAi knockdown worms can be plotted as a percentage of the control worm proportions in X1 and *X2* cell populations.

5. An example of FACS analysis on planarians is the comparison of wild-type to irradiated animals. γ-irradiation severely damages DNA, which kills proliferating cells (Fig. 4). The pASCs begin to be lost immediately and a clear difference can be measured at 1 day following irradiation. The loss of proliferating cells also leads to a depletion of the pool of early post-mitotic progeny (Category 2) many of which are within the *X2* population as they have an undifferentiated morphology (10).

6. Proliferating pASCs can be detected with an antibody against the phosphorylated Serine 10 of histone H3 (anti-H3P). This modification is present in mitotic cells from the beginning of mitosis to the end of telophase. This feature can be used to label proliferating stem cells in planarians by whole-mount immunohistochemistry. Upon injury, the pASCs rapidly begin to proliferate in two mitotic bursts. The first burst occurs at 6–12 h of regeneration, and the second one at 48–72 h of regeneration. The same pattern of proliferation can be observed when studying the FACS profile of X1 (G2/M) cells of regenerating animals. The bursts and dips in proliferation occur earlier in time than in the phosphohistone staining experiment,

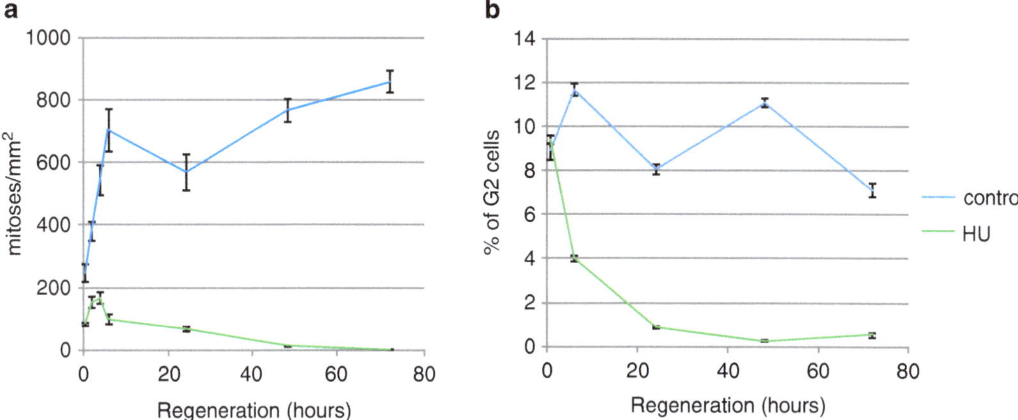

Fig. 5. Effect of HU treatment on dividing cells during planarian regeneration. (**a**) Proliferation curve of wild-type and HU-treated regenerating animals. Each point in the curve is the average number of mitotic cells stained with anti-phosphohistone H3 Serine 10 antibody (anti-H3P) in ten regenerating trunk pieces. (**b**) FACS analysis of the percentage of X1 cells in HU and control regenerating animals. In all cases the organisms were incubated in 20 mM HU for 15 h prior to cutting. The animals were kept in the same solution throughout regeneration, and in the case of FACS analysis HU was added to all the buffers.

which is consistent with the fact that the FACS is measuring the number of G2 cells, as well as those in early M, and the phosphohistone staining is measuring specifically the number of cells in M. Upon hydroxyurea (HU) treatment, pASCs cannot pass through S phase of the cell cycle and are lost. Figure 5 shows the effect of HU treatment on regeneration as analyzed by phosphohistone staining and FACS. This is a simple example of how FACS analysis can be used to successfully determine the effects of certain drugs on the different cell populations of planarians (18).

4. Notes

1. Hydroxyurea (HU) is highly toxic, particularly in its powder form. An eyeshield and mask must be worn.
2. L-cysteine is unbuffered and is thus difficult to pH accurately. A range from pH 7.0–7.3 is generally sufficient for effective mucus removal. Adjustment of the pH should be performed with 2 M NaOH.
3. Instead of CMFHE or 5/8 Holfreter, PBS containing 1 mM EDTA can also be used.
4. Papain can be substituted by other proteinases, such as 0.25%w/v Trypsin (see Subheading 2). Digestion times will vary with different enzymes and their concentrations, so times

should be adjusted accordingly. For example, Trypsin digestion with 30 μg/ml Trypsin inhibitor should be performed for around 1 h, whereas without the Trypsin inhibitor 5 min is sufficient. Note that digestion times will vary with the size of the cut pieces and the batch of enzyme.

5. The digestion solution should be kept frozen in 300 μl aliquots. These take a long time to defrost and should be fully defrosted before starting the dissociation.
6. Hoechst 33342 is toxic when in contact with skin or when swallowed or inhaled. Handle with caution. Prepare stock solution from powder under a fume hood.
7. Propidium iodide is an irritant and a possible mutagen. Handle with care.
8. A square of nylon net melted onto the bottom of a 1.5 ml Eppendorf tube that has had the end sliced off is ideal and can be washed and reused. Alternatively, nylon nets with plastic surrounds are available from cell culture plastics suppliers.
9. Throughout the dissociation protocol it is essential to work fast and accurately to minimize cell death and maximise yield. All the steps must be performed immediately after each other, since the cells only remain alive for a few hours after dissociation.
10. A wash step may be performed at this stage. Many cells will be lost with this wash step, so if performed, it is recommended to increase the starting number of planarians to a minimum of 10.
11. The first time the cell dissociation protocol is performed it is advisable to check the cells under the fluorescent microscope to confirm that cell viability and dissociation are sufficient. If cells show excessive clumping a longer digestion time can be used. Conversely, if cell viability is low indicated by large amounts of PI staining, shorten the digestion period and perform a wash step before staining the cells to remove PI-positive cellular debris (see Note 9).
12. The amount of stain necessary will vary from preparation to preparation. This must be worked out each time and should be the same for all of the samples that are compared against each other.

Acknowledgements

We thank members of Europlannet for help with establishing Planarian FACS analysis in our group.

References

1. Agata K (2003) Regeneration and gene regulation in planarians. Curr Opin Genet Dev 13: 492–6
2. Brockes JP, Kumar A (2008) Comparative aspects of animal regeneration. Annu Rev Cell Dev Biol 24:525–49
3. Reddien PW, Sanchez AA (2004) Fundamentals of planarian regeneration. Annu Rev Cell Dev Biol 20:725–57
4. Newmark PA, Reddien PW, Cebria F, Sanchez AA (2003) Ingestion of bacterially expressed double-stranded RNA inhibits gene expression in planarians. Proc Natl Acad Sci USA 100(Suppl 1):11861–5
5. Reddien PW, Bermange AL, Murfitt KJ, Jennings JR, Sanchez AA (2005) Identification of genes needed for regeneration, stem cell function, and tissue homeostasis by systematic gene perturbation in planaria. Dev Cell 8:635–49
6. Sanchez AA (2004) Regeneration and the need for simpler model organisms. Philos Trans R Soc Lond B Biol Sci 359:759–63
7. Wenemoser D, Reddien PW (2010) Planarian regeneration involves distinct stem cell responses to wounds and tissue absence. Dev Biol 344:979–91
8. Salo E (2006) The power of regeneration and the stem-cell kingdom: freshwater planarians (Platyhelminthes). Bioessays 28:546–59
9. Rossi L, Salvetti A, Marincola FM, Lena A, Deri P et al (2007) Deciphering the molecular machinery of stem cells: a look at the neoblast gene expression profile. Genome Biol 8:R62
10. Eisenhoffer GT, Kang H, Sanchez AA (2008) Molecular analysis of stem cells and their descendants during cell turnover and regeneration in the Planarian Schmidtea mediterranea. Cell Stem Cell 3:327–39
11. Blythe MJ, Kao D, Malla S, Rowsell J, Wilson R et al (2010) A dual platform approach to transcript discovery for the planarian Schmidtea mediterranea to establish RNAseq for stem cell and regeneration biology. PLoS One 5:e15617
12. Asami M, Nakatsuka T, Hayashi T, Kuo K, Kagawa H, Agata K (2002) Cultivation and characterization of planarian neuronal cells isolated by fluorescence activated cell sorting (FACS). Zoolog Sci 19:1257–65
13. Hayashi T, Asami M, Higuchi S, Shibata N, Agata K (2006) Isolation of planarian X-ray-sensitive stem cells by fluorescence-activated cell sorting. Dev Growth Differ 48:371–80
14. Scimone ML, Meisel J, Reddien PW (2010) The Mi-2-like Smed-CHD4 gene is required for stem cell differentiation in the planarian Schmidtea mediterranea. Development 137:1231–41
15. Hayashi T, Shibata N, Okumura R, Kudome T, Nishimura O et al (2010) Single-cell gene profiling of planarian stem cells using fluorescent activated cell sorting and its “index sorting” function for stem cell research. Dev Growth Differ 52:131–44
16. Gonzalez-Estevez C, Arseni V, Thambyrajah RS, Felix DA, Aboobaker AA (2009) Diverse miRNA spatial expression patterns suggest important roles in homeostasis and regeneration in planarians. Int J Dev Biol 53: 493–505
17. Salo E, Baguna J (1984) Regeneration and pattern formation in planarians. I. The pattern of mitosis in anterior and posterior regeneration in Dugesia (G) tigrina, and a new proposal for blastema formation. J Embryol Exp Morphol 83:63–80
18. Evans DJ, Owlarn S, Tejada Romero B, Chen C, Aboobaker AA (2011). Combining classical and molecular approaches elaborates on the complexity of mechanisms underpinning anterior regeneration. PLoS One 6(11): e27927

Chapter 14

Clonal and Lineage Analysis of Melanocyte Stem Cells and Their Progeny in the Zebrafish

Robert C. Tryon and Stephen L. Johnson

Abstract

The study of melanocyte biology in the zebrafish presents a highly tractable system for understanding fundamental principles of developmental biology. Melanocytes are visible in the transparent embryo and in the mature fish following metamorphosis, a physical transformation from the larval to adult form. While early developing larval melanocytes are direct derivatives of the neural crest, the remainder of melanocytes develop from unpigmented precursors, or melanocyte stem cells (MSCs). The Tol2 transposable element has facilitated the construction of stable transgenic lines that label melanocytes. In another application, integration of Tol2 constructs makes possible clonal analysis of melanocyte and MSC lineages. Drugs that block melanin synthesis, ablate melanocytes, and block establishment of MSC populations allow the interrogation of this model system for mechanisms of adult stem cell development and regulation.

Key words: Melanocyte, Stem cell, Clonal analysis, Lineages, Zebrafish, Tol2

1. Introduction

The distinctive pigment pattern of the adult zebrafish quickly led to the identification of spontaneous mutants (*sparse, rose, and leopard*) affecting melanocyte development, allowing initial exploration of the fundamental genes responsible for pigment stripe morphology (1–3). Genetic screens have identified additional alleles that inform our understanding of embryonic melanocyte development (4, 5) and genes involved in adult metamorphosis (6).

The most fundamental technique for evaluating mutants or testing the effect of drugs on melanocyte development and homeostasis is the counting of melanocytes themselves. Melanocytes are first visible by 30 h post-fertilization (hpf) and quickly expand in number over the next 2 days (Fig. 1). Approximately 420 melanocytes are visible by 3 days post-fertilization (dpf), representing ~90% of all larval melanocytes that

Kimberly A. Mace and Kristin M. Braun (eds.), *Progenitor Cells: Methods and Protocols*, Methods in Molecular Biology, vol. 916, DOI 10.1007/978-1-61779-980-8_14,

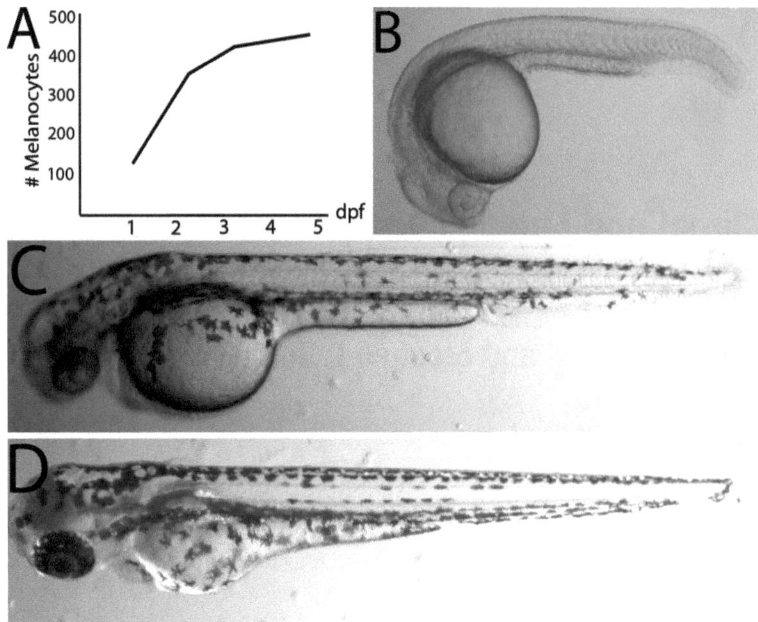

Fig. 1. Melanocyte development in the zebrafish. (**a**) Timeline of melanocyte development in the larvae. (**b**) Zebrafish larvae at 1 dpf just prior to pigmented melanocytes becoming visible. (**c**) At 2 dpf, melanocytes are visible and migrating ventrally. (**d**) At 3 dpf, the ontogenetic, neural crest derived melanocytes have nearly reached the stereotypic dorsal, ventral, yolk, and lateral (*right and left*) stripes.

will develop prior to metamorphosis, which occurs around 14 dpf (7, 8). Wild-type melanocytes frequently abut one another, making accurate counts difficult to obtain. Epinephrine or norepinephrine treatment contracts the melanocytes after 4 dpf (4). Alternatively, the use of the mutant line *mlpha*^{j120} (9), which has constitutively contracted melanocytes, is particularly useful for phenocopying epinephrine treatment in large-scale applications, such as drug screens (10).

One property of some mutants, including *sparse*, is that melanocytes initially differentiate and melanize before ultimately undergoing apoptosis (2). Since melanocytes have a distinct cellular marker (melanin) and often take days to be extruded from the skin, it can be difficult to tell if the melanocytes are living or dead. The transgenic line Tg(fTyrp1 > eGFP)^{j900} expresses a GFP reporter driven by the *Takifugu rubripes* tyrosinase-related protein 1 (Tyrp1) promoter and is specifically expressed in melanocytes (11). Use of this transgenic line allows detection of cell death by looking for extinguishment of the GFP marker and can be combined with the *mlpha*^{j120} mutation to facilitate scoring of living melanocytes.

Zebrafish have a noted capacity for regenerating many tissue types, and properties of melanocyte regeneration were initially explored in fin amputation studies (12, 13). Following amputation of the caudal fin, the blastema regenerates all tissues including bone, artery, skin, and pigment cells, including melanocytes. Drugs that block melanin synthesis, like phenylthiocarbamide (PTU) (14), can be used to test hypotheses about melanocyte precursors in the regenerate. This work has shown that regenerated melanocytes in the fin arise from unpigmented precursors, or melanocyte stem cells (MSCs), rather than existing differentiated melanocytes (13).

Since melanocytes are dispensable in a laboratory setting, they may be ablated via physical (15) or chemical means (11, 16). Chemicals such as 4-hydroxyanisole (4-HA) that molecularly mimic endogenous ligands specifically metabolized in melanocytes produce cytotoxic by-products that ablate melanocytes (16). Following cell death and washout of the drug, MSCs are able to regenerate the ablated melanocyte population and can be studied for their regenerative potential. Consequently, genetic screens using melanocyte ablating drugs have identified novel genes involved in melanocyte regeneration (17).

An ErbB3 mutant that has deficiencies in melanocyte development at metamorphosis has led to the use of ErbB inhibitors such as AG1478 to effect MSC populations (18). Combining AG1478 and 4-HA treatments has made possible the study of MSC establishment during development and regulation of MSCs in the larvae (19). Use of these drugs with PTU has shown that melanocytes found in the early embryo are direct derivatives of the neural crest, developing between 1 and 3 dpf. Following this developmental window, a small proportion (~10%) of embryonic melanocytes are added in a regulative fashion from MSCs (20).

The development of the Tol2 transposable element as a tool for generating transgenic lines in the zebrafish has revolutionized zebrafish genetics (21). This tool can similarly be used to generate clones for studying developmental questions of lineage. Injection of small amounts of transposon harboring a melanocyte-specific reporter construct permits the study of individual precursor cells that give rise to both ontogenetic melanocytes and MSCs. In addition, the use of ubiquitous promoters, such as EF1α (22), allows the study of precursor cells that may give rise to multiple pigment cell types including melanocytes, xanthophores, and iridiphores in the zebrafish (23). An alternative approach for studying clones in the zebrafish was pioneered using gamma-irradiation (24). In a similar fashion, the *albino* mutant has unpigmented melanocytes and can be utilized to study the lineages of melanocytes and MSCs. Disruption or ablation of the *albino*$^{+}$ allele by applying X-irradiation to heterozygous individuals (*alb*^{b4}/+) expressing the fTyrp1 > eGFP transgene results in clonally related albino melanocytes that can be clearly visualized on a pigmented melanocyte background.

2. Materials

2.1. Counting Melanocytes

1. Egg Water: 0.06 ppt scientific grade marine salt (Coralife) in carbon-filtered water. Add 1.2 g of salt to 20 L carbon-filtered water. Store at room temperature (RT).
2. Tricaine methanesulfonate (3-amino benzoic acid ethyl ester or 3-aminobenzoate) (Sigma Aldrich). For stock solution dissolve 400 mg tricaine powder in 97.9 mL DD water and 2.1 mL 1 M Tris (pH 9). Adjust pH to ~7. Store at RT.
3. Epinephrine (Sigma Aldrich) or norepinephrine (Sigma Aldrich). Working solution: 5 mg/mL in H_2O. Can be stored at RT for up to 1 month.
4. TC thumb operated tally counter.
5. Mutant zebrafish line *mlpha*^{j120}, deficient in melanophilin a (9). Available from lab of Steve Johnson (Washington University in St. Louis) and Zebrafish International Research Consortium (ZIRC).
6. Transgenic zebrafish line Tg(fTyrp1 > eGFP)^{j900}, expressing eGFP under the control of the *Takifugu rubripes* tyrosinase-related protein 1 promoter (25). Available from lab of Steve Johnson (Washington University in St. Louis) and ZIRC.

2.2. Reverse Labeling of Melanocytes Using PTU

1. Phenylthiocarbamide (Sigma-Aldrich). Stock solution: 200 mM in ethanol (store at RT). Working solution: Dilute to 200 μM in carbon-filtered water (adults) or egg water (embryos) prior to use.
2. Tricaine methanesulfonate (see Subheading 2.1).
3. Razor blade.

2.3. Birthdating Melanocytes Using fTyrp1 > eGFP Expression and PTU

1. Tg(fTyrp1 > eGFP) ^{j900} (see Subheading 2.1).
2. PTU (see Subheading 2.2).
3. Stereomicroscope with fluorescence filters for GFP detection.

2.4. Drugs for Ablating Melanocytes and Melanocyte Stem Cells

1. Dimethyl sulfoxide (DMSO) (Sigma-Aldrich).
2. 4-hydroxyanisole (4-HA) (Sigma-Aldrich). Stock solution: 10 mg/mL in DMSO. Store at −20°C in 50 μL aliquots. Do not refreeze after thawing for use.
3. AG1478 (4-(3-Chloroanilino)-6,7-dimethoxyquinazoline, Calbiochem), an ErbB kinase inhibitor. Stock solution: 20 mM in DMSO. Store at −20°C in 20 μL aliquots. Do not refreeze after thawing for use.

2.5. Clonal Analysis of Melanocyte Lineages Using Tol2 Transposon Labeling

1. Plasmid DNA containing fTyrp1 > eGFP reporter flanked by Tol2 transposon elements or EF1α > GFP reporter flanked by Tol2 transposon elements.
2. Ambion mMessage mMachine SP6 kit (Ambion, Inc.).
3. Plasmid containing transposase open reading frame.
4. Phenol red (Sigma Aldrich); 1% solution in sterile milliQ water.
5. Sterile milliQ water.
6. Sutter P-87 Micropipette puller (Sutter Instrument Co.).
7. Glass thin walled capillary with filament;1.0 mm O.D, 0.75 mm I.D., 4 in. length (World Precision Instruments, Inc.).
8. Glass plate or slide covered with paraffin film.
9. Razor blade.
10. Dissecting scope (40× magnification) with reticle.
11. Sterile 30 G1 precision glide needle (Becton Dickinson).
12. Glass syringe, 25 μl capacity.
13. MPPI-3 Pressure Injector with micropipette holder kit (Applied Scientific Instrumentation, Inc.).
14. Tank with compressed N_2.
15. Micro-manipulator (World Precision Instruments, Inc.).
16. Mineral oil (Sigma Aldrich) in Petri dish.
17. Grooved silicon pad for holding fertilized eggs.
18. Stereomicroscope with fluorescence filters for GFP detection.

2.6. Lineage Analysis Using X-Ray Induced Clones

1. *Albino* mutant zebrafish line *alb*$^{b4/b4}$ (ZIRC).
2. Tg(fTyrp1 > eGFP)^{j900} (see Subheading 2.1).
3. X-ray machine (Faxitron Cabinet X-Ray System—Model 43855D) (see Note 1).
4. Stereomicroscope with fluorescence filters for GFP detection.

3. Methods

3.1. Counting Melanocytes

1. To anesthetize fish, use 1 mL stock tricaine solution per 25 mL egg water.
2. Once embryos are anesthetized, transfer five fish to a clean Petri dish and remove excess egg water. Add 100 μl of 5 mg/mL solution of epinephrine or norepinephrine to the embryos and wait 5–10 min for melanocytes to contract (Fig. 2a, b).
3. Once melanocytes are confirmed to be contracted, transfer embryos to fresh egg water containing tricaine so fish stay

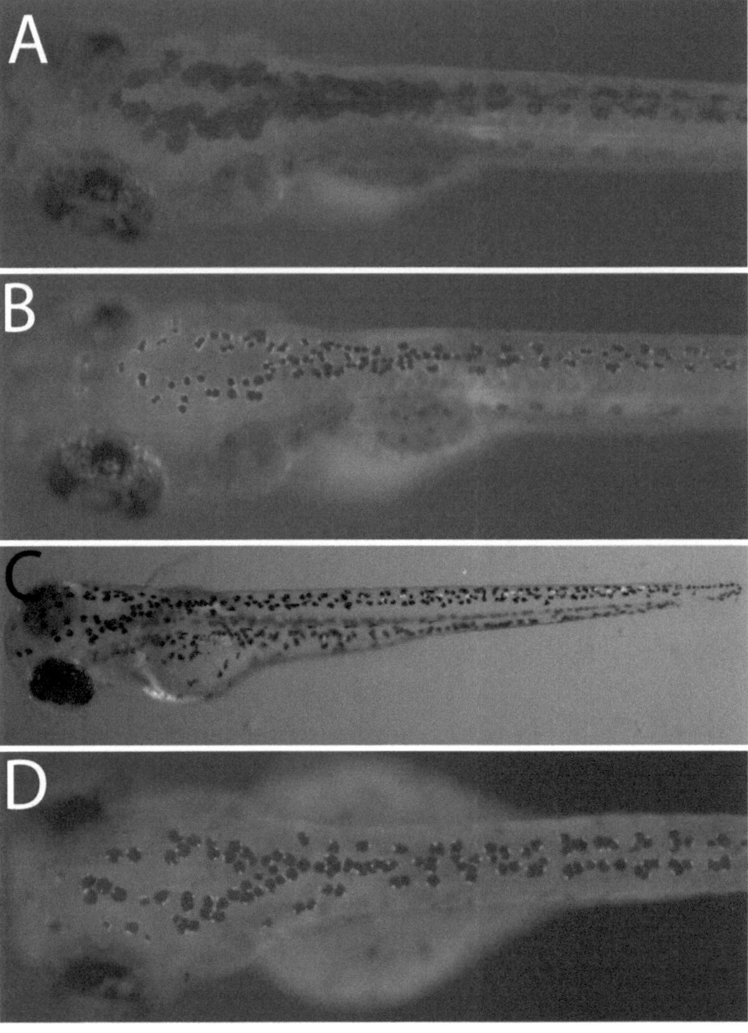

Fig. 2. Techniques for visualizing individual melanocytes. (**a**) Wild-type zebrafish at 5 dpf expressing fTyrp1 > eGFP transgene. (**b**) 10 min epinephrine treatment (5 mg/mL) of fish shown in (**a**), which facilitates counting cells and clearly shows cells expressing GFP. (**c**) The *mlpha*^{j120} mutant line has constitutively contracted melanocytes that facilitate counting cells. (**d**) The fTyrp1 > eGFP transgene on a *mlpha*^{j120} background allows unambiguous identification of GFP + melanocytes.

immobilized (see Note 2). Count melanocytes under a dissecting microscope with a TC thumb counter.

4. To facilitate counting, focus on the five embryonic melanocyte stripes (one dorsal, two lateral, one ventral, one yolk) one at a time.
5. Adult zebrafish can be anesthetized in a one-half strength tricaine solution (0. 5 mL per 25 mL) and epinephrine treated (5 mg/mL) concurrently. By the time fish are anesthetized

melanocytes are typically contracted. If not fully anesthetized by time of melanocyte contraction, add a small additional amount of tricaine.

6. Transfer fish from tricaine/epinephrine solution to a Petri dish with a slotted spoon and quickly proceed to count melanocytes on a dissecting microscope. If fish begins to wake from anesthetic, return to water until immobile and proceed counting.
7. The *mlpha*^{j120} line (Fig. 2c) is particularly useful for counting melanocytes without the need for epinephrine treatment. In combination with fTyrp1 > eGFP expression (Fig 2d) melanocytes can be both counted and determined if alive when evaluating mutant phenotypes or effects of drugs on melanocyte development.

3.2. Reverse Labeling of Melanocytes Using PTU

1. Anesthetize adult fish in tricaine.
2. Once fish are unresponsive, blot the fish several times on a paper towel to remove excess water, and place on a Petri dish.
3. Using a sharp razor blade, firmly press down on the caudal fin approximately one-half the distance between the distal tip of the fin and where the fin emerges from the body. Detach excised fin tissue and discard.
4. Place fish back into water to recover from anesthetic.
5. Following recovery, transfer fish to 200 μM solution of PTU to block melanin synthesis in newly developing melanocytes (Fig. 3b). PTU solution should be replaced every third day to ensure that melanin synthesis remains blocked.
6. To reveal pigmentation in regenerated melanocytes, remove fish from PTU and allow several days for melanin synthesis to resume and pigmentation to develop (Fig. 3c).

3.3. Birthdating Melanocytes Using Tg(fTyrp1 > eGFP)^{j900} and PTU

1. Breed Tg(fTyrp1 > eGFP)^{j900} adults and rear embryos (26).
2. By 2 dpf, hundreds of ontogenetic melanocytes are visible and embryos can be sorted for those that inherited the transgenic GFP marker (see Note 3).
3. By 3 dpf, the majority (~90%) of ontogenetic embryonic melanocytes have developed. At this point, transfer embryos to a solution of egg water containing 200 μM PTU to block melanin synthesis in remaining melanocytes yet to differentiate.
4. Change egg water with PTU every 2 days.
5. At 7 dpf, screen embryos for GFP+, unmelanized pigment cells (Fig. 4).

3.4. Drugs for Ablating Melanocytes and Melanocyte Stem Cells

1. Embryos can be produced either by in vitro fertilization or natural breeding (26).

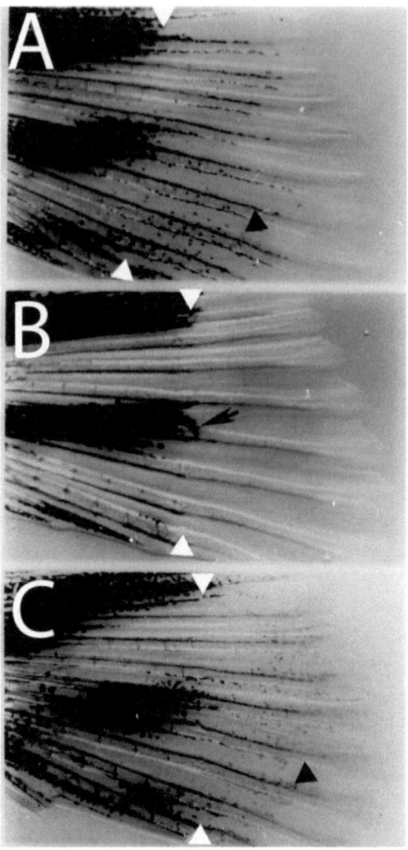

Fig. 3. Reverse labeling of melanocytes using PTU. (**a**) Fin regeneration in an untreated caudal fin. *White arrowheads* indicate amputation plane and *black arrowhead* indicates regenerated pigmented melanocyte. (**b**) Fin regeneration in a PTU treated caudal fin. Note the lack of pigmented melanocytes in regenerate, except for a few melanocytes immediately distal to the amputation plane that preceded the amputation (*black arrow*). (**c**) Washout of PTU following fin regeneration (similar to **b**) reveals the presence of melanocytes (*black arrowhead*) that develop from unpigmented precursors.

2. Dilute stock solution of AG1478 to 3 μM in egg water prior to application. AG1478 treatment is optimal between 8 and 48 hpf (Fig. 5a). While melanocytes that are derived directly from the neural crest develop normally, later developing melanocytes that require a MSC intermediate fail to develop (Fig. 5e).
3. Dilute stock solution of 4-HA to 4 μg/mL in egg water prior to application. To block or kill ontogenetic melanocytes prior to their development, apply 4-HA at 1 dpf for 2 days (Fig. 5a, c). To kill previously differentiated melanocytes, apply 4-HA for 2 days (Fig. 5a) until melanocytes appear small and punctate and signs of detritus are evident.

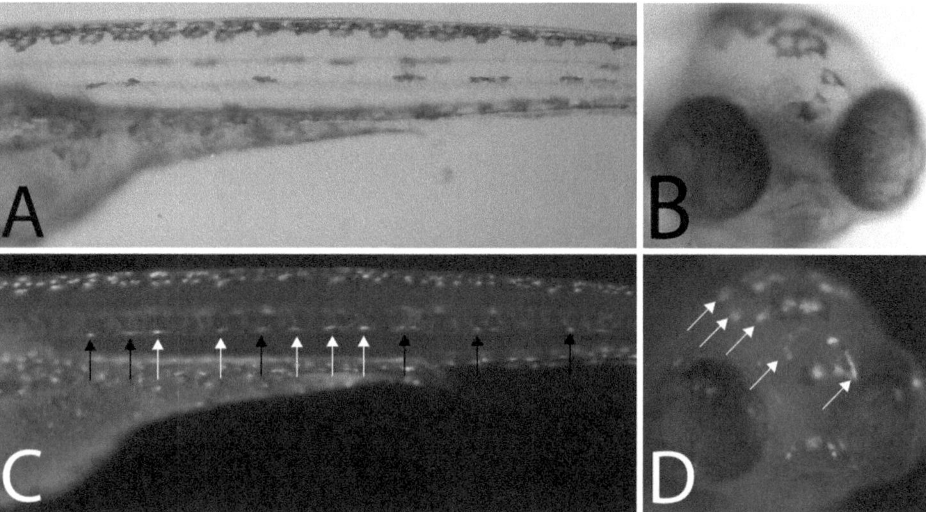

Fig. 4. Birthdating melanocytes using Tg(fTyrp1 > eGFP)[j900] in conjunction with PTU (**a**) Trunk and (**b**) head of 5 dpf zebrafish larvae treated with PTU at 2.5 dpf, blocking melanin synthesis in MSC derived embryonic melanocytes in the lateral stripe. (**c**, **d**) Epifluorescence of (**a**, **b**) reveals additional melanocytes that are not visible in bright field. *Black arrows* are previously differentiated melanized, GFP + melanocytes. *White arrows* are more recently differentiated unmelanized, GFP + melanocytes.

4. To regenerate melanocytes from MSCs, washout 4-HA. Between 1 and 2 days post-washout, new melanocytes should be visible in all canonical embryonic zebrafish stripes, except for the yolk stripe which fails to regenerate (Fig. 5d) (16).
5. AG1478 and 4-HA treatments can be combined in the larvae for testing models of MSC regulation (Fig. 5f) (19).

3.5. Clonal Analysis of Melanocyte Lineages Using Tol2 Transposon Labeling

1. Prepare capped transposase mRNA using Ambion mMessege mMachine SP6 kit. Dilute to 75 ng/μL and store at –70°C in 5 μL aliquots.
2. Prepare injection needles by pulling 4 in. 1 mm thin capillary needles under the following conditions: [H = 329, P = 150, V = 100, T = 200, Psi = 150], resulting in two usable injection needles per pull (see Note 4). Place pulled needle on a glass slide covered with paraffin and view under a dissecting microscope. At full magnification (40×), align sharp tip of pulled needle with reticle markings and cut at a 45° angle with a razor blade so that the sharp, cut end is approximately one reticle tick wide.
3. Prepare injection solution consisting of: 1 μL fTyrp1 > eGFP: Tol2 plasmid (15 ng/μL), 1 μL transposase mRNA (75 ng/μL), 1 μL phenol red, and 12 μL sterile milliQ water (see Note 5).
4. Load 5–7 μL of injection solution into a glass syringe with a sterile 30 gauge needle. Insert 30 gauge needle into the blunt

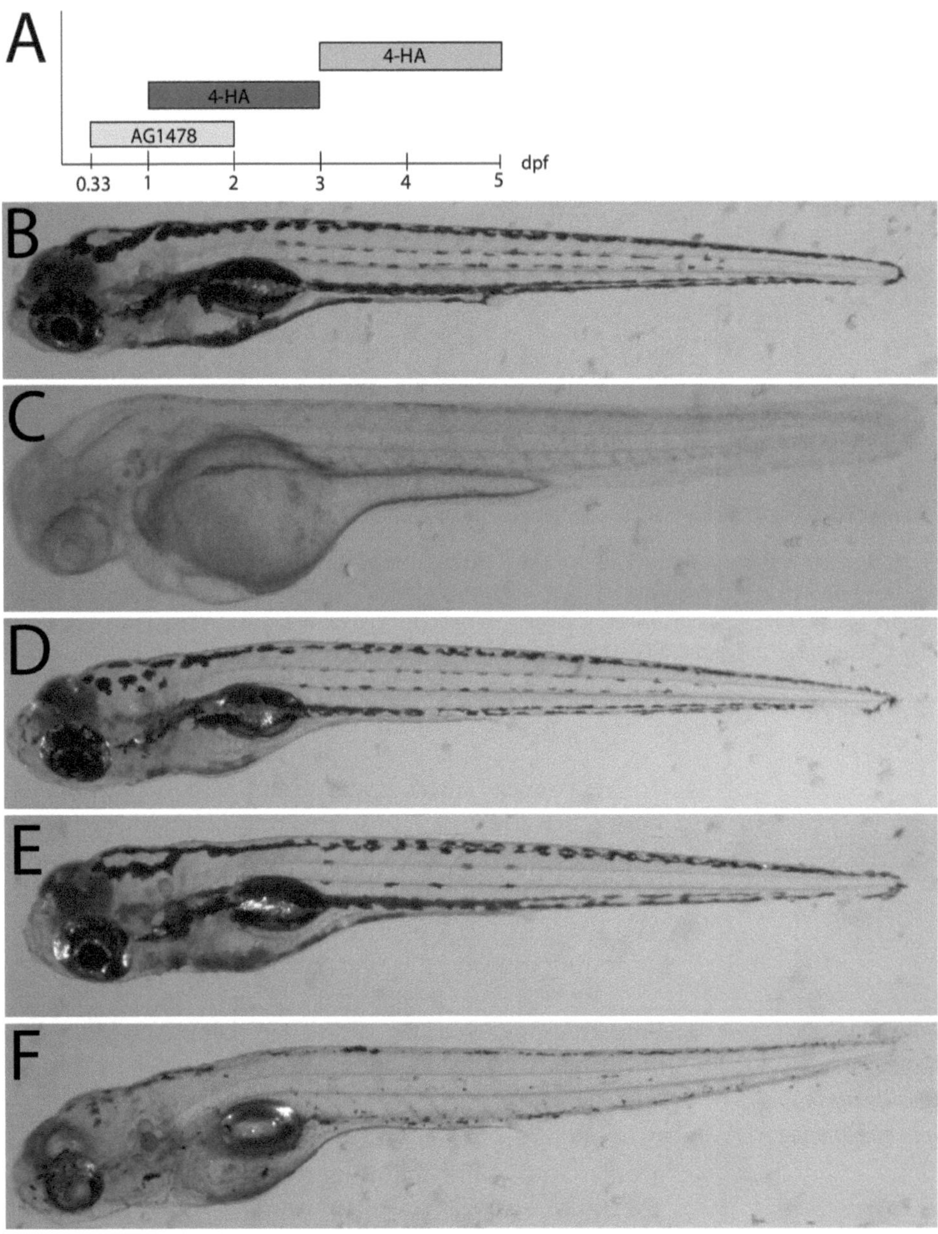

Fig. 5. Drugs that ablate melanocytes and MSCs (**a**) Timeline of drug applications. Early 4-HA treatment ablates melanocytes prior to their melanization. Late 4-HA treatment ablates differentiated melanocytes. AG1478 treatment from 8 to 48 hpf blocks establishment of MSCs. (**b**) 8 dpf untreated zebrafish. (**c**) 2 dpf zebrafish treated with 4-HA beginning at 1 dpf. (**d**) 8 dpf zebrafish treated with 4-HA between 1 and 3 dpf fully regenerates. (**e**) 8 dpf zebrafish treated with AG1478 has normal development of neural-crest derived ontogenetic melanocytes but fails to fill in the lateral stripes as seen in (**b**) and (**d**). (**f**) 8 dpf zebrafish treated with AG1478 and 4-HA largely fails to regenerate melanocytes. The majority of pigment visible is detritus that has not yet cleared from the embryo.

side of prepared injection needle and transfer injection solution. Solution should wick to the sharp, pulled point of the injection needle.

5. Insert injection needle into micropipette holder and tighten firmly without breaking the needle.
6. Open tank of compressed N_2 and set pressure to 30 psi. Turn on pressure injector unit.
7. Adjust micro-manipulator to ~45 ° angle, and place tip of needle into a Petri dish one-half full of mineral oil, just below the surface. Center the tip of the injection needle immediately next to reticle and pulse one time to release a single injection of labeled solution. Adjust pressure on pressure injector unit until desired volume, approximately five reticle ticks wide at 40× magnification, is consistently released on multiple pulses. Phenol red makes the injection solution clearly visible as it is injected into the mineral oil. Leave the injection needle submerged in mineral oil to avoid drying out and clogging the tip of needle when waiting for embryos to be bred.
8. Produce one clutch of zebrafish embryos via in vitro fertilization to ensure consistent, uniform timing of development. 15–20 min following application of sperm to eggs, cytoplasmic flow of the yolk into the one-cell embryo should be visible.
9. Transfer fertilized embryos to a grooved silicone pad lying in a Petri dish with a Pasteur pipet and remove excess egg water to reduce excessive rolling and movement of embryos while attempting to inject.
10. Carefully pierce the chorion of the one cell embryo with the injection needle and continue to puncture into the yolk, aiming for a location just underneath the visible single cell. Apply a single pulse to inject plasmid-transposase solution. Carefully remove the needle from the embryo (see Note 6).
11. Continue to inject remainder of eggs from the clutch until done (see Note 7).
12. Transfer injected embryos back into a Petri dish with egg water and place at 28.5° (standard temperature).
13. By 6–8 hpf, clean up embryos and remove all damaged, dead, or dying embryos. Allow embryos to grow at ~50 per Petri dish in egg water under standard conditions.
14. At 3 dpf, screen through injected embryos with a stereomicroscope and set aside all fish with GFP + melanocytes (Fig. 6a–c).
15. Procedure for EF1α > GFP:Tol2 plasmid is identical. Clones generated with this transposon are not limited to melanocytes since the promoter is ubiquitously expressed (Fig. 6d–f).

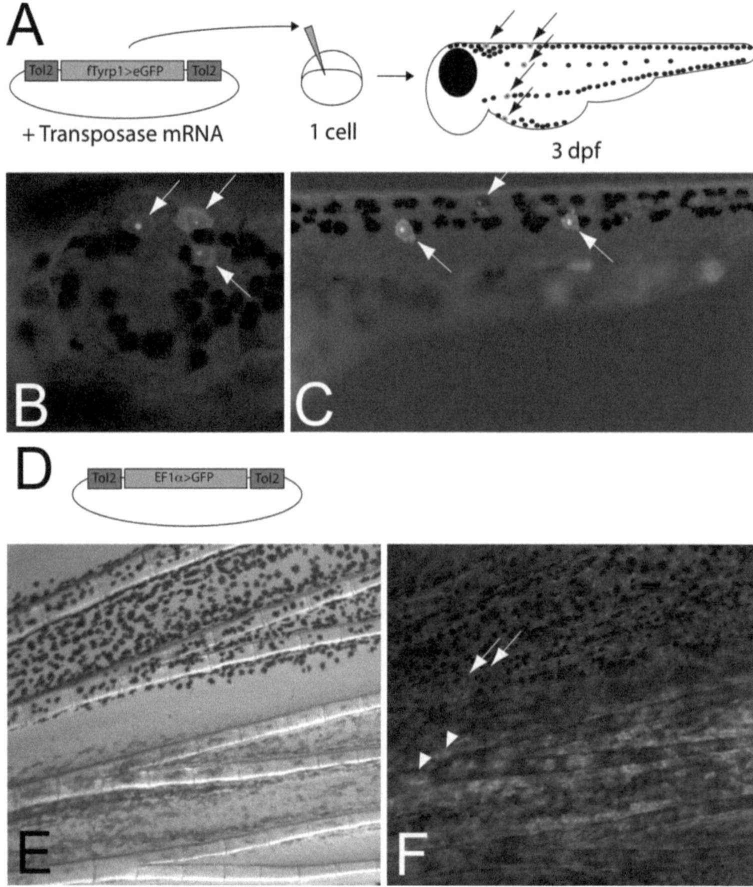

Fig. 6. Clonal analysis of melanocyte lineages using Tol2 transposon labeling (**a**) Protocol for generating clones with the fTyrp1 > eGFP expressing melanocytes. Plasmid and transposase mRNA are coinjected into 1-cell embryo and screened at 3 dpf for GFP + melanocytes. (**b**) Epifluorescence of 3 GFP + melanocytes in the head and (**c**) trunk. (**d**) EF1α > GFP transposon lineage construct used for generating clones that can be visualized in either melanocytes or other cells. (**e**) Brightfield image of adult caudal fin. (**f**) Epifluorsecence image of fin. This experiment reveals that melanocytes (*arrows*) and xanthophores (*arrowheads*) develop from the same precursor.

3.6. Lineage Analysis Using X-Ray Induced Clones

1. Breed *albino* homozygotes (*alb*$^{b4/b4}$) with Tg(fTyrp1 > eGFP)^{j900} (see Note 8).
2. Sort embryos using a dissecting microscope and Pasteur pipet, removing unfertilized eggs and confirming approximately equivalent developmental staging of fertilized embryos.
3. Prepare X-ray cabinet by warming up X-ray tube incrementally.
4. When working with embryos, place up to 500 embryos in the center of a Petri dish and place in the X-ray cabinet directly underneath X-ray source (Fig. 7a). When working with adults, anesthetize fish in tricaine, quickly blot dry on paper towels,

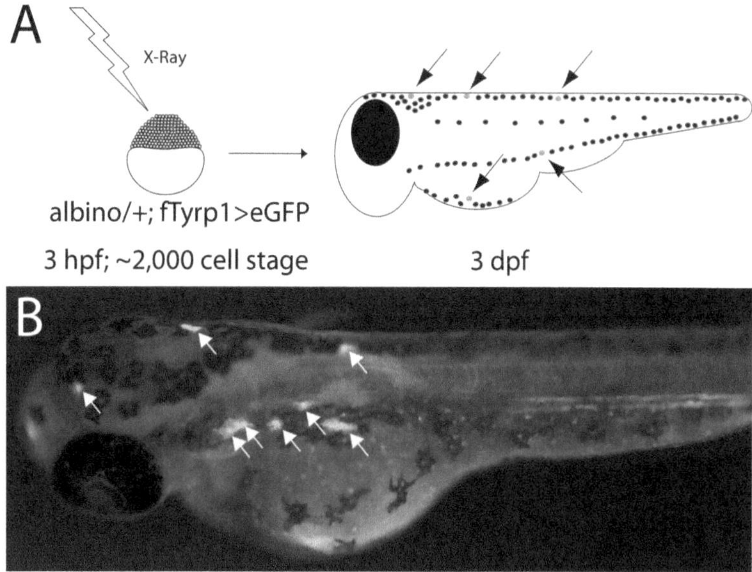

Fig. 7. Lineage analysis using X-ray induced clones on albino heterozygous background. (**a**) Protocol for inducing albino melanocytes with X-rays. X-rays can be applied with temporal precision at various time points in development and screened after 3 dpf following completion of melanocyte differentiation. (**b**) Epifluorescence image of 3 dpf zebrafish following X-irradition to generate albino clones. fTyrp1 > eGFP reveals albino melanocytes (*white arrows*).

and transfer to Petri dish with a slotted spoon just prior to application of X-rays.

5. For low rates of clone production, embryos are placed ~30 cm from the X-ray source and exposed to 2 min of 120 kVp, equivalent to ~787 Rads (see Note 9).
6. At 2 dpf, once melanocytes have begun clearly expressing the fTyrp1 > eGFP marker, remove all fish lacking GFP expression (see Note 3).
7. At 3 dpf, sort through GFP-positive embryos for GFP$^+$, *albino* melanocytes lacking pigment (Fig. 7b) (see Note 10).

4. Notes

1. Any source of ionizing radiation, such as gamma rays from a cesium source, will suffice to generate double strand breaks for the purpose of generating hemizygous clones.
2. Larval fish will die or visibly begin to degrade if left in epinephrine or norepinephrine for periods longer than 10 min. Adult fish can typically withstand up to 30 min in epinephrine without negative effects.

3. In embryos with wild-type melanocytes, sorting after 2 dpf makes identifying GFP expression more difficult as melanin levels increase. Treatment with epinephrine (see Subheading 3.1) to contract melanocytes can facilitate sorting in older embryos.
4. *H* = heat; *P* = pull; *V* = velocity; *T* = time, Psi = Pressure.
5. The amount or concentration of plasmid and transposase can be varied to increase or decrease the number of clones generated per injected clutch. The concentrations indicated result in approximately 5–10% of injected fish containing melanocyte clones.
6. If properly injected, the phenol red in the injection solution will cause a small area of yolk to be distinctly dyed red. If dye quickly wicks away and does not stain the yolk, the needle was not sufficiently inserted and the transposon will likely fail to integrate.
7. We find that a single investigator can perform the in vitro fertilization and transposon injection using 3–4 clutches and labeling 500–1,000 embryos per morning.
8. If application of X-rays at specific early developmental time points is essential, in vitro fertilization is the optimal method for breeding (26). Alternatively, if natural breeding is used, barriers separating males and females should be removed in the morning to avoid rare, accidental late night breeding.
9. Amount of X-rays applied can be varied depending on the number of precursors to be induced per fish. Empirical testing of X-ray dosage in relation to the number of clones generated should be done.
10. While melanocytes in the dorsal and lateral stripes are particularly easy to identify, care should be taken when examining the ventral strip near the yolk, which is more difficult to screen.

References

1. Johnson SL, Africa D, Walker C, Weston JA (1995) Genetic control of adult pigment stripe development in Zebrafish. Dev Biol 167: 27–33
2. Parichy DM, Rawls JF, Pratt SJ, Whitfield TT, Johnson SL (1999) Zebrafish *sparse* corresponds to an orthologue of c-kit and is required for the morphogenesis of a subpopulation of melanocytes, but is not essential for hematopoiesis or primordial germ development. Development 126:3425–36
3. Parichy DM, Mellgren EM, Rawls JF, Lopes SS, Kelsh RN, Johnson SL (2000) Mutational analysis of endothelin receptor b1 (*rose*) during neural crest and pigment pattern development in the zebrafish *Danio rerio.* Dev Biol 227:294–306
4. Rawls JF, Johnson SL (2003) Temporal and molecular separation of the kit receptor tyrosine kinase's roles in melanocyte migration and survival. Dev Biol 262:152–61
5. Mellgren EM, Johnson SL (2004) A requirement for kit in embryonic zebrafish melanocyte differentiation is revealed by melanoblast delay. Dev Gene Evol 214:493–502
6. Parichy DM, Ransom DG, Paw B, Zon LI, Johnson SL (2000) An orthologue of the kit-related gene fms is required for development

of neural crest-derived xanthophores and a sub- population of adult melanocytes in the zebrafish, Danio rerio. Development 127: 3031–3044

7. Kimmel CB, Ballard WW, Kimmel SR, Ullmann B, Schilling TF (1995) Stages of embryonic development of the zebrafish. Dev Dyn 203:253–310
8. Parichy DM, Elizondo MR, Mills MG, Gordon TN, Engeszer RE (2009) Normal table of postembryonic zebrafish development: staging by externally visible anatomy of the living fish. Dev Dyn 238:2975–3015
9. Sheets L, Ransom DG, Mellgren EM, Johnson SL, Schnapp BJ (2007) Zebrafish melanophilin facilitates melanosome dispersion by regulating dynein. Curr Biol 17:1721–34
10. Hultman KA, Scott AW, Johnson SL (2008) Small molecule modifier screen for kit-dependent functions in Zebrafish embryonic melanocytes. Zebrafish 5:279–287
11. O'Reilly-Pol T, Johnson SL (2008) Neocuproine ablates melanocytes in adult Zebrafish. Zebrafish 5:257–264
12. Rawls JF, Johnson SL (2000) Zebrafish kit mutation reveals primary and secondary regulation of melanocyte development during fin stripe regeneration. Development 127:3715–24
13. Rawls JF, Johnson SL (2001) Requirements for the kit receptor tyrosine kinase during regeneration of zebrafish fin melanocytes. Development 128:1943–9
14. Milos N, Dingle A (1978) Dynamics of pigment pattern formation in the zebrafish, Brachydanio rerio: I. Establishment and regulation of the lateral line melanophore stripe during the first eight days of development. J Exp Zool 205:205–16
15. Yang CT, Sengelmann RD, Johnson SL (2004) Larval melanocyte regeneration following laser ablation in zebrafish. J Invest Dermatol 123: 924–29
16. Yang CT, Johnson SL (2006) Small molecule-induced ablation and subsequent regeneration of larval zebrafish melanocytes. Development 133:3563–73
17. Yang CT, Hindes A, Hultman KA, Johnson SL (2007) Mutations in gfpt1 and skiv2l2 cause distinct stage-specific defects in larval melanocyte regeneration in zebrafish. PLoS Genet 3:e88
18. Budi EH, Patterson LB, Parichy DM (2008) Embryonic requirements for ErbB signaling in neural crest development and adult pigment pattern formation. Development 135:2603–14
19. Hultman K, Budi E, Teasley D, Gottlieb A, Parichy D, Johnson SL (2009) Defects in ErbB-dependent establishment of adult melanocyte stem cells reveals independent origins for embryonic and regeneration melanocytes. PLoS Genet 5:e1000544
20. Hultman KA, Johnson SL (2010) Differential contribution of direct-developing and stem cell-derived melanocytes to the zebrafish larval pigment pattern. Dev Biol 337:425–3
21. Kawakami K, Koga A, Hori H, Shima A (1998) Excision of the tol2 transposable element of the medaka fish, Oryzias latipes, in zebrafish, *Danio rerio*. Gene 225:17–22
22. Johnson AD, Krieg PA (1995) A Xenopus laevis gene encoding EF-1 alpha S, the somatic form of elongation factor 1 alpha: sequence, structure, and identification of regulatory elements required for embryonic transcription. Dev Genet 17:280–90
23. Tu S, Johnson SL (2010) Clonal analyses reveal roles of organ founding stem cells, melanocyte stem cells, and melanoblasts in establishment, growth, and regeneration of the adult zebrafish fin. Development 137:3931–3939
24. Streisinger G, Coale F, Taggart C, Walker C, Grunwald DJ (1989) Clonal origins of cells in the pigmented retina of the zebrafish eye. Dev Biol 131:60–69
25. Zou J, Beermann F, Wang J, Kawakami K, Wei X (2006) The Fugu TYRP11 promoter directs specific GFP expression in zebrafish: tools to study the RPE and the neural crest derived melanophores. Pigment Cell Res 19:615–627
26. Westerfeld M (2000) The zebrafish book. A guide for the laboratory use of zebrafish (Danio rerio), 4th edn. University of Oregon Press, Eugene, OR

Chapter 15

Reconstitution of the Central Nervous System During Salamander Tail Regeneration from the Implanted Neurospheres

Levan Mchedlishvili, Vladimir Mazurov, and Elly M. Tanaka

Abstract

Urodele amphibians such as axolotl are well known for their regenerative potential of the damaged central nervous system structures. Upon tail amputation, neural stem cells behind the amputation plane undergo self-renewing divisions and contribute to the functional spinal cord in the newly formed regenerate. The neural stem cells, harboring this potential, can be isolated from the animal and cultured under the suspension conditions. After 2–3 weeks in vitro they will proliferate and form the floating aggregates of the spherical shape, so-called neurospheres. Reimplanted back into the animal, the neurospheres can efficiently integrate in the spinal cord lesion and contribute to the following spinal cord regeneration events. Here we demonstrate the unique method of the axolotl tail spinal cord regeneration from the implanted neurosphere.

Key words: Axolotl, Neurospheres, Spinal cord injury, CNS regeneration, Neural stem cells

1. Introduction

Salamanders are well known for the remarkable capacity to regenerate the central nervous system (1). Tail amputation and regeneration represent a convenient experimental context to study this problem. We have been studying the neural stem cells that are responsible for regenerating the spinal cord. As shown previously, the stem cells appear to be the GFAP radial glial cells lining the central canal of the spinal cord (2). These spinal cord radial glia from within 500 μm behind the tail amputation contribute to the regenerating spinal cord where they undergo self-renewing divisions for approximately 2 weeks before the onset of neural differentiation in a rostral to caudal sequence (3). Our lineage mapping

Kimberly A. Mace and Kristin M. Braun (eds.), *Progenitor Cells: Methods and Protocols*, Methods in Molecular Biology, vol. 916, DOI 10.1007/978-1-61779-980-8_15,

experiments suggest that at least some of the cells represent multipotent progenitors that will contribute to different regions (and hence neuronal subtypes) in the regenerating spinal cord (3). We have also been studying the spinal cord neural stem cells by isolating them in culture, and reimplanting them back into the animal, where they efficiently integrate into the lesioned spinal cord and can contribute to even the majority of cells in the regenerating spinal cord.

Here we describe the method for culturing Axolotl neurospheres, and reimplanting them back into the spinal cord before tail amputation to induce regeneration. Neural stem cells of the axolotl central nervous system can be isolated. Cultured under cell suspension conditions they will proliferate and form floating aggregates of the spherical shape, so-called neurospheres. Axolotl neurospheres can be obtained as a pool of neural stem and progenitor cells. Implanted back into the spinal cord lesion they integrate in the regenerated host tissue architecture. Following tail amputation close to the implant, the integrated cells undergo proliferation and contribute to a large extent to the outgrowing tail spinal cord.

2. Materials

1. 0.01% Ethyl p-aminobenzoate dissolved in tap water.
2. Antibiotics (Penicillin, Streptomycin).
3. PBS–pen/strep (1× PBS, 100 units/ml penicillin, 100 μg/ml streptomycin).
4. Dissection tools (forceps, tungsten needles, etc.).
5. Petri dishes (3.5 and 10 cm diameter).
6. Stereo microscope.
7. Papain-mix (L15 medium containing 30 U/ml papain, 0.24 mg/ml cysteine, and 40 μg/ml DNAsa), sterile filtered.
8. Ovomucoid-mix (1 mg/ml trypsin inhibitor, 0.5 mg/ml BSA, 40 μg/ml DNAsa in L15 medium), sterile filtered.
9. Tissue digestion solution (1:1 volume of Papain-mix and Ovomucoid-mix).
10. Neurosphere medium (DMEM/F12 + Glutamax® supplemented by 2% B27, 20 ng/ml human recombinant FGF-2).
11. 25 cm^2 cell culture flask.

3. Methods

3.1. Axolotl Tail Spinal Cord Dissection and Dissociation

1. Anesthetize Axolotls in 0.01% Ethyl p-aminobenzoate (see Note 1).
2. Amputate the tail and rinse in 1× PBS containing antibiotics (see Note 2). All steps from here should be done under the sterile conditions, using sterile tools and solutions.
3. Transfer the tail to 10 cm Petri dish with PBS–pen/strep.
4. Dissect the spinal cord out of the tail by disrupting the surrounding tissue with sharp tools (e.g., tungsten needles and forceps). The dissection should be done under the stereo microscope in PBS–pen/strep (see Note 3).
5. Transfer the isolated spinal cord piece into a 3.5 cm Petri dish with PBS–pen/strep and cut it in smaller (3–5 mm) pieces (see Note 4).
6. Prepare the tissue digestion solution by mixing equal volumes of Papain- and Ovomucoid-mix (1:1); incubate 5 min at room temperature.
7. Pick up the spinal cord pieces gently (e.g., with forceps) and transfer them into the 1.5 ml Eppendorf tube containing 300 μl tissue digestion solution.
8. Incubate the tissue pieces in tissue digestion solution for 1 h at room temperature. Shake the tube gently every 15 min.
9. In order to inhibit the digestion reaction, add now an equal volume (300 μl) of the Ovomucoid-mix to the reaction tube.
10. Dissociate the spinal cord pieces by pipetting them up and down about 1–2 min in the tube. Longer or too brutal dissociation procedure can damage the cells and will decrease the efficiency of the neurosphere formation.
11. Wash the dissociated cells by adding them to the 9.4 ml DMEM/F12 medium containing pen/strep.
12. Optional: in order to get rid of the undissociated tissue pieces the cell suspension can be filtered through a sieve.
13. Centrifuge the cells 3 min at $80\times g$.
14. Remove the supernatant (see Note 5).
15. Resuspend the cells in 4 ml neurosphere medium and transfer them to the 25 cm^2 cell culture flask. The optimal cell density for the neurosphere formation is about 5,000–10,000 cells/ml (see Note 6).

3.2. Neurosphere Culture

1. Culture the cell suspension in the incubator setup at 25°C supplied with 2% CO_2.
2. The neurosphere medium should be changed once a week, by centrifuging the suspension and replacing the supernatant with the fresh medium (see Steps 13–15 of Subheading 3.1.)
3. After 2–3 weeks of culturing, neurospheres of different size (100–300 μm) will appear in the flask (see Note 7).

3.3. Neurosphere Implantation in the Tail Spinal Cord

1. For the neurosphere implantation into the tail spinal cord, 2.5–4 cm long juvenile axolotls are of optimal size.
2. Anesthetize the host animals in 0.01% Ethyl p-aminobenzoate (about 15 min) and check that the animal is no longer responsive to pinching of the tail with forceps.
3. Place the anesthetized axolotl in the 10 cm silicon-coated Petri dish filled with the animal dissection solution (0.01% Ethyl p-aminobenzoate in 1× PBS). All steps of the animal dissection and neurosphere implantation should be done in the animal dissection solution. Conditions for this procedure do not need to be sterile.
4. The spinal cord dissection should be done from the dorsal side. Immobilize the tail in the perpendicular position so that dorsal is up, e.g., by pinning the animal to silicon layer incision with the small needles (see Note 8).
5. The lateral view of the uninjured tail is shown in Fig. 1a. Dissect 2–4 mm long portion of the tail fin with two cuts from the top and one cut from lateral, just above the myotomes, and flip it to the side opposite the lateral cut in order to expose the tissue layers under the fin (Fig. 1b). The spinal cord should be visible as a whitish rod between the myotomes.
6. Remove approximately 1 mm section of the spinal cord by clipping the spinal cord in two locations with iridectomy scissors, and gently freeing the spinal cord segment from its surroundings with forceps (Fig. 1c).
7. Implant one or several 100–200 μm size neurospheres into the spinal cord lesion (Fig. 1c). Put the neurospheres in with forceps, or tungsten needles (see Note 9).
8. Cover the lesion by flipping the tail fin back to the perpendicular position (Fig. 1d).
9. Let the animals heal for 1–2 weeks.

3.4. Spinal Cord Regeneration from the Implanted Neurospheres

1. Using a scalpel, amputate the tail with a straight cut close to the integrated neurosphere (Fig. 1e). The implant should be maintained within a 500 μm spinal cord zone behind the amputation plane (3) (see Note 10).
2. Let the animal regenerate the tail for 3–6 weeks (Fig. 1f) (see Note 11).

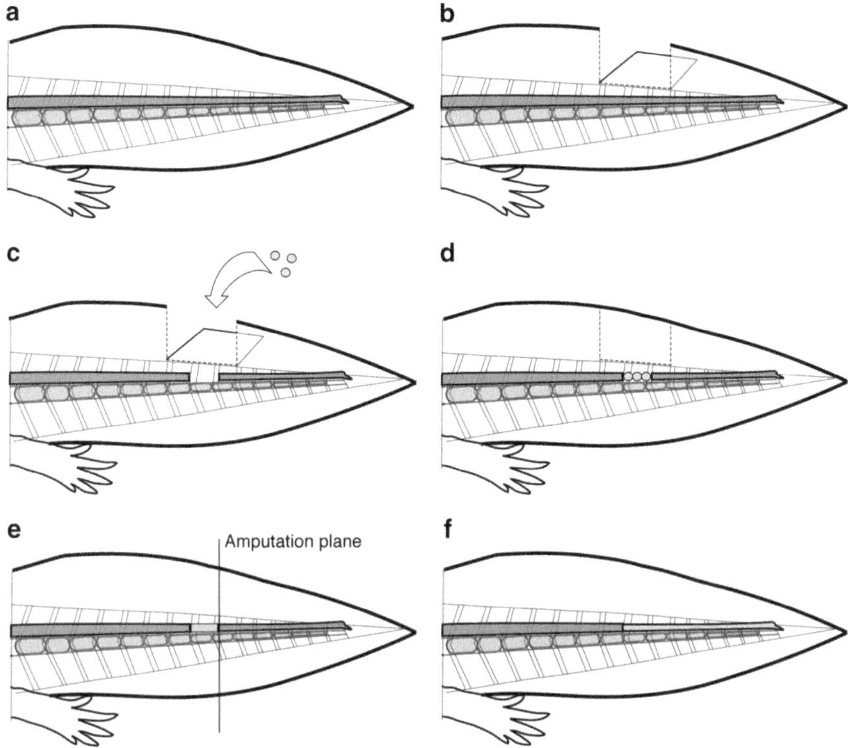

Fig. 1. (**a**) Lateral view of the intact axolotl tail; (**b**) the tail fin is dissected with two incisions from the top and one from lateral and flipped to the side opposite to the lateral cut to access the spinal cord; (**c**) a piece of the spinal cord is removed from the tail and the neurospheres are implanted into the lesion; (**d**) the tail fin is flipped back to original position in order to close the wound and support healing; (**e**) the tail is amputated close to the integrated neurospheres; (**f**) tail spinal cord regenerates from the implanted neurospheres.

4. Notes

1. In order to distinguish the implanted cells from the host, the neurospheres can be cultured from eGFP + transgenic axolotls (4) and implanted into the white hosts (d/d alleles).
2. Approximately half of the tail, measured from the cloacae to the tail tip, should be amputated for the spinal cord isolation. For the optimal neurosphere culture cell density usually two adult animal tail spinal cords are enough. Accordingly, more tails of the smaller animal size are needed for the equal cell density (e.g., four tails of the 8–10 cm size animals).
3. In order to avoid the bacterial or fungal contaminations sterile conditions are very important during the spinal card isolation and dissociation procedure. It helps for the sterility to clean well the dissection place with 70% ethanol and flame the dissection tools with phosphor flame time to time, or wash them in 70% ethanol.

4. After isolating the spinal cord out of the tail other tissues can stick to the explant and contaminate the future cell culture. In order to obtain a maximally pure neural stem cell culture the layer of meninges, that is, visible as a transparent film covering the spinal cord, can be removed with the sharp tools.
5. After centrifuging the cell suspension the cell palette is usually not detectable in the falcon tube. It is better to leave 100–200 μl liquid in the tube after aspirating the supernatant. The leftover in the tube can be resuspended then in the neurosphere medium and the cells can be viewed under the microscope.
6. Cell counting is recommended for standardizing the conditions.
7. It is recommended to shake the flask containing the cells once a day. Otherwise, the forming neurospheres can stick to the plastic surface and become useless for the following implantations.
8. The donor animal receiving the neurosphere implant should be fixed well on the silicone surface. The tail should not move during the dissection procedure. That insures precision of the implantation.
9. Make sure that the neurospheres are placed in one line with the spinal cord, by checking the implant from the side after finishing the implantation procedure.
10. It is possible that not all of the implants give rise to the regenerating spinal cord to the same extent upon the tail amputation. If so, the tails can be re-amputated after 1–2 weeks close to the visible implant. Try to maintain the implanted cells within 500 μm behind the amputation plane (3). In order to get the full extent of spinal cord regeneration from the implant, other amputation distances can also be tried out (e.g., 300 or 700 μm).
11. The spinal cord regenerate from the implanted neurospheres can be viewed and imaged under UV-light based on the eGFP fluorescence using a standard fluorescence dissecting microscope. Histological sections of the tail can be performed for a detailed view of the regenerate (3).

References

1. Butler EG, Ward MB (1967) Reconstitution of the spinal cord after ablation in adult Triturus. Dev Biol 15:464–86
2. O'Hara CM, Egar MW et al (1992) Reorganization of the ependyma during axolotl spinal cord regeneration: changes in intermediate filament and fibronectin expression. Dev Dyn 193:103–15
3. Mchedlishvili L, Epperlein HH, Telzerow A, Tanaka EM (2007) A clonal analysis of neural progenitors during axolotl spinal cord regeneration reveals evidence for both spatially restricted and multipotent progenitors. Development 134:2083–93
4. Sobkow L, Epperlein HH, Herklotz S, Straube WL, Tanaka EM (2006) A germline GFP transgenic axolotl and its use to track cell fate: dual origin of the fin mesenchyme during development and the fate of blood cells during regeneration. Dev Biol 290:386–97

Chapter 16

Following the Fate of Neural Progenitors by Homotopic/ Homochronic Grafts in *Xenopus* Embryos

Raphaël Thuret and Nancy Papalopulu

Abstract

The neural plate consists of neuroepithelial cells that serve as progenitors for the mature central nervous system. The neural plate is a highly regionalized structure, harboring neural progenitors with different programs of differentiation, due to signaling or intrinsic differences in their differentiation potential. In the frog neural plate, neural progenitors located in the deep or superficial layer differ in their ability to contribute to early (primary) neurogenesis but intercalate during neurulation. In order to understand the origins and mechanisms of this progenitor heterogeneity, it is necessary to be able to follow directly the fate of different progenitors. Here, we describe a fate mapping method, which is based on homotopic and homochronic grafts of labeled tissue to unlabeled, or differentially labeled, hosts. This method can be combined with immunohistochemical analysis with cell type specific markers, thus allowing one to determine the contribution that each early progenitor type makes to the differentiated nervous system. Such labeling can also be used to examine the morphogenetic movements that take place during neurulation.

Key words: *Xenopus*, Neural progenitors, Graft, Fate map, Morphogenesis

1. Introduction

The frog *Xenopus* is an excellent model system for the study of neural progenitors in vivo. The external development of the embryo enables the observation of the earliest stages of neural development. Furthermore, it is possible to chemically, molecularly, and surgically manipulate *Xenopus* embryos. Finally, its different life phases, from embryo to adult through larval stage undergoing metamorphosis, make it a unique model to understand the establishment and the maintenance of neural progenitors in vertebrates during embryonic and postembryonic development.

The *Xenopus* central nervous system is gradually set up during the life of the animal. A first wave of neurogenesis is identified at neural plate stage, when a set of neuroepithelial cells differentiates

Kimberly A. Mace and Kristin M. Braun (eds.), *Progenitor Cells: Methods and Protocols*, Methods in Molecular Biology, vol. 916, DOI 10.1007/978-1-61779-980-8_16, © Springer Science+Business Media, LLC 2012

into primary neurons (Nieuwkoop and Faber stage 12–20 (1, 2)). Other neuroepithelial cells remain as progenitors and gradually give rise to neurons throughout the life of the animal. The *Xenopus* neural plate is composed of two layers of neuroepithelial cells unlike the neural plates of other vertebrates which are thought to be single layered epithelia. However, recent reports showed that the zebrafish neural plate is also multilayered (3). These two cell layers composing the *Xenopus* neural plate are already set up by the onset of gastrulation (NF stage 9, (4)). During neurulation, they interdigitate to form a single layered closed neural tube (by NF stage 25). The superficial layer is composed of apico-basally polarized epithelial cells that sit on top of a layer of nonpolar deep cells. Primary neurons arise from nonpolar deep layer cells, whereas progenitors that are maintained derive from both deep and superficial layers (1). At neural plate stage, superficial cells have been shown to be unresponsive to overexpression of neuronal differentiation factors (5). Further work has shown that molecules that set up the apico-basal polarity, notably the atypical kinase aPKC, are specifically enriched in superficial cells, conferring them resistance to differentiation (6, 7). Therefore, three cell types can then be identified in *Xenopus* neural plate: primary neurons expressing differentiation markers, deep precursors that can be experimentally pushed toward differentiation, and superficial polarized progenitors that are refractory to primary neurogenesis. After neural tube closure, these two types of progenitors interdigitate and cannot be distinguished based on morphology or position. It is not known whether progenitors originating from the deep or superficial layers maintain differentiation differences later during development and/or express different genes. Thus, methods allowing the tracking of these two cell layers are of primary interest to provide insight on the establishment of the long-term fate of these two progenitor populations, as well as to understand how the neural tube is formed.

Different methods can be used to follow the fate of superficial cells in *Xenopus*, including single cell injection (1) or biotinylation of the whole embryo (8). The method we describe here consists of homotopic and homochronic grafting of a lineage labeled part of either the superficial or the deep layer of the prospective neuroepithelium, carried out at early gastrula stage. Unlike single cell injection, this technique enables following of a cohort of cells rather than individual cells. It can be combined with immunohistochemistry, and therefore, it is a powerful method to understand tissue movements, notably, cell intercalation during neural tube formation and cell fate during development. This technique has been used to analyze morphogenetic movements occurring during gastrulation and neurulation (8, 9) to investigate the inductive properties of the ectoderm (10) and, more recently, to study the role of polarity molecules during primary neurogenesis (11). Additionally, the persistence of the lineage tracer until late stages of development also allows us to study the fate of the graft, on a long-term basis.

2. Materials

2.1. Obtaining Xenopus laevis Embryos (See Note 1)

1. Pregnant mare serum gonadotropin (PMSG): 100 U/mL PMSG (P.G.600®, Intervet). Dissolve powder in water, make 5 mL aliquots, and store at -20°C.
2. Human chorionic gonadotropin (HCG): 1,000 U/mL HCG (Chorulon®, Intervet). Dissolve powder in water and store at 4°C for not more than a week.
3. 1 mL syringe and Gauge 27 needles.
4. 0.1× Marc's Modified Ringer (MMR): 10 mM NaCl, 0.2 mM KCl, 0.1 mM $MgCl_2$, 0.2 mM $CaCl_2$, 0.5 mM HEPES, pH 7.5. Prepare a 10× stock solution, and adjust the pH with NaOH to 7.5. Sterilize the 10× solution by autoclaving.
5. 1× MMR (High salt solution).
6. 40% Tricaine methanesulfonate (MS222, aminobenzoic acid ethyl ester, Sigma). Dissolve in distilled deionized water. Make 400 μL aliquots and store at -20°C.
7. A pair of small scissors and a pair of No 4 forceps to dissect the male.
8. L-15 Leibovitz medium (Sigma).
9. 2% Cysteine hydrochloride 1-hydrate (Sigma) in 0.1× MMR (adjust to pH 8.0 with 10 N NaOH). Make up fresh 200 mL solution and use during the day.
10. 90 mm Petri dishes.

2.2. Injecting Tracer in Fertilized Embryos

1. Micro-ruby 2.5% (Invitrogen) diluted in distilled deionized water. Make 5 μL aliquots and store at -20°C in the dark (see Note 2).
2. Micro-emerald 2.5% (Invitrogen) diluted in distilled deionized water. Make 5 μL aliquots and store at -20°C in the dark.
3. 1% Ficoll in 0.1× MMR (Injection Buffer: embryos will be injected in this buffer).
4. Injection needles: GC100TF-10 glass capillaries (Harvard Apparatus) are pulled with a Flaming/Brown micropipet puller Model P-87 (Sutter Instruments Co.). Conditions used are temp = ramp test, pull = 50, vel = 100, and time = 5 (see Note 3). Once set up, the end tip of the needle should be approximately 10 μm.
5. Air-pulse based nano-injector type Pico-Spritzer (Intracel).
6. Microladder tips to load injection needles with dextran.
7. Forceps size 5 (Dumont).

2.3. Grafting Superficial or Deep Layer of the Neuroepithelium

1. 14 V power supply with crocodile clips fitted on anode and cathode.
2. 5 N NaOH.
3. Carbon rod.
4. Tungsten wire (0.1 mm diameter, Goodfellow) with holder (see Note 4 for etching technique).
5. Forceps size 5 and 55 (Dumont) for vitelline membrane removal.
6. Fire-polished Pasteur pipette.
7. 1% Ficoll, 0.5× MMR buffer with 5 μg/mL of gentamicin (Sigma).
8. Dishes coated with 1% Agarose in 0.1× MMR with wells the size of an embryo, at blastula stage (approximately 1.3–1.5 mm diameter).
9. 24-well dishes coated with 1% Agarose in 0.1× MMR.
10. 0.1× MMR with 5 μg/mL gentamicin.
11. Forceps size 5 and 55 (Dumont).

2.4. Analyzing the Fate of the Grafted Embryos by Immunohistochemistry

2.4.1. Fixing Samples

1. Wheaton glass scintillation vials (20 mL, VWR).
2. MEMFA: 0.1 M MOPS pH 7.4, 2 mM EGTA, 1 mM $MgSO_4$, 3.7% Formaldehyde. Prepare fresh and use on the day.
3. 100% methanol.

2.4.2. Sectioning Samples

1. 15% fish gelatine (Sigma), 15% sucrose dissolved in distilled deionized water for embryos up to NF stage 30. Use within a week of preparation.
2. 25% fish gelatine (Sigma), 15% sucrose dissolved in distilled deionized water for embryos from NF stage 30 onwards. Use within a week of preparation.
3. Cryostat (Leica).
4. Superfrost plus slides (VWR).
5. OCT compound (Lamb).
6. Cryomolds 15×15×5 mm (Fisher).

2.4.3. Immunostaining on Sections

1. 1× PBS 0.1% Triton X-100 (Sigma).
2. Heat Treated Lamb Serum (HTLS). Incubate Lamb Serum at 60°C for 1 h, aliquot in 50 mL Falcon tubes, and store at −20°C.
3. 1× PBS 0.1% Triton X-100 + 5% HTLS.
4. Primary and secondary antibodies (such as anti-PH3, anti-acetylated tubulin, anti-Myt1 (7), and anti-Sox3 (12)).

5. Mowiol mounting media: mix 6 mL of glycerol, 2.4 g of Mowiol 4-88 (Calbiochem), and 6 mL of distilled deionized water. Mix at room temperature for 2 h. Add 12 mL of 200 mM Tris–HCl pH 8.5 and incubate at 50°C with occasional shaking until the Mowiol has completely dissolved. Filter on 0.45-μm membrane filter and store aliquots at −20°C.
6. DAPI diluted at 10 mg/mL in distilled deionized water (Sigma). Make 50 μL aliquots and keep at −20°C.
7. Glass coverslips (22 × 50, thickness N°1, VWR).

3. Methods

Layer grafting is performed between NF stage 9 and 10, just before the appearance of the blastopore lip and the process of gastrulation (see Fig. 1a for a general overview of the method). Studies of fate map at blastula stages reveal that most of the central nervous system derives from blastomeres A1, A2, B1, B2, and C2 (13). In late blastula-early gastrula stages, this region corresponds to the dorsal and lateral region of the animal hemisphere as has been shown by vital dye marking experiments ((14), Fig. 1b). Prior to the appearance of the blastopore lip, the dorsal region of the ectoderm is identifiable by a lighter pigmentation than the ventral side (Fig. 1c).

At this stage, mechanical separation of deep and superficial layers of the ectoderm is relatively easy, making the process of grafting particularly successful. This is a transient phenomenon and separation of deep from superficial layer is very difficult either before or after this stage (14). Therefore accurately staging embryos at NF stage 9 is very important. In order to get the embryos at the right stage, it is recommended to start removing the vitelline membrane as soon as NF stage 8 and then try to separate the layers until this process can be done easily. Ideally, the best moment to start the dissection is when bottle cells, which will lead the gastrulation movements, are appearing on the vegetal dorsal side. This is identified by the appearance of pigment constriction in the involuting marginal zone (Fig. 1d). On this step reach, one will be able to easily peel the superficial from the deep layer (as schematized in Fig. 1e).

3.1. Obtaining *Xenopus laevis* Embryos (See Note 1)

1. Prime two female adult *X. laevis* 3–5 days before HCG injection with 50 U PMSG. Inject them with 500 U HCG, 12–15 h before they are needed.
2. Place each frog into 5 L of 1× MMR and let them lay eggs.
3. Sacrifice a male by injecting 400 μL of MS222 solution in dorsal lymph sac and dissect testis. Keep testes on ice in L-15 (testes can be kept for a week in L-15 at 4°C).

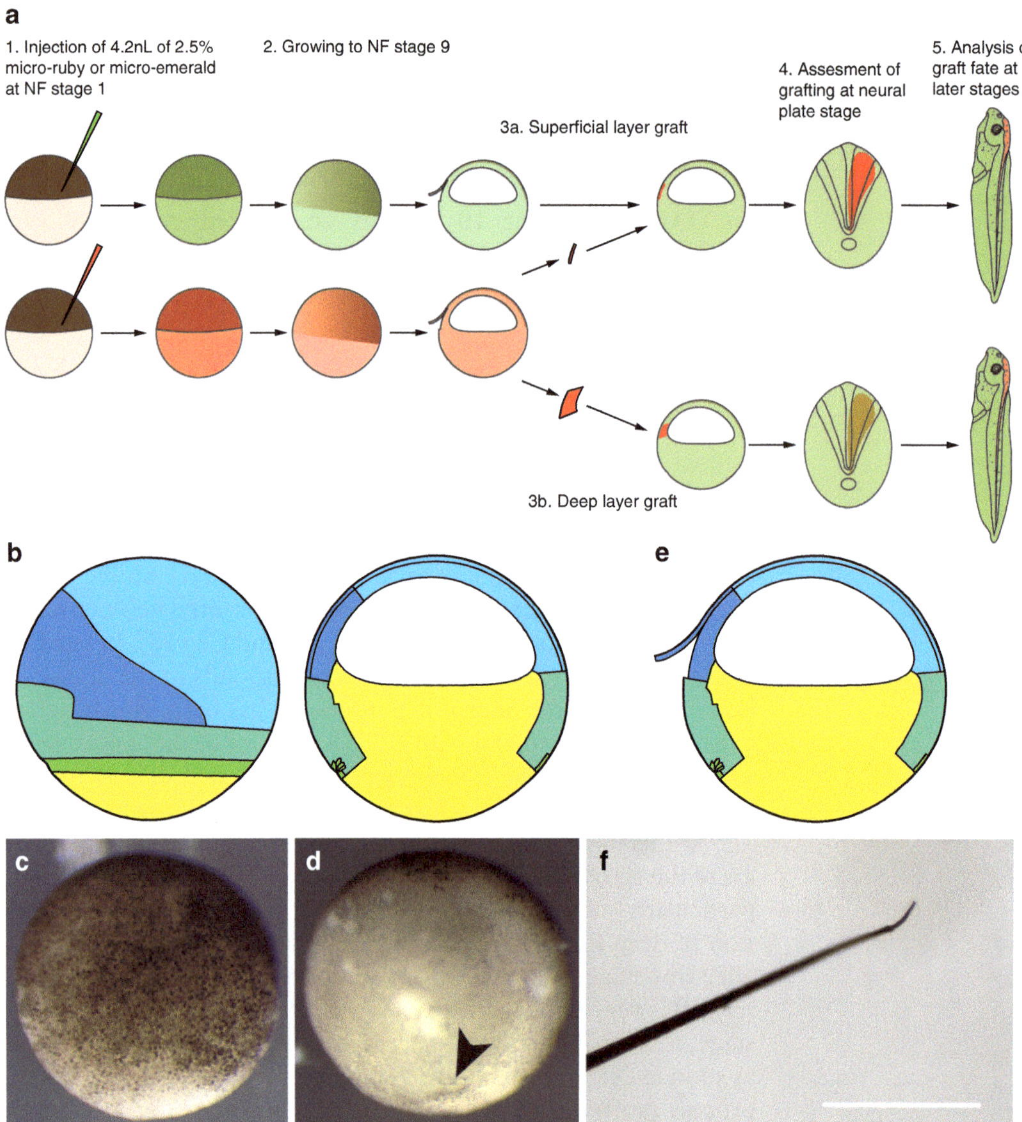

Fig. 1. (**a**) Overview of the grafting procedure. (**b**) Fate map at NF stage 9 on whole embryo and on a sagittal section. Animal pole is up; vegetal pole is down. *Light blue*: epidermal ectoderm, *dark blue*: neuroectoderm, *green*: mesoderm, *light green*: bottle cells, *yellow*: endoderm. NF stage 9 is identified by the initiation of the formation of the bottom cells. (**c**) View of a NF stage 9 embryo from the animal pole. Note the lighter pigmentation in the dorsal region, allowing the identification of the presumptive neuroectoderm. (**d**) View of a NF stage 9 embryo from the animal pole. *Arrowhead* indicates the appearance of the contraction of the bottle cells, identifiable by a condensation of pigmentation. (**e**) Schematic of a NF stage 9 embryos showing the peeling of the superficial layer of the presumptive neuro-ectoderm. (**f**) Picture of the tip of a dissecting tungsten wire after electrochemical etching. Magnification ×40. *Scale bar* = 500 μm.

4. Harvest laid eggs with a 25 mL pipette to a 90 mm Petri. Remove most of the liquid and proceed to fertilization. A dish full of a monolayer of eggs will be fertilized with 1/3 of a testis. To do so, cut onethird of testis, slightly dilacerate it, and

gently pass it over the eggs for 1–2 min. Let it stand for 10 min and cover eggs with 0.1× MMR after 10 min. After 30 min, fertilized embryos will rotate to orient themselves with animal (pigmented) pole facing up. This allows one to assess the quality of the fertilization.

5. Remove liquid and dejelly embryos with 2% Cysteine in 0.1× MMR for 4–8 min until complete disappearance of the jelly. Rinse three times and finally keep embryos in a dish of 0.1× MMR.

3.2. Injecting Tracers in Fertilized Embryos

1. Backfill an injection needle with 5 μL of micro-ruby solution using a microladder tip. Install the needle on the holder, break the needle tip under the microscope with a pair of forceps, and set up the injector in order to obtain a 4.2 nL bubble (if injecting at one-cell stage) or 2 nL (if injecting in each blastomere of a two-cell-stage embryo) using an ocular micrometer for measurement.
2. Inject embryos in 0.1× MMR 1% Ficoll using forceps to hold them gently in place (see Note 5).
3. Let embryos recover for an hour in injection buffer. Then grow them in 0.1× MMR until they reach NF stage 9 to perform homotopic and homochronic grafts. These embryos are the donors; therefore, at the same time, grow embryos to serve as hosts (recipients) of the graft (see Note 6).

3.3. Grafting Superficial or Deep Layer of the Neuroepithelium

1. Remove vitelline membrane from donor and recipient embryos in 0.1× MMR on an agarose-coated dish.
2. Very carefully dispense embryos in 0.5× MMR, 1% Ficoll, 5 μg/mL gentamicin in the agarose dishes with wells sized to embryos (see Note 7).
3. With the help of a tungsten wire, gently cut superficial cells on a square of 20 × 20 cells approximately from the donor embryo. This should be done in the less pigmented area of the ectoderm (the future neuroectoderm), above the nascent blastopore lip (see Note 8). It is easier to start from the most posterior part since the deep layer is thicker there. Then cut gently on both sides of the initial slit going anteriorly. From there, tease the superficial and deep layer apart by swiping with caution the tungsten wire in between the deep and superficial layers of the ectoderm in order to dissociate them. The superficial layer should detach fairly easily, if the embryos are at the right stage. Do so until reaching the side uncut and then finish to cut the bit of superficial layer still using the tungsten needle. Save the piece of tissue next to your recipient embryo (see Note 9).
4. Use the same procedure to peel the superficial layer from the recipient embryo. To ensure that the graft is homotopic, aim

to place the graft in the same general area of the host. Try to remove a slightly bigger piece of tissue in order to make it easier to fit the grafted tissue. Once the piece of tissue is removed, delicately place the donor piece of tissue with forceps, taking care not to damage it. Lightly press it and let it heal for an hour (see Note 10).

5. Transfer the grafted embryo with a fire-polished glass Pasteur pipette in 0.1× MMR gentamicin and let it grow up to NF stage 18 at 14°C (see Note 11).
6. Check for the position of the graft under a fluorescent microscope by NF stage 18 (Fig. 2a, b, superficial grafts; Fig. 2c, d, deep grafts).
7. Let embryos grow until the desired stage in 0.1× MMR gentamicin.

3.4. Analyzing the Fate of the Grafted Embryos by Immuno histochemistry

3.4.1. Fixing Samples

1. Once embryos reach the desired developmental stage, fix them in MEMFA for 1 h at room temperature in glass scintillation vials (see Note 12).
2. Rinse three times in 100% Methanol and store at least overnight in 100% Methanol at –20°C. Samples can be stored up to 6 months before being processed for immunostaining.

3.4.2. Sectioning Samples

1. Freshly prepare some Fish Gelatine (15% for embryos up to NF stage 30, 25% for later stages). For example, 5 mL of Fish Gelatine will be enough for 10 NF stage 30 embryos.
2. Gradually rehydrate your fixed samples in 1× PBS, before putting them in fish gelatine. Let them embed at least overnight at room temperature on a rocker.
3. Mount them in embedding molds of the appropriate size. First put a bed of fish gelatine and let it set on dry ice. Then place the sample and orient it with forceps under the dissecting microscope quickly enough to keep the fish gelatine bed frozen.

Fig. 2. (**a, b**) NF stage 16 live embryos injected with micro-emerald and grafted at NF stage 9 with superficial layer cells injected with micro-ruby under bright field (**a**) or fluorescent light (**b**). (**c, d**) NF stage 16 embryos injected with micro-emerald and grafted at NF stage 9 with deep layer cells injected with micro-ruby under bright field (**c**) or fluorescent light (**d**). Note the position of the micro-ruby grafts (*red*) in the neural plate area, around the midline (**b** and **d**, see Note 8). (**e, f**) 12 μm sections of NF stage 16 embryos injected with micro-emerald and grafted at NF stage 9 with micro-ruby injected superficial (**e**) or deep (**f**) layer cells. Grafts are successfully restricted to superficial and deep layers (*arrowheads*). Contaminating red labeled cells can be seen in both case; nonetheless, they are not located in neural tissue. *Red*: micro-ruby, *green*: micro-emerald, *blue*: DAPI. Magnification ×40, *scale bars* 100 μm. (**g, j**) Examples of graft progeny recovery in the hindbrain (**g** and **i**) and in the spinal cord (**h** and **j**) of NF stage 42 tadpoles grafted with superficial (**g** and **h**) or deep cell (**i** and **j**). *Red*: micro-ruby, *blue*: DAPI. Note that most of the cells recovered after superficial grafting are located close to the ventricle (progenitors), whereas red cells are found in the differentiated neuron compartment in the case of deep graft. 12 μm sections, magnification ×40, *scale bars* 100 μm.

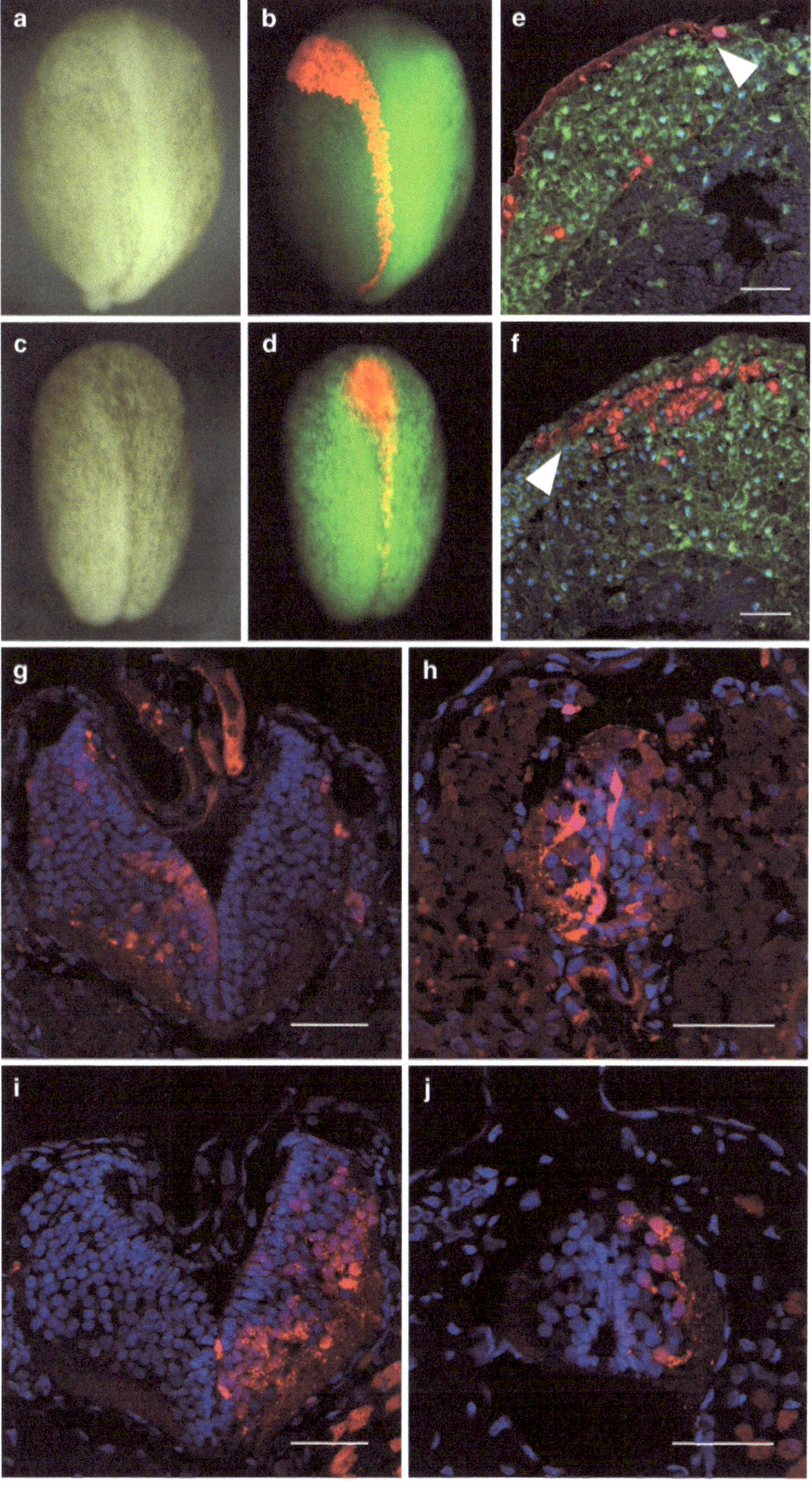
a
b
e
c
d
f
g
h
i
j

Mark the mold in order to locate the sample, top up with fish gelatine, and let it freeze on dry ice.

4. Once set, mark the position of the sample on the block on the dry ice-cold block with a pencil and allow the block to reach the cryostat temperature for at least 10 min in the cryostat chamber. Mount it on the chunk with OCT compound and let it set. Set up your block on the cryostat head (see Note 13), let it cool down to object temperature for 5 min, and start sectioning. First trim between 30 and 50 μm until the sample is reached. Section at 12 μm thickness and gather sections on Superfrost slides.
5. Keep slides in –80°C for at least 1 h before processing for immunostaining. Slides can be stored for at least 3 years at –80°C.

3.4.3. Immunostaining on Sections

1. Take slides out of the –80°C freezer and let them dry under a fume hood on a tray for 1 h.
2. Dip the slide in acetone for 2 min and rehydrate them in 1× PBS 0.1% Triton X-100 in a Coplin jar type slide tank.
3. Draw a boundary with a PAP pen and block with 1× PBS 0.1% Triton X-100, 5% HTLS for 30 min in a humid chamber at room temperature.
4. Incubate with primary antibody for 2 h at room temperature or overnight at 4°C in a humid chamber.
5. Wash three times 10 min in 1× PBS 0.1% Triton X-100; then block with 1× PBS 0.1% Triton X-100, 5% HTLS for 30 min in a humid chamber.
6. Incubate with secondary antibodies for an hour at room temperature. Usually, secondary antibodies will be used from 1:300 to 1:500.
7. Wash three times 10 min in 1× PBS 0.1% Triton X-100.
8. Incubate with DAPI diluted at 5 μg/mL in PBS for 10 min and wash three times for 10 min in 1× PBS 0.1% Triton X-100.
9. Take slides out of the washing tank and partially dry them before mounting them in Mowiol. Seal it with nail varnish. Image with a fluorescent microscope (see Fig. 2g–j for representative examples of graft recovery at NF stage 42).

4. Notes

1. The methods presented here have been used in *Xenopus laevis*. *Xenopus tropicalis* species could be used as well, although the embryos are smaller (0.7 mm for *X. tropicalis*, 1.3 mm for *X. laevis*).

2. The length of time that the cells can be followed is limited by the long-term stability of the lineage marker in the embryo. In our hands, using micro-ruby as an injectable dye coupled to dextran, we have been able to follow cells reliably up to NF stage 45. This kind of dye-coupled dextran is lysine fixable and is stable after aldehyde-based fixation.
3. According to the type of glass used, the ramp test will vary. Follow the manufacturer's instructions regarding the ramp test.
4. Tungsten wires are electrochemically sharpened in 5 N NaOH. In a small beaker, put 5 N NaOH and dip the carbon rod connected to the anode of the power supply. The cathode is then connected to a to a 2–3 cm long piece of tungsten. Etch it by dipping it in the NaOH solution, aiming for a long and fine edge (Fig. 1f). This edge can be later shaped like a hook, facilitating the dissection of the layers.
5. In order to facilitate visualization of neural progenitor interdigitation or the origin of neural cells, the recipient embryos can also be injected with a similar fluorescent dextran like micro-emerald, which is a green equivalent of micro-ruby. Micro-ruby and micro-emerald are detectable on live and fixed samples under fluorescent light, respectively, absorbing at 555 nm and 494 nm and emitting at 580 nm and 518 nm.
6. From the dejellying step, all the embryos (or a subset of them) should be kept at the same temperature in order to limit developmental lag affecting homochronic grafts. Nonetheless, after injection, subsets of donor and recipient embryos can be grown in different incubators in order to allow the grafting procedure on a longer time frame. For example, embryos kept at 18°C from the two-cell stage will reach NF stage 9 in 9 h (15). If those same embryos are kept at 16°C, they will reach NF stage 9 in 14 h. Varying the temperature and development speed will expand your ability to perform multiple grafts on a single day.
7. A "grafting dish" can be set up where several donor and recipient embryos had their vitelline membranes removed. Once the membrane is removed, embryos are very fragile and disintegrate at the contact of the air. To manipulate and move them, you should use a fire-polished glass Pasteur pipette and avoid any contact of the embryo with the air–water interface.
8. Such dorsal grafts tend to label midline tissue (see Fig. 2b, d). Depending on the experimental question, one can graft more laterally, but note that it is not known whether the deep and superficial layers extend equally to the lateral side of the gastrula. At the neural plate, the width of superficial and deep layers is different according to the expression of neural progenitor markers (e.g., Sox3, (5)). Therefore, it may be different early on as well, although this has not been established.

9. The donor piece of tissue tends to curl up after dissection and the host tissue heals very quickly after opening. The complete grafting procedure will be done without complication if all the steps described are performed in a time frame of 5 min. It is important to note that we use 0.5× MMR in the grafting buffer to slow down the healing process and thus to facilitate the grafting procedure. Note as well that the bigger the piece of superficial layer is, the slower the curling up will happen.
10. This part explains how to graft labeled superficial layer. In order to perform a deep graft, proceed the same way to peel the superficial layer from the donor embryo. This time, cut a piece of deep layer instead of the superficial and save it in order to graft it on the recipient embryos. Peel the superficial layer of the recipient embryo, without cutting the fourth edge. Instead, fold it and dilacerate the deep layer underneath. Tuck in the donor piece of tissue, fold back the superficial layer, and let it heal for an hour.
11. At this point, cleanliness of the grafting procedure can be checked by fixing and sectioning a couple of embryos. This should be done to ensure reproducibly clean dissection of each layer. Grafts should just be restricted to the layer targeted (Fig. 2e and f).
12. Tadpoles from NF stage 50 are fixed overnight at 4°C. The CNS can be subsequently dissected in order to facilitate further processing by reducing the size of the sample.
13. Sectioning on the cryostat carries risk of injury; adhere to local safety rules.

References

1. Hartenstein V (1989) Early neurogenesis in Xenopus: the spatio-temporal pattern of proliferation and cell lineages in the embryonic spinal cord. Neuron 3:399–411
2. Lamborghini JE (1980) Rohon-beard cells and other large neurons in Xenopus embryos originate during gastrulation. J Comp Neurol 189: 323–333
3. Hong E, Brewster R (2006) N-cadherin is required for the polarized cell behaviors that drive neurulation in the zebrafish. Development 133:3895–3905
4. Chalmers AD, Strauss B, Papalopulu N (2003) Oriented cell divisions asymmetrically segregate aPKC and generate cell fate diversity in the early Xenopus embryo. Development 130: 2657–2668
5. Chalmers AD, Welchman D, Papalopulu N (2002) Intrinsic differences between the superficial and deep layers of the Xenopus ectoderm control primary neuronal differentiation. Dev Cell 2:171–182
6. Chalmers AD, Pambos M, Mason J, Lang S, Wylie C, Papalopulu N (2005) aPKC, Crumbs3 and Lgl2 control apicobasal polarity in early vertebrate development. Development 132: 977–986
7. Sabherwal N, Tsutsui A, Hodge S, Wei J, Chalmers AD, Papalopulu N (2009) The apicobasal polarity kinase aPKC functions as a nuclear determinant and regulates cell proliferation and fate during Xenopus primary neurogenesis. Development 136:2767–2777
8. Minsuk SB, Keller RE (1997) Surface mesoderm in Xenopus: a revision if stgae 10 fate map. Dev Genes Evol 207:389–401
9. Keller R, Danilchik M (1988) Regional expression, pattern and timing of convergence and

extension during gastrulation of Xenopus laevis. Development 103:193–209
10. Shih J, Keller R (1992) The epithelium of the dorsal marginal zone of Xenopus has organizer properties. Development 116:887–899
11. Tabler JM, Yamanaka H, Green JB (2010) PAR-1 promotes primary neurogenesis and asymmetric cell divisions via control of spindle orientation. Development 137:2501–2505
12. Zhang C, Basta T, Jensen ED, Klymkowsky MW (2003) The beta-catenin/VegT-regulated early zygotic gene Xnr5 is a direct target of SOX3 regulation. Development 130:5609–5624
13. Dale L, Slack JM (1987) Fate map for the 32-cell stage of Xenopus laevis. Development 99:527–551
14. Keller R (1991) Early embryonic development of Xenopus laevis. Methods Cell Biol 36: 61–113
15. Khokha MK, Chung C, Bustamante EL, Gaw LW, Trott KA, Yeh J, Lim N, Lin JC, Taverner N, Amaya E, Papalopulu N, Smith JC, Zorn AM, Harland RM, Grammer TC (2002) Techniques and probes for the study of Xenopus tropicalis development. Dev Dyn 225: 499–510

Part III

Mammalian Model Systems

Chapter 17

Analyzing the Angiogenic Potential of Gr-1⁺CD11b⁺ Immature Myeloid Cells from Murine Wounds

Elahe Mahdipour and Kimberly A. Mace

Abstract

Analysis of tissue repair and regeneration in a variety of organisms has demonstrated that stem and progenitor cells play a critical role in the healing and regenerative response. In particular, during cutaneous wound healing bone marrow-derived cells are recruited to the site of injury in large numbers, often comprising over 50% of the cells within the wound milieu. These bone marrow-derived cells are comprised mostly of a heterogeneous mix of myeloid cells. In the early stages of wound healing, the most prominent subtypes are Gr-1^{+}CD11b^{+} cells that consist of progenitor cells and more differentiated granulocytes. Under certain conditions, these cells have the potential to strongly promote angiogenesis, and thus tissue repair and regeneration. This chapter provides methods by which one can isolate these cells from wound tissue and assess their pro-angiogenic capacity via gene expression analyses and functional in vivo angiogenesis assays.

Key words: Gr-1+CD11b+ cells, Myeloid progenitor cells, Neovascularisation, Wound healing, In vivo angiogenesis assay, Tissue dissociation

1. Introduction

Inflammation is one of the earliest responses following cutaneous injury. In addition to the protective function of inflammatory cells in which pathogens and debris are neutralized and phagocytized, inflammatory cells also contribute to the proliferative phase of wound healing. During the proliferative phase, regenerative processes such as neovascularization take place, coordinated by wound resident cells and recruited progenitor cells and inflammatory cells that produce angiogenic cytokines such as vascular endothelial growth factor (Vegf) (1). Different cell types are involved in this stage of wound healing, including fibroblasts, endothelial cells, and bone marrow-derived progenitor cells. A variety of progenitor cell types including CD34^{+} cells, CD14^{+} cells, and Gr-1^{+}CD11b^{+}

Kimberly A. Mace and Kristin M. Braun (eds.), *Progenitor Cells: Methods and Protocols*, Methods in Molecular Biology, vol. 916, DOI 10.1007/978-1-61779-980-8_17, © Springer Science+Business Media, LLC 2012

cells participate in neovascularization via direct (differentiation into endothelial cells) and indirect (paracrine induction of angiogenesis) mechanisms (2, 3).

Gr-1⁺CD11b⁺ cells are a heterogeneous population of myeloid progenitor cells with the potential to fully differentiate into monocytic or granulocytic lineages and more differentiated polymorphonuclear cells. Environmental signals may regulate the inflammatory or angiogenic behavior of these cells (4). Gr-1⁺CD11b⁺ cells are a source of angiogenic factors such as matrix metalloproteinase-9 (Mmp-9), Vegf, and Prokineticin 2 (Bv8) (5, 6). They also have been shown to differentiate into endothelial cells and incorporate into growing vessels in tumors (5). We have recently reported that during the first several days following injury Gr-1⁺CD11b⁺ cells accumulate in wound tissue and reach their maximum level by the end of the inflammatory phase (days 3–4). Shortly after the start of angiogenesis (days 4–5), Gr-1⁺CD11b⁺ cell numbers decline in the wound tissue, and are at relatively low levels by day 7 (7), suggesting that these cells play a role in the initiation of angiogenesis, but are not required for its maintenance, which proceeds until day 10. However, in impaired wound healing conditions such as diabetes, which are characterized by poor angiogenesis, Gr-1⁺CD11b⁺ cell numbers are significantly higher and remain at high levels far longer than their counterparts in nondiabetic animals. In addition, Gr-1⁺CD11b⁺ progenitor cells derived from diabetic animals have aberrant differentiation potential, producing a much higher percentage of monocytes than granulocytes compared to nondiabetic-derived equivalents (7). Moreover, these cells have a defective potential to produce angiogenic factors, such as Vegf and transforming growth factor β (Tgf-β), and fail to promote angiogenesis in vivo (7).

In order to better understand the role of these and other myeloid progenitor cells in tissue repair and regeneration, and to develop methods to manipulate their behavior in order to enhance this process, it is important to be able to isolate these cells from bone marrow, circulation, and injured tissue. It is also critical to ultimately test the effects of this manipulation in in vivo angiogenesis assays. This chapter presents methods to evaluate the angiogenic potential of Gr-1⁺CD11b⁺ cells in vitro via analyzing their angiogenic gene expression profile and in vivo, by injecting them to the wound site to assay their effects on angiogenesis.

2. Materials

2.1. Cutaneous Injury Model

1. Mice, 6–14 weeks of age (see Note 1).
2. Surgical scissors or punch biopsy, sterilized.

3. 6-mm wound template, sterilized. This can be made by cutting a precisely measured circle, with a diameter of 6 mm, out of the center of a 4-cm square piece of any flexible clear material, such as developed X-ray film or thin clear plastic sheet.
4. Buprenorphine (0.3 mg/ml) (see Note 2).
5. 29–30 gauge needles with 0.3–1.0 cc syringes.
6. Anesthesia unit to provide isoflurane + oxygen mixture to animal.
7. Electric warming blanket.
8. Sterile drape.
9. Marking pen.
10. 50 ml Falcon tube containing 100% ethanol.
11. 15 ml Falcon tube containing 9.5 ml sterile normal saline or phosphate-buffered saline (PBS).
12. Electric shavers/clippers for small animals.
13. Small scale for weighing animals.
14. Ethanol wipes.

2.2. Isolation of Gr-1+CD11b+ Cells from Wound

1. Surgical scissors.
2. Razor blades.
3. 19 gauge needle and 5 ml syringe.
4. 70 μm cell strainers (BD Biosciences).
5. Shaking incubator (37°C).
6. PBS.
7. Hanks' balanced salt solution (HBSS).
8. Fetal bovine serum (FBS).
9. Dispase I.
10. G418 powder (cell culture grade).
11. Collagenase D.
12. DNase I.

2.3. Antibody Staining for Fluorescence-Activated Cell Sorting

1. PBS.
2. FBS.
3. PE-Cy7-conjugated anti-mouse Ly-6 G (Gr-1) antibody (or other suitable fluorophore).
4. Alexa 647-conjugated anti-mouse CD11b antibody (or other suitable fluorophore).
5. RPMI 1640.
6. 12 × 75 mm, 5 ml round-bottomed polystyrene tubes for FACSAria cell sorter (e.g., Falcon #2051).
7. Centrifuge.

2.4. RNA Extraction from Gr-1+CD11b+ Cells

1. Sterile, RNase-free pipet tips (1–20, 20–200 and 200–1,000 μl).
2. Microcentrifuge.
3. 96–100% ethanol.
4. 70% ethanol.
5. Cell pellet, prewashed with RNase-free PBS.
6. RNeasy Kit (Qiagen)—contains spin columns, RLT buffer (lysis buffer), RW1 buffer (wash buffer), RPE buffer (final wash buffer, dilute the RPE buffer with pure 100% ethanol as recommended), RNase-free water.
7. QIAshredder homogenizer columns (Qiagen).
8. 2-Mercaptoethanol.
9. RNase-free DNase set (Qiagen).

2.5. Analysis of Angiogenic Gene Expression

1. Reverse transcription reagents (SABiosciences RT2 First Strand Kit is optimal).
2. Mouse angiogenesis RT2 profiler arrays, at least 6 (SABiosciences, array platform is specific depending on which qPCR instrument you use).
3. Quantitative PCR machine (e.g., StepOnePlus, Applied Biosystems).
4. RT2 qPCR master mix, for at least six arrays (SABiosciences, mix is specific for qPCR instrument).
5. Nuclease-free water.

2.6. In Vivo Angiogenesis Assay

1. Wounded mice, 6–14 weeks of age (see Note 1).
2. Gr-1+CD11b+ cells [from fluorescence-activated cell sorting (FACS)].
3. PBS.
4. 29–30 gauge needles with 0.3–1.0 cc syringes.
5. Anesthesia unit to provide isoflurane + oxygen mixture to animal.

3. Methods

3.1. Isolation of Gr-1+CD11b+ Cells

Gr-1+CD11b+ cells can be isolated from bone marrow, peripheral blood, or the peripheral tissues to which they are recruited (i.e., wounds or tumors). Further analyses can then be conducted to investigate potential mechanisms underlying their role in processes such as injury-induced neovascularization. Here we present the technique for isolating these cells from wound tissue from one mouse (presumably of a particular background, e.g., wild type, mutant, or transgenic) to transplant into the wound tissue of another mouse (presumably of a different background), as described below, or, alternatively to carry out in vitro analyses, such as angiogenic gene expression profiling, as discussed further below.

3.1.1. Animal Wounding and Tissue Harvest

1. Set up surgical area with warming blanket set to 37°C, covered by sterile drape. Sterile surgical instruments (biopsy punch or surgical scissors and tissue forceps with teeth) should be kept soaking in 100% ethanol but should be laid out to air dry on sterile drape prior to use. Anesthesia unit with isoflurane and oxygen should be arranged such that there is a holding container to initially anesthetize the animal, and once under, transfer the animal to nose cone delivering anesthesia to animal with minimal gas escape. Buprenorphine stocks generally come at 0.3 mg/ml and should be diluted 1:20 into sterile PBS or normal saline (15 μg/ml final concentration).
2. Anesthetize mouse with 2–3% isoflurane in oxygen at 2 l/min, quickly weigh animal, and record weight.
3. Shave hair (in the opposite direction to its growth) in a ~3-cm square region on the dorsum, two thirds of the way from the rostral to the caudal end, using clippers (Fig. 1a).
4. Remove excess hair and sterilize mouse dorsum using ethanol wipes and inject buprenorphine subcutaneously at 0.05 μg–0.2 μg/g mouse (e.g., for a 25 g mouse inject 100 μl).
5. Using sterile wound template and permanent marker, draw a dotted circular wound "target" on skin (Fig. 1b).
6. Using 6 mm biopsy punch, excise piece of targeted skin by pressing firmly and rotating biopsy punch (Fig. 1c). Alternatively, using tissue forceps and curved surgical scissors, tent targeted skin and cut along dotted line to excise skin. Wound should appear as shown in Fig. 1d.
7. House animal in fresh cage alone in quiet, darkened, warm (30°C) area until fully recovered from anesthesia (about 30 min). Return to normal housing area (but keep separate from other animals) and monitor behavior on a daily basis (see Note 3).
8. At day 2 following wounding (or whatever time point you desire), harvest wound tissue from mouse. This can be performed immediately after sacrificing animal, or under 2–3% isoflurane in oxygen at 2 l/min. The entire wound area including a 2-mm region outside the wound edge as well as granulation tissue under the wound should be removed using surgical scissors. The animal should be sacrificed at the end of this procedure.
9. The harvested wound should be kept in a clean microfuge tube on ice. Proceed to digestion step as soon as possible.

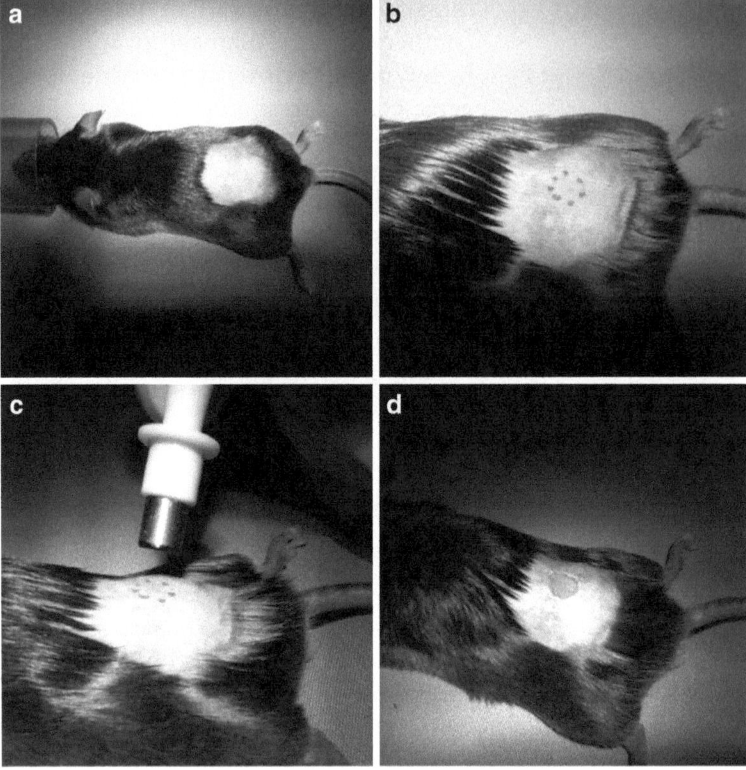

Fig. 1. *Creating a 6-mm diameter excisional wound.* (**a**) Example of mouse under light anesthesia with 3-cm square region shaved on dorsum. (**b**) 6-mm template drawn on mouse dorsum. (**c**) 6-mm biopsy punch used to excise skin. (**d**) Appearance of wound immediately after excision.

3.1.2. Wound Digestion

1. Weigh the freshly harvested wound and cut it into 2 mm pieces using a sharp blade.
2. Incubate the wound pieces in HBSS (40 μl/mg wound tissue) containing 1 mg/ml dispase I, 3% FBS, and 10 mg/ml G418 overnight at 4°C, on a rotating platform.
3. Transfer the tissue pieces to HBSS (80 μl/mg wound tissue) containing 1 mg/ml collagenase D, 100 U/ml DNase I, and 5 mg/ml G418. Set aside the overnight tubes containing the HBSS/dispase I mix on ice.
4. Incubate the tissue pieces in the HBSS/collagenase D mix for 2 h at 37°C in a shaking incubator (300 rpm).
5. After 2 h incubation, mix the cell suspension from steps 2–4 and pass through a 19 gauge needle several times to break up cell clumps.
6. Pass cell suspension through 70 μm cell strainer into a 50 ml Falcon tube on ice.

7. Count cells and centrifuge at $350 \times g$ for 5 min. Pour off supernatant and resuspend cell pellet in 1 ml PBS + 5% FBS.
8. Transfer cells to 1.5 ml tube and centrifuge to pellet the cells at $350 \times g$ for 5 min. Aspirate supernatant.
9. Resuspend cells in appropriate volume of PBS + 5% FBS for antibody staining (see below). Set aside 10^4–10^5 cells for unstained or isotype control.

3.1.3. Antibody Staining for FACS Sorting

1. Add 0.06 μg each of anti-Ly6G (Gr-1) and anti-CD11b per 10^6 cells in 100 μl of staining buffer (PBS + 5% FBS). Scale up volume for larger numbers (e.g., for 10^7 cells, use 500 μl staining buffer).
2. For unstained/isotype control, leave cells in staining buffer or add appropriate amount of each isotype control. This control is needed to measure false background signals in the flow cytometer.
3. Incubate cells on ice for 60–90 min.
4. After incubation, wash cells three times with staining buffer.
5. After final wash, resuspend cells in RPMI medium containing penicillin (100 U/ml) and streptomycin (100 μg/ml).
6. Sort cells using a FACS machine, obtaining sample positive for both antibodies (see Note 4).
7. Transfer collected Gr-1$^+$CD11b$^+$ cells to a 1.5 ml Eppendorf tube, being careful to rinse the walls of the collection tube to avoid any loss of cells.
8. Pellet the sorted cells at $400 \times g$ for 5 min and wash them once with RNase-free sterile PBS before proceeding to RNA extraction or injection step.
9. Count viable cells using a haemocytometer.

3.2. RNA Extraction

These instructions assume the use of Qiagen RNeasy Mini kit for purification of total RNA from animal cells. This method is based on the use of high-salt buffer which allows RNAs larger than 200 base pairs to bind to silica membrane. The other contaminants will be washed away using the ethanol-based wash provided; then the purified RNA will be eluted using RNase-free water.

3.2.1. Preparation of a Homogenized Cell Lysate

1. Following the RNAse-free PBS wash of the Gr-1$^+$CD11b$^+$ cells, pellet again as before and remove the supernatant completely.
2. Add 350 μl of RLT buffer for $<5 \times 10^6$ cells, or 600 μl of buffer for 5×10^6 to 1×10^7 cells.
3. Resuspend the pellet completely for efficient cell lysis.

4. Transfer mixture to a QIAshredder spin column to homogenize the lysate.
   Note: Incomplete homogenization results in losing a significant amount of RNA.
5. Centrifuge column for 2 min at full speed in a microfuge (>15,000 × *g*) and either transfer the homogenized lysate to an RNase-free Eppendorf tube and snap-freeze or proceed to extraction step immediately.

3.2.2. RNA Extraction

1. Mix the homogenized lysate with 1 volume of RNase-free 70% ethanol.
2. Transfer the sample to an RNeasy spin column and follow the manufacturer's instructions, including performing the DNase I on-column digestion step.
3. Measure the concentration of eluted RNA and store at −80°C until further use.

3.3. Angiogenesis Array

1. Perform reverse transcription reactions using between 10 and 1,000 ng RNA according to the reverse transcriptase manufacturer's instructions. Heat RNA for 2–3 min at 80°C and snap-chill on ice just prior to reverse transcription.
2. Heat RT reactions after transcription to 95°C for 5 min, snap-chill on ice, and repeat if desired (recommended for small starting amounts of RNA).
3. Dilute reverse transcription reactions with nuclease-free water up to a final volume of 102 μl. Store dilute cDNA at −20°C until ready to perform qPCR.
4. Perform qPCR using diluted cDNA, angiogenesis array (containing gene-specific primers and controls), RT2 qPCR master mix (containing buffer, enzyme, dNTPs, and SYBR green), and nuclease-free water according to the manufacturer's instructions for the specific instrument you are using.
5. Repeat qPCR on a separate array plate for each sample.
6. Analyze the data using the ΔΔCt method, either on the SABiosciences free web portal (http://www.SABiosciences.com/pcrarraydataanalysis.php) or using software of your choice.

3.4. In Vivo Angiogenesis Assay

Cells should be injected into wounds at day 2 post wounding. A minimum number of 2.5×10^5 cells are needed per animal. Animals can be injected all on the same day (preferable as this will reduce experimental variation) or in pairs (one control + one experimental on same day).

1. Prepare cells for injection by diluting 2.5×10^5 cells/50 μl of sterile PBS per animal in a sterile microtube (see Note 5).

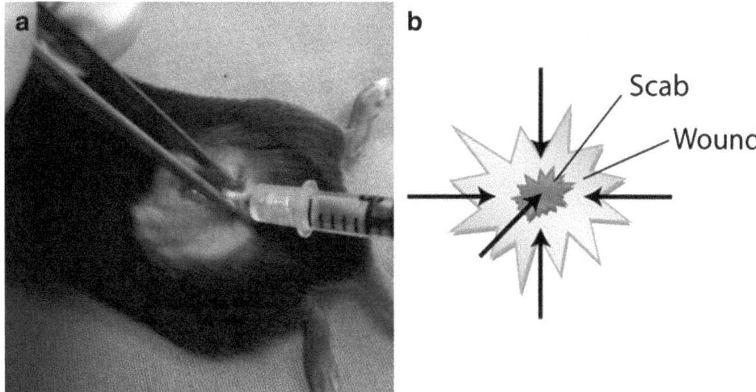

Fig. 2. *Injection of progenitor cells into wound tissue at day 2 following wounding.* (**a**) Inject cells in 50 microliters of sterile PBS using a 45° angle towards center of wound. (**b**) Schematic representation of injection to the wound. *Black arrows* show the five suitable sites for injecting the cells beneath the wound and scab.

2. Keep cells on ice until just prior to injection; then warm to 37°C.
3. For injection into wound, animal should be under anesthesia as before (2% isoflurane in oxygen at 2 l/min).
4. When the animal is ready, using forceps tent skin at the perimeter of the wound and using a 0.3–1.0 cc insulin syringe inject 50 μl of diluted cells (2.5×10^5 cells in total) into the wound at a 45° angle, splitting volume over four separate sites around the perimeter as well as just under the middle of the forming scab tissue (Fig. 2).
5. Injection of control cells and/or vehicle only (PBS) to wounds should be used as the control.
6. After injection, return animals to their cages until day 7 post wounding.
7. To analyze the effect of injected experimental cells on angiogenesis, the whole wound, including a 2-mm perimeter outside the visible scab, should be harvested at day 7 post wounding and the tissue should be dissociated for flow cytometry as described above, or embedded in OCT for frozen sections and//or paraffin sections following fixation for 24 h in formalin. It is prudent to divide the wound tissue for a combination of these assays.
8. The tissue is then analyzed for new microvessel formation in the granulation tissue using antibodies such as anti-CD31 or anti-CD34 (as described elsewhere, e.g., (7)). An increase in microvessels indicates enhanced angiogenesis.

4. Notes

1. All animal work must be performed in accordance with local rules and regulations as well as national and international laws governing the humane treatment of animals in scientific experimentation. The minimum number of animals for each experiment should be used, thus, where possible, control treated and experimentally treated wounds should be made on the same animal. However, no more than four wounds should be made on the same animal. A power analysis (power = 80%, alpha = 0.05) has determined that the assay described in this chapter requires a group size of six (i.e., six controls vs. six experimental). Only surgically trained individuals should perform this procedure. Donor mice from which the Gr-1^+CD$11b^+$ cells are obtained can be wild type or a mutant/transgenic depending on the question to be answered, and the recipients can likewise be wild type or some particular background. As a baseline control, one can use wild-type donor into wild-type recipient, which shows no measurable effect on angiogenesis compared to PBS (7).
2. Buprenorphine is a controlled substance and must be obtained from and used by licensed individuals only. Animal facility veterinary staff can provide further information for your particular state/country/facility.
3. Following wounding, mice should not show any digression from normal behavior, including nesting, playing, eating, and drinking. If abnormal behavior is observed, seek veterinary advice immediately.
4. Collect cells into a FACS tube that has been completely coated in 100% FBS, as cells may adhere to dry tube walls as they come out of the FACS machine. Collect cells into 1 ml of 100% FBS or PBS + FBS (from 5 to 50%). Collection medium MUST be sterile.
5. All steps of preparation should be done using sterile materials and laminar flow cabinet.

Acknowledgements

The authors would like to thank Mike Jackson for his help with developing the flow cytometry methods described here. This work was funded by Iranian Ministry of Health and Medical Education (E. M), and the Healing Foundation (K. A. M.).

References

1. Singer AJ, Clark RA (1999) Cutaneous wound healing. N Engl J Med 341:738–746
2. Ferrara N (2010) Role of myeloid cells in vascular endothelial growth factor-independent tumor angiogenesis. Curr Opin Hematol 17:219–224
3. Schatteman GC, Dunnwald M, Jiao C (2007) Biology of bone marrow-derived endothelial cell precursors. Am J Physiol Heart Circ Physiol 292:H1–18
4. Ostrand-Rosenberg S, Sinha P (2009) Myeloid-derived suppressor cells: linking inflammation and cancer. J Immunol 182: 4499–4506
5. Yang L, DeBusk LM, Fukuda K, Fingleton B, Green-Jarvis B, Shyr Y, Matrisian LM, Carbone DP, Lin PC (2004) Expansion of myeloid immune suppressor Gr+CD11b+cells in tumor-bearing host directly promotes tumor angiogenesis. Cancer Cell 6:409–421
6. Shojaei F, Ferrara N (2008) Refractoriness to antivascular endothelial growth factor treatment: role of myeloid cells. Cancer Res 68:5501–5504
7. Mahdipour E, Charnock JC, Mace KA (2011) Hoxa3 promotes the differentiation of hematopoietic progenitor cells into proangiogenic Gr-1+CD11b+myeloid cells. Blood 117:815–826

Chapter 18

In Vivo Imaging of Hematopoietic Stem Cells in the Bone Marrow Niche

Oliver Barrett, Roberta Sottocornola, and Cristina Lo Celso

Abstract

Even though hematopoietic stem cells (HSC) are amongst the first somatic stem cells exploited for therapeutic purposes, their application is still limited by the inability to expand them ex vivo without impairing their function. Moreover, it has recently emerged that several types of leukemia develop and relapse through complex interactions with bone marrow (BM) components and may directly affect the HSC and their niche. Increasing attention has therefore been dedicated to the BM microenvironment the HSC reside in, with the view that a better understanding of the molecular regulators of HSC-niche interaction in vivo will allow improving HSC mobilization, collection and transplantation and provide clues for the development of innovative leukemia treatments. This chapter focuses on a recently established technique for the visualization of transplanted hematopoietic stem and progenitor cells (HSPC) within the calvarium bone marrow of live mice (Lo Celso et al. Nature 457:92–96, 2007). Intravital microscopy is a rapidly developing field, driven by constant improvement in both detection technologies (i.e., spatial resolution, depth of penetration, spectral definition) and probe availability (i.e., increasingly sophisticated genetic and chemical reporter systems). We therefore discuss the current limitations and challenges related to intravital microscopy of the HSC niche and introduce a number of potential imaging approaches, which could be promising candidates for future development of this technique.

Key words: Intravital microscopy, Hematopoietic stem cells, Stem cell niche, Mouse, Bone marrow, Calvarium, fluorescent probes, Red-shifted fluorophores, Confocal microscopy, Two-photon microscopy

1. Introduction

Despite a long research tradition in the field of hematopoiesis, direct observation of the hematopoietic stem cell (HSC) niche has been traditionally difficult and therefore its nature remains elusive. The development of an intravital microscopy setup allowing analysis of transplanted hematopoietic stem and progenitor cells (HSPC) in the calvarium bone marrow of anesthetized mice is a first step

Kimberly A. Mace and Kristin M. Braun (eds.), *Progenitor Cells: Methods and Protocols*, Methods in Molecular Biology, vol. 916, DOI 10.1007/978-1-61779-980-8_18, © Springer Science+Business Media, LLC 2012

towards the unequivocal identification of the cellular and molecular components HSC reside in.

Bone marrow from the femur has been traditionally used for hematopoietic research; however calvarium bone marrow shows identical HSC frequency and function (1). Imaging approaches reaching femoral marrow rely on dissection (2), fiber optic insertion (3), or mechanical bone thinning (4) to overcome the obstacle posed by the thickness of compact bone. Calvarium intravital microscopy has the key advantages of being minimally invasive for the mouse and of allowing observation of virtually undisrupted bone marrow microenvironment, while maintaining cellular resolution (5).

Penetration depth is one of the major challenges associated with in vivo imaging. Techniques such as functional photoacoustic microscopy (fPAM) (6, 7) and bioluminescence (8, 9) are promising deep imaging modalities; however their achievable resolution diminishes drastically with depth, making it impossible to obtain cellular resolution within any bone marrow cavity. Approximately 150 μm is the current maximum observable depth by confocal/two-photon microscopy, provided that the skull surface is thoroughly cleaned and free of any scar tissue (1).

There is an enormous range of fluorescent probes, both chemical and genetically encoded, that could be adapted for use in intravital microscopy. Chemical probes currently tend to be brighter (for example, DiD labeling of HSPC provides signal several orders of magnitude brighter than that of GFP expression driven by most conventional promoters). However, the use of the appropriate promoters, such as the *Col2.3* promoter for osteoblastic cells (1) or the nestin promoter for mesenchymal progenitors (10), does lead to transgenic reporter expression sufficiently elevated for deep detection. The probes would also have to conform to other parameters necessary for the particular experiment, such as maturation rate, stability, and non-toxicity (11). Deeper penetration can be reached through the use of red and far red-shifted probes, whose signal is not efficiently quenched by the water and hemoglobin present within live tissues. Recently developed fluorescent proteins, such as mCherry, plum, and Neptune (12, 13) have excitation and emission spectra previously exclusively achievable through the use of chemical probes. The use of such fluorescent reporters may increase not only imaging depth, but also spatial resolution, leading to subcellular definition and in vivo analysis of parameters currently observable in detail only in cultured cells, such as protein and RNA distribution (14), protein–protein interactions (15–17), caspase (18, 19), and GTPase activity (20, 21), as well as cell cycle stage (22), calcium concentration (18, 23), redox potential (24, 25), membrane potential (26), and pH (27).

As the HSC niche is a complex and dynamic system, its efficient visualization requires probing multiple components over time. The generation of scar tissue following scalp suturing is the major limiting factor for efficient HSPC tracking over a time window of a few days. While we are still far from in vivo long term time-lapse imaging of the HSC niche, it is relatively easier to tackle the complexity of the BM microenvironment. The number of parameters observed can be drastically increased by working with probes distributed as widely as possible along the light spectrum (5), by choosing fluorophores with tight emission spectra, such as quantum dots (28, 29), and by taking advantage of spectral imaging (30, 31). We have conducted simultaneous imaging of bone, osteoblasts, vasculature, autofluorescent bone marrow cells, and DiD-labeled HSPC (1). Spectral imaging has allowed identification of neuronal interactions based on randomized expression of a few fluorescent proteins (30). Depending on the availability of cell type-specific promoters, a simplified version of this approach based on the combinatorial, predefined expression of fluorophores could allow identification of a greater number of cell types by using a small number of fluorescent probes. This chapter focuses specifically on the harvest, DiD labeling, and in vivo imaging of transplanted HSPC in mouse calvarium bone marrow as a basic protocol, amenable to a wide range of modifications in terms of transplanted cell populations, labeling dyes, and microenvironmental components analyzed.

2. Materials

2.1. Harvesting of Bone Marrow Mononuclear Cells

1. Donor wild-type mice (C57BL/6, 6–12 weeks old) can be purchased from Harlan (UK) or Jackson Laboratories (USA).
2. Harvest medium: Hanks' balanced salt solution (HBSS; Stemcell Technologies) supplemented with 2% fetal bovine serum (FBS; Stemcell Technologies).
3. Cell strainer, 70 μM (BD).
4. Dissection and surgery tools (scissors, forceps).
5. Mortar and pestle (VWR).

2.2. Isolation of Long-Term Repopulating Hematopoietic Stem and Progenitor Cells (LT-HSPC) by Lineage Depletion and Cell Sorting

1. Lineage antibodies cocktail: Biotin-conjugated anti-mouse CD3e (clone 145-2C11, eBioscience); Biotin-conjugated anti-mouse CD11b (clone M1/70, eBioscience); Biotin-conjugated anti-mouse Ly-6 G (clone RB6-8 C5, eBioscience); Biotin-conjugated anti-mouse CD8a (clone 53-6.7, eBioscience); Biotin-conjugated anti-mouse CD4 (clone GK1.5, eBioscience); Biotin-conjugated anti-mouse TER-119 (clone Ter119, eBioscience); Biotin-conjugated anti-mouse CD45R (clone Ra3-6B2, eBioscience). Mix the antibodies at 1:1:1:1:1:1:1 ratio. The cocktail can be prepared in advance and stored at 4°C.
2. Streptavidin-conjugated magnetic beads (Miltenyi Biotec).
3. PBS.

4. Steriflip filter unit (Millipore).
5. LD columns (Miltenyi Biotec).
6. QuadroMACS (Miltenyi Biotec) (see Note 1).
7. Hanks' balanced salt solution (HBSS) (Stemcell Technologies).
8. Fetal bovine serum (FBS) (Stemcell Technologies).
9. Anti-Mouse CD117 (c-Kit) antibody, APC conjugated (clone 2B8, eBioscience).
10. Anti-Mouse CD34 antibody, FITC conjugated (clone RAM34, eBioscience).
11. Anti-mouse Ly-6A/E (Sca-1) antibody, Pacific Blue™ conjugated (clone D7, Biolegend).
12. Anti-Mouse CD135 antibody, PE conjugated (clone A2F10.1, BD Bioscience).
13. PE-Cy™7-conjugated Streptavidin (BD Bioscience).
14. Compensation beads (BD™ CompBead, BD Bioscience).
15. Cell sorter (e.g., BD FACS ARIA) (see Note 2).
16. Sorting collection medium: StemSpan® Serum-Free Expansion Medium (Stemcell Technologies) supplemented with 10% FBS (Stemcell Technologies). It can be stored in aliquots at –20°C indefinitely or at 4°C for up to 2 months.

2.3. Irradiation of Recipient Mice

1. Recipient wild-type mice (C57BL/6, 8–14 weeks old) can be purchased from Harlan (UK) or Jackson laboratories (USA).
2. γ-Irradiator (see Note 3).
3. Mouse holder fitting into the irradiator chamber (see Note 3).

2.4. LT-HSPC Labeling and Injection

1. PBS.
2. DiD (Vybrant® DiD cell-labeling solution, Invitrogen) (see Note 4).
3. Insulin syringes (Terumo).
4. Heated box (VetTech Solutions).
5. Mouse restrainer (VetTech Solutions).

2.5. In Vivo Imaging

1. Hypnorm (VetaPharma).
2. Hypnovel (Roche).
3. Homeothermic blanket system for rodents (Stoelting).
4. Leica SP5 MP/FLIM upright confocal microscope, with infrared laser (Spectra Physics Mai Tai 690-1020) for multiphoton imaging (see Note 5).
5. Leica 25× water immersion objective 0.95 numerical aperture (see Note 6).

3. Methods

In order to isolate LT-HSPC from the whole bone marrow, differentiated cells are first eliminated using the MACS™ separation system. Total bone marrow is stained with a cocktail of biotin-conjugated antibodies (lineage antibody cocktail) against differentiated cells and with streptavidin-conjugated magnetic beads and finally loaded onto a column positioned besides a strong magnet. Differentiated cells are retained in the column, while undifferentiated cells flow through the column (32). The cell suspension obtained is then stained with specific markers to allow the isolation of LT-HSPC by cell sorting. PE-Cy™7-conjugated streptavidin is also added to eliminate remaining differentiated cells. LT-HSPC are identified as Lineage$^-$, c-kit$^+$, Sca-1$^+$, CD34$^-$, Flk-2$^-$. Different antibody cocktails can be used to purify different HSPC populations (33). Sorted cells are then labeled with DiD (a lipophilic dye that efficiently binds the plasma membrane lipids (34), injected intravenously in irradiated recipient mice and visualized by confocal microscopy. Two-photon microscopy allows visualization of the calvarium bone through second harmonic generation signal from the collagen, one of the main components of calcified bone (35). Depending on the Investigator's interest, different components of the bone marrow microenvironment can be visualized through confocal/two-photon microscopy by using injectable dyes or fluorescent reporter transgenic mouse strains. Table 1 summarizes published probes and detection settings described for in vivo imaging of the HSC niche. In vivo immunofluorescence labeling of bone marrow has also been achieved (3, 34).

3.1. Total Bone Marrow Cells Harvest

1. Sacrifice two donor mice by cervical dislocation
2. Dissect all hind limb bones (tibias, femurs, and hip bones) and the spine. Put them in PBS.
3. Crush the dissected bones in harvest medium with mortar and pestle (see Note 7).
4. Filter the cell suspension obtained through a cell strainer and collect the total bone marrow from each mouse in a 50 ml Falcon tube.

3.2. Purification of LT-HSPC by Magnetic Lineage Depletion and Cell Sorting

1. Centrifuge the cell suspension at 500 × *g* for 5 min at 4°C, discard the supernatant, and resuspend the cell pellet in 1 ml of harvest medium.
2. Add 50 μl of Lineage antibody cocktail to each tube, mix briefly, and incubate at 4°C for 15 min.
3. During this incubation, prepare degassed PBS. Prepare ~40 ml of PBS in a 50 ml Falcon tube, connect it to a Steriflip filter,

Table 1
Bone marrow in vivo imaging setup

Signal source	Niche component	Microscopy	Excitation	Emission filter	Reference
SHG	Collagen (bone)	Two-photon	840–900	420–480	(35)
DiD	Transplanted HSPC	Confocal	He–Ne 633	660–690	(1)
Autofluorescence	AF BM cells	Confocal	He–Ne 543	560–620	(1)
GFP	Osteoblasts, Nestin + cells	Confocal	Arg-488	500–550	(1, 10)
GFP	Osteoblasts	Two-photon	900–960	500–550	(1)
Qdot800	Vasculature	Confocal	He–Ne 543	>795	(1)
Qdot800	Vasculature	Two-photon	960	750–850	(1)

HSPC and bone marrow components can be observed simultaneously using the settings indicated. GFP reporter mice have been successfully used to highlight both osteoblasts and nestin-positive cells. GFP and Quantum dots (Qdots) can be detected with either confocal or two-photon microscopy, while Second Harmonic Generation signal (SHG) can only be detected by two-photon microscopy and DiD by confocal microscopy. Autofluorescence is more easily detected with confocal microscopy

and keep it under vacuum for a few minutes without filtering it. Prepare fresh each time.

4. To wash the unbound antibodies, fill each tube with PBS and spin at 500 × *g* for 5 min at 4°C.
5. Resuspend the cell pellet in 1 ml of degassed PBS, removing all cell clumps.
6. Add 30 μl of streptavidin-conjugated magnetic beads to each tube, mix briefly, and incubate at 4°C for 15 min.
7. During this incubation, position two LD columns on the magnet and load them with 3 ml of degassed PBS each. Allow the PBS to run entirely through the column and discard.
8. Add 2 ml of degassed PBS to each tube and load the cell suspension onto a column through a cell strainer to eliminate potential clumps, which otherwise may cause blockage of the column. Collect lineage depleted cells in a 15 ml Falcon tube.
9. Add 2 ml of degassed PBS to each 50 ml tube to collect any residual cells. Once the cell suspension has entirely run through the column, load the additional 2 ml onto the column using the same cell strainer (see Note 8).
10. Combine the two cell suspensions in one 15 ml conical tube and spin at 500 × *g* for 5 min at 4°C.
11. Resuspend the cell pellet in 500 μl of harvest medium. Set aside 50 μl of cell suspension as unstained control for cell sorting. To label LT-HSPC, add 5 μl (1:100 dilution) of c-Kit,

Sca-1, CD34, and Flk-2 antibodies and 1 μl (1:500 dilution) of PE–streptavidin and incubate at 4°C for 20 min. For compensation purposes, at the same time prepare single-color control for each antibody by mixing one drop (50 μl) of positive and one drop of negative compensation beads with 1 μl of antibody (1:100 dilution). The compensation beads are ready to use without any further steps (see Note 9).

12. Fill the tube with PBS and spin at 500×g for 5 min at 4°C.
13. Resuspend the cell pellet in 3 ml of harvest medium and sort at room temperature. LT-HSPC are collected in a 1.5 ml Eppendorf tube prefilled with 1 ml of sorting collection medium. In average, 5-10,000 LT-HSPC should be obtained from two donor C57BL/6 mice.

3.3. Irradiation of Recipient Mice

1. Administer two doses of 5.5 Gy 3 h apart to the recipient mice. Recipient mice must be injected within 4–24 h from the last dose of irradiation.

3.4. LT-HSPC Labeling and Injection

1. Spin the sorted cells at 500×g for 5 min at 4°C and aspirate the supernatant without disturbing the cell pellet.
2. To remove the serum, which will decrease the efficiency of the staining, fill the tube with PBS and spin the cell suspension at 500×g for 5 min at 4°C. Aspirate the supernatant and resuspend the cell pellet in 100 μl of PBS.
3. Add 0.5 μl of DiD, vortex immediately, and then incubate at 37°C for 10 min in a tissue culture incubator (see Note 10).
4. Fill the tube with PBS and spin the cell suspension at 500×g for 5 min at 4°C. Resuspend the cell pellet in 300 μl of PBS.
5. Inject the labeled cells in the tail vein of irradiated recipient mice using insulin syringes to minimize cell loss (insulin syringes have no "dead" space near the needle). To better visualize the tail vein, keep the mice at 37°C in a heated box for about 10–15 min. A number of different restrainers can be equally efficient to immobilize the mouse.

3.5. In Vivo Imaging

1. Prepare the Hypnorm/Hypnovel cocktail. Mix 1 ml of Hypnorm with 2 ml of water in a test tube. Add 1 ml of Hypnovel to the Hypnorm/water mix in the test tube. Prepare fresh each time.
2. Anesthetize the mouse by injecting 6 ml/kg of Hypnorm/Hypnovel cocktail intraperitoneally (see Note 11).
   - Clip the scalp hair using scissors and remove the hair fragments using a wet tissue.
3. Assess the pedal reflex (toe pinch) to determine the depth of anesthesia. When the pedal reflex is absent, make a longitudinal

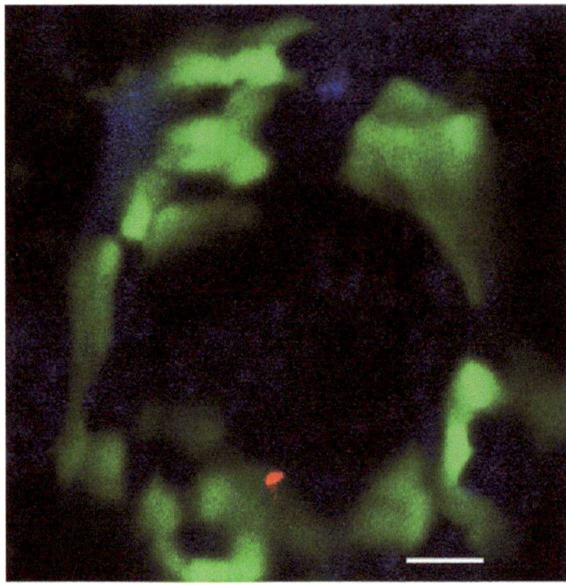

Fig. 1. A mixture of hematopoietic stem and progenitor cells (Lineage low, c-Kit+, Sca1+) was purified, labeled with DiD and injected in Col2.3GFP osteoblast reporter mice (1), and visualized as described. *Blue* is second harmonic generation signal (bone collagen), *green* is GFP (osteoblastic cells), and *red* is DiD signal (LKS cell). *Scale bar* : 50 μm.

or a C-shaped incision onto the scalp and separate the skin flaps to expose the skull (36).

4. Apply a drop of warm PBS on the skull to avoid drying.
5. Switch on the microscope lasers and allow them to warm up for 10–20 min.
6. Place the mouse on the microscope stage, keeping the calvarium area containing the frontal bones as horizontal as possible. Position the objective directly on the PBS-covered calvarium.
7. Arrange the excitation/emission setup as indicated in Table 1.
8. For orientation purposes, identify the central and coronal sutures by observing the bone through the second harmonic generation (SHG) signal of the collagen (36).
9. Scan the whole bone marrow area contained within the frontal bones and identify injected cells by comparing DiD and autofluorescence signals (Fig. 1). The signal from a DiD-labeled cell should be at least twofold more intense than the autofluorescence signal from the same cell of interest (non-DiD-labeled, autofluorescent cells will present similar signal intensity in both channels. DiD-labeled cells, however, have weak or no autofluorescent signal compared with DiD signal).
10. At the end of the imaging session, euthanize the mouse using an approved procedure, or suture the scalp incision if further

imaging or long-term follow-up of the mouse is required. Suturing needs to be performed very carefully if planning to image the mouse again, because the formation of any amount of scar tissue between the surgical suture and the skull will dramatically impair signal detection. The second imaging session procedure will be identical to the first one.

4. Notes

1. The quadromacs magnet holds four columns. If fewer columns are going to be routinely used, single magnets can be purchased instead (MidiMACS™ separator, Miltenyi Biotec cat. no. 130-042-302).
2. The described fluorophore combination works well when using a FACS Aria cell sorter. If working with a different sorter, we recommend optimizing the antibody (and fluorophores) cocktail in advance as different sorters may be equipped with different lasers and filter sets.
3. Irradiator installation is strictly regulated by national and federal laws and it is usually dealt with by each hospital/university. Different types and models of irradiators are available and Cesium 137 is the most commonly used source for gamma rays. All irradiators have a chamber inside which a container holding the mice can be positioned. A plastic container does not shield gamma irradiation and allows thorough cleaning after use. If the gamma irradiator is located outside the animal facility, we recommend housing the irradiated mice in a different location than pre-irradiation to minimize the risk of contamination of long-term mouse stocks.
4. DiD belongs to the family of "Vybrant" lipophilic dyes. It can be substituted with DiI, DiO, or DiR, depending on the available lasers and emission filter sets and on the overlap with other fluorophores used within the same experiment. In our hands, DiD is the brightest dye; therefore we recommend using the other dyes at slightly higher concentration.
5. Confocal and two-photon microscopes for in vivo imaging are in constant development. A number of commercial and custom/self-made setups offer slightly different performances but are equally suitable for calvarium bone marrow live imaging. We describe the microscope we are currently using, but any alternative system should work, provided it is equipped with similar lasers, filters, and objective. The main difference between an in-house built microscope and a commercial one is in the servicing of the machine. Self-built machines tend to have "open table" form and are easily accessible for the

expert user to resolve day-to-day alignment and optics issues. Commercial systems are more user friendly and therefore preferable for most biologists; however usually they can only be serviced by the authorized engineers and therefore their maintenance and optimization can be time consuming.

6. Lower magnification objectives tend not to offer sufficient resolution for single cell imaging. Higher magnification objectives tend not to allow a sufficient amount of light to collect any non-superficial signal.
7. Bone crushing can be time consuming and might require some optimization in order to minimize the amount of time required for the harvest and, as a consequence, stem cell death. Two rounds of crushing for the hind limb bones and three rounds of crushing for the spine should be sufficient to collect all hematopoietic marrow, leaving only white bone chips. Occasionally some red patches remain associated with the bone chips, but unless they are very large it is best to avoid further rounds of crushing.
8. The LD columns do not run dry and therefore it is not a problem if they are left unattended for longer than it takes for the cell suspension to flow through. Normally, the flow rate is about 1 ml/5 min, and the main reason not to wait longer than necessary is to reduce the overall time required for the harvest and purification, which has an impact on the viability of HSPC.
9. Another widely recognized LT-HSPC population is Lineage$^-$, c-kit$^+$, Sca-1$^+$, CD48$^-$, CD150$^+$ (37, 38). CD34-Fitc and Flk2-PE antibodies can be substituted by CD48-Fitc (Biolegend, cat. no. 103404) and CD150-PE (Biolegend, cat. no. 115904) antibodies, diluted 1:1,000 and 1:100, respectively. Generally, variation is observed between different batches and aliquots of antibodies and we do recommend titrating each antibody and adjusting the working dilution accordingly.
10. DiD is used at 5 μM. For simplicity of use, any amount of HSPC up to 100,000 cells can be resuspended in 100 μl PBS and stained with 0.5 μl of DiD. For larger amounts of cells, the ideal concentration of the cell suspension is 10^6 cells/ml. Efficiency and toxicity of the staining may vary depending on the cell population used. We therefore recommend validating the staining procedure for the cell population of interest using both FACS analysis and bone marrow transplantation assay (for example, 1:1 competitive transplantation using stained versus unstained cells (1)).
11. The indicated dose is recommended for C57BL/6 mice and may need to be adjusted if working with different mouse strains.

Acknowledgments

We are grateful to Mark Scott and Dr. Martin Spitaler for discussion and input on the setup and subsequent description of in vivo confocal and two-photon microscopy with a commercially available system.

References

1. Lo Celso C, Fleming HE, Wu JW, Zhao CX, Miake-Lye S, Fujisaki J, Cote D, Rowe DW, Lin CP, Scadden DT (2009) Live-animal tracking of individual haematopoietic stem/progenitor cells in their niche. Nature 457:92–96
2. Xie Y, Yin T, Wiegraebe W, He XC, Miller D, Stark D, Perko K, Alexander R, Schwartz J, Grindley JC, Park J, Haug JS, Wunderlich JP, Li H, Zhang S, Johnson T, Feldman RA, Li L (2009) Detection of functional haematopoietic stem cell niche using real-time imaging. Nature 457:97–101
3. Lewandowski D, Barroca V, Duconge F, Bayer J, Van Nhieu JT, Pestourie C, Fouchet P, Tavitian B, Romeo PH (2010) In vivo cellular imaging pinpoints the role of reactive oxygen species in the early steps of adult hematopoietic reconstitution. Blood 115:443–452
4. Kohler A, Schmithorst V, Filippi MD, Ryan MA, Daria D, Gunzer M, Geiger H (2009) Altered cellular dynamics and endosteal location of aged early hematopoietic progenitor cells revealed by time-lapse intravital imaging in long bones. Blood 114:290–298
5. Lo Celso C, Wu JW, Lin CP (2009) In vivo imaging of hematopoietic stem cells and their microenvironment. J Biophotonics 2:619–631
6. Maslov K, Zhang HF, Hu S, Wang LV (2008) Optical-resolution photoacoustic microscopy for in vivo imaging of single capillaries. Opt Lett 33:929–931
7. Zhang HF, Maslov K, Stoica G, Wang LV (2006) Functional photoacoustic microscopy for high-resolution and noninvasive in vivo imaging. Nat Biotechnol 24:848–851
8. Branchini BR, Ablamsky DM, Davis AL, Southworth TL, Butler B, Fan F, Jathoul AP, Pule MA (2010) Red-emitting luciferases for bioluminescence reporter and imaging applications. Anal Biochem 396:290–297
9. Nakajima K, Komiyama Y, Hojo H, Ohba S, Yano F, Nishikawa N, Ihara S, Aburatani H, Takato T, Chung UI (2010) Enhancement of bone formation ex vivo and in vivo by a helioxanthin-derivative. Biochem Biophys Res Commun 395:502–508
10. Mendez-Ferrer S, Michurina TV, Ferraro F, Mazloom AR, Macarthur BD, Lira SA, Scadden DT, Ma'ayan A, Enikolopov GN, Frenette PS (2010) Mesenchymal and haematopoietic stem cells form a unique bone marrow niche. Nature 466:829–834
11. Chudakov DM, Matz MV, Lukyanov S, Lukyanov KA (2010) Fluorescent proteins and their applications in imaging living cells and tissues. Physiol Rev 90:1103–1163
12. Lin MZ, McKeown MR, Ng HL, Aguilera TA, Shaner NC, Campbell RE, Adams SR, Gross LA, Ma W, Alber T, Tsien RY (2009) Autofluorescent proteins with excitation in the optical window for intravital imaging in mammals. Chem Biol 16:1169–1179
13. Shaner NC, Campbell RE, Steinbach PA, Giepmans BN, Palmer AE, Tsien RY (2004) Improved monomeric red, orange and yellow fluorescent proteins derived from Discosoma sp. red fluorescent protein. Nat Biotechnol 22:1567–1572
14. Lange S, Katayama Y, Schmid M, Burkacky O, Brauchle C, Lamb DC, Jansen RP (2008) Simultaneous transport of different localized mRNA species revealed by live-cell imaging. Traffic 9:1256–1267
15. Ai HW, Hazelwood KL, Davidson MW, Campbell RE (2008) Fluorescent protein FRET pairs for ratiometric imaging of dual biosensors. Nat Methods 5:401–403
16. Chu J, Zhang Z, Zheng Y, Yang J, Qin L, Lu J, Huang ZL, Zeng S, Luo Q (2009) A novel far-red bimolecular fluorescence complementation system that allows for efficient visualization of protein interactions under physiological conditions. Biosens Bioelectron 25:234–239
17. Shyu YJ, Suarez CD, Hu CD (2008) Visualization of ternary complexes in living cells by using a BiFC-based FRET assay. Nat Protoc 3:1693–1702
18. Tomosugi W, Matsuda T, Tani T, Nemoto T, Kotera I, Saito K, Horikawa K, Nagai T (2009) An ultramarine fluorescent protein with increased photostability and pH insensitivity. Nat Methods 6:351–353

19. Xu X, Gerard AL, Huang BC, Anderson DC, Payan DG, Luo Y (1998) Detection of programmed cell death using fluorescence energy transfer. Nucleic Acids Res 26:2034–2035
20. Itoh RE, Kurokawa K, Ohba Y, Yoshizaki H, Mochizuki N, Matsuda M (2002) Activation of rac and cdc42 video imaged by fluorescent resonance energy transfer-based single-molecule probes in the membrane of living cells. Mol Cell Biol 22:6582–6591
21. Mochizuki N, Yamashita S, Kurokawa K, Ohba Y, Nagai T, Miyawaki A, Matsuda M (2001) Spatio-temporal images of growth-factor-induced activation of Ras and Rap1. Nature 411:1065–1068
22. Sakaue-Sawano A, Ohtawa K, Hama H, Kawano M, Ogawa M, Miyawaki A (2008) Tracing the silhouette of individual cells in S/G2/M phases with fluorescence. Chem Biol 15:1243–1248
23. Palmer AE, Tsien RY (2006) Measuring calcium signaling using genetically targetable fluorescent indicators. Nat Protoc 1:1057–1065
24. Dooley CT, Dore TM, Hanson GT, Jackson WC, Remington SJ, Tsien RY (2004) Imaging dynamic redox changes in mammalian cells with green fluorescent protein indicators. J Biol Chem 279:22284–22293
25. Hanson GT, Aggeler R, Oglesbee D, Cannon M, Capaldi RA, Tsien RY, Remington SJ (2004) Investigating mitochondrial redox potential with redox-sensitive green fluorescent protein indicators. J Biol Chem 279:13044–13053
26. Knopfel T, Tomita K, Shimazaki R, Sakai R (2003) Optical recordings of membrane potential using genetically targeted voltage-sensitive fluorescent proteins. Methods 30:42–48
27. Jankowski A, Kim JH, Collins RF, Daneman R, Walton P, Grinstein S (2001) In situ measurements of the pH of mammalian peroxisomes using the fluorescent protein pHluorin. J Biol Chem 276:48748–48753
28. Medintz IL, Uyeda HT, Goldman ER, Mattoussi H (2005) Quantum dot bioconjugates for imaging, labelling and sensing. Nat Mater 4:435–446
29. Texier I, Josser V (2009) In vivo imaging of quantum dots. Methods Mol Biol 544: 393–406
30. Livet J, Weissman TA, Kang H, Draft RW, Lu J, Bennis RA, Sanes JR, Lichtman JW (2007) Transgenic strategies for combinatorial expression of fluorescent proteins in the nervous system. Nature 450:56–62
31. Phan TG, Bullen A (2010) Practical intravital two-photon microscopy for immunological research: faster, brighter, deeper. Immunol Cell Biol 88:438–444
32. Lo Celso C, Scadden D (2007) Isolation and transplantation of hematopoietic stem cells (HSCs). J Vis Exp (2) 157
33. Bryder D, Rossi DJ, Weissman IL (2006) Hematopoietic stem cells: the paradigmatic tissue-specific stem cell. Am J Pathol 169: 338–346
34. Sipkins DA, Wei X, Wu JW, Runnels JM, Cote D, Means TK, Luster AD, Scadden DT, Lin CP (2005) In vivo imaging of specialized bone marrow endothelial microdomains for tumour engraftment. Nature 435:969–973
35. Zipfel WR, Williams RM, Webb WW (2003) Nonlinear magic: multiphoton microscopy in the biosciences. Nat Biotechnol 21: 1369–1377
36. Lo Celso C, Lin CP, Scadden DT (2011) In vivo imaging of transplanted hematopoietic stem and progenitor cells in mouse calvarium bone marrow. Nat Protoc 6:1–14
37. Foudi A, Hochedlinger K, Van Buren D, Schindler JW, Jaenisch R, Carey V, Hock H (2009) Analysis of histone 2B-GFP retention reveals slowly cycling hematopoietic stem cells. Nat Biotechnol 27:84–90
38. Kiel MJ, Yilmaz OH, Iwashita T, Yilmaz OH, Terhorst C, Morrison SJ (2005) SLAM family receptors distinguish hematopoietic stem and progenitor cells and reveal endothelial niches for stem cells. Cell 121:1109–1121

Chapter 19

Characterizing the Phenotype of Murine Epidermal Progenitor Cells: Complementary Whole-Mount Visualization and Flow Cytometry Strategies

Korinna D. Henseleit, Ann P. Wheeler, Gary Warnes, and Kristin M. Braun

Abstract

The epidermis and its appendages, the hair follicle and sebaceous gland, have the capacity to constantly regenerate throughout adult life. Postnatal hair follicles undergo a cyclic mode of tissue homeostasis, defined by periods of growth, degeneration, and rest. A multipotent population of stem cells residing within the hair follicle bulge not only generates the hair lineages during each hair cycle, but also transiently contributes to the repair of epidermis following wounding. In this chapter, we provide methods for identifying epidermal stem cells and investigating their proliferative and apoptotic characteristics. We introduce whole-mount and flow cytometry techniques, which complement each other by permitting visualization of the epidermal stem cell compartment in situ and assessment of the phenotype of purified cells. These techniques can easily be adapted to characterize novel putative epidermal stem or progenitor cell populations. By applying whole-mount and flow cytometry techniques to characterize normal and genetically modified mice with skin defects, we expect to learn more about the factors that regulate stem cell self-renewal and differentiation.

Key words: Epidermis, Stem cells, Progenitor cells, Hair follicles, Bulge, Wholemounts, Cell cycle analysis, Flow cytometry

1. Introduction

Mammalian epidermis consists of a stratified squamous epithelium (interfollicular epidermis) with associated hair follicles and glandular structures which form a protective waterproof barrier essential for survival. The epidermis is maintained by a process of tissue homeostasis, a fine balance which ensures that dead or damaged cells are continually replaced throughout adult life. There is significant evidence that homeostasis is maintained by epidermal stem cells which reside in the basal layer of the interfollicular

Kimberly A. Mace and Kristin M. Braun (eds.), *Progenitor Cells: Methods and Protocols*, Methods in Molecular Biology, vol. 916, DOI 10.1007/978-1-61779-980-8_19,

epidermis (IFE), the hair follicle (HF), and the sebaceous gland (1–4). In contrast to the interfollicular epidermis which is constantly renewed, postnatal hair follicles undergo a cyclic mode of tissue homeostasis, defined by periods of growth (anagen), degeneration (catagen), and rest (telogen) (5, 6). The stem cell population is activated to proliferate during anagen, whilst keratinocyte apoptosis is an important element of hair follicle regression during catagen.

In murine hair follicles, both nucleotide and H2B-GFP tracing studies have identified epidermal label-retaining cells (LRC) in a specialized region of the outer root sheath known as the bulge (7–10). While label retention merely reflects the proliferative history of a cell, several lines of evidence suggest that this infrequently cycling subpopulation has attributes of epidermal stem cells (10, 11). It is now well established that the murine hair follicle bulge contains a multipotent stem cell population, capable of giving rise to all hair lineages during the growth phase of the hair cycle and contributing to acute repair of interfollicular epidermis following tissue damage (1, 2, 4, 7, 12–15). However, the issue of hair follicle stem cell identity may not be as straightforward as previously thought, since cells with stem or progenitor properties recently have been reported in other regions of the hair follicle (16–20).

The availability of promoters permitting conditional overexpression and gene deletion in epidermal subcompartments has provided excellent tools to investigate the molecules regulating epidermal homeostasis and wound healing. To characterize the role of stem and progenitor cells in these processes, it is essential to use complementary experimental approaches, facilitating both the evaluation of purified populations of cells and assessment of their behavior in situ. Bulge stem cells have been purified by using fluorescence-activated cell sorting (FACS), enriching for labeled cells isolated from genetically modified mice (9, 14) and, more commonly, by using antibodies against the cell surface proteins CD34 and α6-integrin (CD49f) (21, 22) to purify epidermal stem cells from any mouse stain. In this protocol, we describe flow cytometry techniques that can be applied to assess the cell cycle and apoptotic characteristics of classical bulge keratinocytes or novel putative stem/progenitor cells. In contrast to assessing the phenotype of individual FACS-purified cells, we also describe a method of whole-mount labeling that permits visualization of large expanses of epidermis. This technique has been widely used to reveal novel insights regarding patterning of stem cells, cycling cells, and differentiating cells in both human and mouse epidermis (23–25). Taken together, these flow cytometry and whole-mount techniques have proven useful for assessing the characteristics of stem and progenitor cells in normal mice and in mice with skin abnormalities.

2. Materials

2.1. Preparation of Epidermal Whole Mounts from Murine Tail

1. Any standard laboratory mouse strain (see Note 1).
2. Optional: Bromodeoxyuridine (BrdU) solution 12.5 mg/mL in PBS. Warm the solution to get the BrdU to fully dissolve. Sterilize by using a 10 mL syringe to pass the solution through a syringe filter with 0.22 μm pores. Aliquots of the sterile BrdU solution may be stored at −20°C until required.
3. Insulin syringes, with needle (VWR).
4. Disposable scalpels.
5. Forceps (2×).
6. Phosphate-buffered saline (PBS).
7. 5 mM EDTA in PBS.
8. 10% Neutral Buffered Formalin (10% strong formalin in H_2O; 0.4% (w/v) sodium dihydrogen phosphate, monohydrate; 0.65% (w/v) disodium hydrogen phosphate, anhydrous).
9. Sodium azide (Sigma) (see Note 2).
10. 50 mL conical tubes (e.g., Falcon™; BD Biosciences).

2.2. Immunolabeling of Epidermal Whole Mounts

1. 96-well tissue culture plates.
2. Primary antibodies (as required).
3. Alexa Fluor® secondary antibodies (species-specific, as required) (Life Technologies).
4. Sodium Buffer (0.9% NaCl, 20 mM HEPES, pH 7.2).
5. Blocking Buffer (0.5% skim milk powder, 0.25% fish skin gelatin, and 0.5% Triton X-100 in Sodium Buffer). Store at 4°C for up to 2 days.
6. Laboratory rocker.
7. PBS containing 0.2% Tween 20.
8. DAPI (Sigma-Aldrich). Prepare a 1 mg/mL stock in double-distilled H_2O. Store at −20°C.
9. Aluminum foil.
10. Microscope slides.
11. Coverslips.
12. Glycerol.
13. MOWIOL® 4-88 (Calbiochem).
14. Double-distilled water (ddH_2O).
15. Tris–HCl pH 8.5.
16. DABCO (Sigma).
17. Laboratory bench-top centrifuge.

2.3. Confocal Imaging of Epidermal Whole Mounts

1. Zeiss LSM Meta 510 point scanning confocal (or similar machine).
2. Lens cleaning tissue (Whatman).
3. 100% Ethanol.
4. Imaris version 7.2 (Bitplane Scientific Software).

2.4. Disaggregation of Mouse Epidermis

1. Contura cordless clipper (Harvard Apparatus).
2. Sterile dissection tools (scissors, forceps, scalpel).
3. 100 mm cell culture dishes.
4. Ice.
5. Cell culture hood.
6. Betadine.
7. 70% Ethanol.
8. Trypsin (0.25%).
9. FAD—Calcium basal media. Mix three parts high glucose DMEM and one part Ham's F12 medium (F12), from which calcium chloride has been omitted (custom-made by PAA). Supplement with 1.8×10^{-4} M adenine, penicillin (100 IU mL^{-1}), and streptomycin (100 μg mL^{-1}). Media can be stored for several weeks at 4°C. Add fresh glutamine (450 mg L^{-1}) immediately prior to use.
10. FAD low calcium complete medium. Supplement FAD—Calcium basal medium with 10% fetal bovine serum (FBS). The serum must be batch-tested for its ability to support growth of keratinocytes. Finally, add hydrocortisone (0.5 μg mL^{-1}), cholera enterotoxin (10^{-10} M) (Enzo Life Sciences), EGF (10 ng mL^{-1}), and insulin (5 μg mL^{-1}) (all final concentrations). Completed medium can be stored at 4°C for 1 week prior to use.
11. 10 mL serological pipettes.
12. 70 μm cell strainer.
13. Laboratory bench-top centrifuge with refrigeration.
14. Trypan blue solution (0.4%).
15. Hemocytometer.

2.5. Immunofluorescence Identification and Cell Cycle Analysis of Epidermal Stem Cells

1. Minimal Essential Medium (MEM) without magnesium and calcium (Invitrogen), complemented with 2% (final concentration) FBS (PAA Laboratories Ltd.).
2. Phosphate-buffered saline (PBS) without magnesium and calcium.
3. FACS buffer: 2% FBS in 1× PBS.
4. Pasteur pipettes (polyethylene).
5. 40 μm nylon strainers (BD Biosciences).

6. Round bottom tubes with snap cap 12×75 mm (BD Biosciences).
7. Compensation beads (BD Biosciences).
8. FITC-conjugated rat anti-CD49f antibody, clone GoH3 (BD Biosciences).
9. FITC-conjugated isotype rat IgG2a (BD Biosciences).
10. Alexa Flour® 647 anti-mouse CD34, Clone RAM34 (eBioscience).
11. Alexa Flour® 647 isotype rat IgG1 (eBioscience).
12. 70% ice-cold Ethanol.
13. RNAseA (Sigma-Aldrich, UK) Prepare a 100 g/mL stock in 1× PBS and store at 4°C.
14. LSRII flow cytometer system (BD Biosciences) (or a similar machine).

2.6. Analysis of Apoptosis in Epidermal Stem Cells by Annexin V Staining

1. Annexin V Binding Buffer (BD Biosciences).
2. Annexin V-PE (Invitrogen).

3. Methods

3.1. Preparation of Epidermal Whole Mounts from Murine Tail

By testing epidermis isolated from several anatomical locations, we determined that mouse tail epidermis was most informative (10, 24), as robust whole mounts could be generated from any stage of the hair cycle. In addition, tail skin contains large hair follicles with clear anatomical features, which are ideal for microscopic visualization. In contrast, dorsal epidermis is more fragile and contains a much higher density of hair follicles, both factors which prevent consistent generation of high-quality epidermal sheets. Hair follicles in the tail grow as triplets and undergo cycles of growth with similar timing to dorsal follicles, although the degree of synchronization is less tight (26, 27).

1. Any wild-type or genetically modified mouse strain may be used for this procedure; however the age of the mice will determine the stage of the hair cycle (see Note 3).
2. *Optional*: Injection of BrdU into the intraperitoneal cavity may be used to generate BrdU label-retaining cells (see Note 4) or to achieve short-term labeling to identify cells undergoing DNA synthesis (see Note 5).
3. Kill the mouse using an approved technique. Remove the mouse tail by slicing with a scalpel at the point of attachment to the body.

4. Slit the mouse tail lengthwise with a scalpel and use a pair of forceps to peel the skin from the underlying connective tissue.
5. Use scalpel to cut tail skin into small pieces (e.g., 5×5 mm^2).
6. Place the skin in a 50 mL conical tube and incubate the pieces of skin in 25 mL of 5 mM EDTA in PBS at 37°C for approximately 4 h.
7. Remove the epidermis from the dermis. Use one pair of forceps to hold the tissue on the dermal side and a second pair of forceps to gently peel the intact epidermal sheet starting from one corner (see Note 6).
8. Fix epidermal whole mounts in 10% Neutral Buffered Formalin for 2 h at room temperature.
9. Wash 3× with PBS.
10. Store tissues in PBS containing 0.2% (w/v) sodium azide at 4°C. Whole mounts can be stored for several weeks to months prior to immunolabeling with most antibodies.

3.2. Immunolabeling Epidermal Whole Mounts

This is a general protocol which can be adapted to investigate different markers of interest, including putative epidermal progenitor/stem cell subpopulations (see Subheading 1). Some examples of specific applications are shown in Table 1 and Fig. 1 (see Note 7).

Table 1
Primary antibodies and dyes used to fluorescently label the whole-mount staining panels shown in Fig. 1

Structure/epitope labeled	Source of primary antibody or labeling agent	Species	Dye used to visualize	Excitation wavelengths (nm)	Laser line used (nm)
DNA	Invitrogen	N/A	DAPI	340–410	UV (361 or 405)
Fragmented DNA of apoptotic cells	Promega	N/A	Fluorescein	494–521	488
BrdU CD34 Keratin 15 Lrig	Oxford Biotech eBioscience Cancer Research UK R&D Systems	Rat Mc Rat Mc Ms Mc Gt Pc	Alexa Fluor® 488 secondary antibody	470–500	488
CD49f Keratin 14 Ki67	BD Biosciences Covance Novacastra	Rat Mc Rb Pc Rb Pc	Alexa Fluor® 568 secondary antibody	530–580	543/561

Also indicates the wavelengths which can be used to excite the dyes to facilitate selection of an appropriate laser on a confocal system to acquire high resolution images. *Mc* monoclonal *Pc* polyclonal, *Ms* Mouse, *Gt* Goat, *Rb* Rabbit

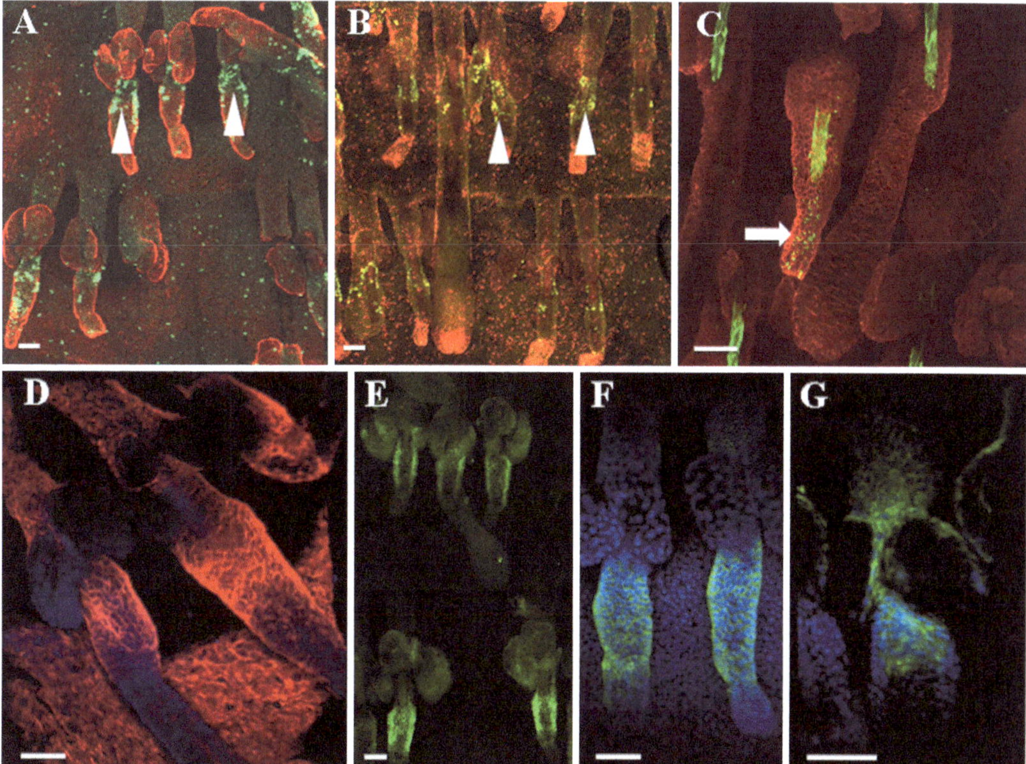

Fig. 1. Immunofluorescence of mouse tail whole mounts. Epidermal sheets were labeled and excited as indicated in Table 1. (**a**, **b**) BrdU injection (see Notes 4 and 8) was used to generate label-retaining cells. Mice were maintained for at least 70 days postinjection. (**a**) BrdU (*green*) and Keratin 14 (*red*); (**b**) BrdU (*green*) and the proliferative marker Ki67 (*red*); (**c**) Apoptotic cells (*green*; see Note 9) and Keratin 14 (*red*); (**d**) CD49f (*red*); (**e**) CD34 (*green*); (**f**) Keratin 15 (*green*) and DAPI (*blue*); (**g**) Lrig (*green*) and DAPI (*blue*). Size bars indicate 50 μm. *Arrowheads*, BrdU label-retaining cells; *Arrow*, apoptotic cells.

1. Place each epidermal sheet in an individual well of a 96-well tissue culture plate.
2. Block and permeabilize epidermal sheets for 30 min at room temperature in Blocking Buffer. A volume of 100 μl of solution should be sufficient to cover the tissue.
3. Aspirate Blocking Buffer.
4. *Optional*: BrdU detection (see Note 8) or detection of apoptotic cells (see Note 9).
5. Dilute primary antibodies in Blocking Buffer and incubate epidermal sheets overnight with gentle agitation at room temperature. The concentration of all antibodies needs to be determined by testing several dilutions (see Note 10).
6. Aspirate primary antibody solution. Wash epidermal sheets in PBS containing 0.2% Tween 20, changing solution 4× over a minimal 4 h time period.

7. Incubate with species-specific secondary antibodies diluted in Blocking Buffer overnight with gentle agitation at room temperature. DAPI may also be added at a final concentration of 1 μg/mL to label cell nuclei. Secondary antibodies must be conjugated to appropriate fluorescent dyes to permit confocal detection and avoid spectral overlap (for examples see Table 1; Fig. 1). Tissue culture plates are covered with aluminum foil for the remainder of the protocol as the dyes are light-sensitive.
8. Aspirate secondary antibody and wash epidermal sheets in PBS containing 0.2% Tween 20, changing the solution 4× over a minimal 4 h period.
9. If required (see Step 10), prepare MOWIOL/DABCO mounting media. Mix 6 g glycerol and 2.4 g MOWIOL® 4-88 in a 50 mL conical tube and vortex. Add 6 mL distilled water and leave at room temperature for 1.5 h. Add 12.5 mL 200 mM Tris–HCl pH 8.5 and vortex the solution. Heat to 50°C for 10 min followed by vortexing, repeating this process three times. Place the solution on a laboratory rotator overnight at room temperature. Add DABCO anti-fade to a final concentration of 2.5% (w/v). Rotate several more hours to dissolve DABCO. The solution is then centrifuged at 5,000×*g* for 10 min at room temperature. The supernatant is removed and stored in aliquots at −20° C. The pellet should be discarded as it contains undissolved crystals.
10. Mount the epidermal sheets on a microscope slide with the basal surface facing the coverslip using MOWIOL-DABCO (or commercially available mounting media). Add enough mounting media so that the tissue is completely covered.
11. Store slides containing epidermal whole mounts in the dark at 4°C. Tissues will retain fluorescence signal for several months.

3.3. Confocal Imaging of Epidermal Whole Mounts

One advantage of confocal microscopy over conventional optical microscopy is the ability of the instrument to collect serial optical sections from thick specimens, such as tail epidermal whole mounts which measure approximately 100 μm in depth. In addition, the confocal instrument is able to exclude secondary fluorescence emitted by the specimen away from the region of interest, resulting in images of higher resolution. We collect images using a Zeiss LSM 510 Meta point scanning confocal, typically using a 20× 0.5NA objective. However the methodology outlined below can be adapted for confocal microscopes produced by any manufacturer. We have included a detailed protocol to aid inexperienced users, also explaining some of the theoretical aspects of confocal microscopy.

1. Prior to data collection the microscope objective lens is cleaned thoroughly with 100% ethanol and lens cleaning tissue.

The glass coverslip covering the specimen is also cleaned using lens cleaning tissue, double-distilled water (ddH_2O), and then 100% ethanol.

2. Initially, samples are visualized by eye and a central focal plane is selected where the antibody staining is brightest and most in focus.
3. Lasers on the confocal are then selected based on the excitation maxima of the dye (examples in Table 1).
4. In a typical study, multiple structures are fluorescently labeled (generally nuclei and one or two protein epitopes). Therefore, the confocal microscope is configured to image each fluorescent label in the sample in separate acquisition channels, i.e., sending this signal from each dye to a separate detector.
5. Next the confocal is set up to scan through the channels sequentially, in order to prevent bleed-through of signals from different dyes in the final image.
6. Initially, the sample is visualized live using a high scan speed (e.g., pixel dwell time >0.8 μs $pixel^{-1}$). This frequently updating, noisy image is used to quickly improve the focus and detector (see Note 11) settings with minimal photo-damage to the sample. To optimize the detectors for sensitivity and image contrast the following steps (7–11) are carried out using high speed scanning.
7. First, the brightest point in the whole-mount sample is brought into focus to ensure the field of view with highest dynamic range is used for optimizing the detectors.
8. To acquire a true confocal image the pinhole of the detector is set to 1 Airy unit for the dye detector channel of the epitope of interest. *N.B.* If two epitopes equally of interest are to be imaged, the pinhole of the channel imaging the dye with the longest wavelength is set to 1 Airy unit. The other channel is then set so the pinhole size gives an optical slice identical to the first channel. For visualization of nuclei, the pinhole for the channel containing DAPI is consistently set to 1 Airy unit.
9. The detectors are then set into a mode to reveal underexposed (pseudo-colored blue) and overexposed (pseudo-colored red) pixels in the image.
10. The digital offset of the detectors is altered to ensure that there are no underexposed pixels in the image. The offset is then corrected so that the background region in the image is as low intensity as possible without being underexposed.
11. The detector gain is adjusted to ensure there are no overexposed pixels in the image. The offset is then corrected to ensure that the brightest point in the image is as high as possible without becoming overexposed (see Note 12).

12. The confocal scan speed is reduced to 1.6–2 μs pixel^{-1} and the scan average increased to 4. Reducing confocal scan speed allows the pixel to be imaged to receive more light, thereby maximizing signal. Increasing the scan average causes each pixel in the image to be scanned multiple times and the averaged result of the multiple scans is then displayed as the final image (see Note 13).
13. When acquiring the image, the bit depth of data is set to 16 bit to allow a maximum contrast image to be recorded (see Note 14).
14. To collect the *Z* stack the confocal microscope is set in high scan speed mode and focussed to the bottom of the sample and then a small amount beyond. The position of the *Z* axis is then recorded into the software. This process is repeated with the top of the sample. The focal planes slightly above and below the sample are captured to allow high-quality three-dimensional image rendering of the whole sample.
15. The step size of the *Z* stack is determined by Nyquist Criterion; this states that the optimal *Z* interval must be equal to half of the optical slice thickness (see Note 15).
16. Once imaging is complete, data are saved in both the manufacturers' proprietory data format (.lsm) and also in a nonproprietary (.tif) format to ensure it can be rendered in all image analysis programs. Image rendering is carried out using Imaris version 7.2 software.

3.4. Disaggregation of Mouse Epidermis

Isolation of murine keratinocytes from telogen or anagen skin results in a heterogeneous population of differentiated and undifferentiated cell types. To achieve a high-quality single-cell preparation, it is important to handle cells gently and to incubate them on ice as much as possible to maintain cell viability.

1. Kill mice using an approved protocol (see Note 1) and use the electric clippers to remove the hair from the dorsal side of the skin. Cut against the direction of growth to remove as much hair as possible, but try not to nick the skin.
2. Use dissection tools to remove the dorsal skin in one piece.
3. Place the skin in a 100 mm cell culture dish. Intact tissue can be stored for several hours chilled on wet ice.
4. *Optional*: If viable cells are required (e.g., for tissue culture), work in a cell culture hood to sterilize skin. Use sterile instruments, plastics, and glassware throughout all procedures. Sequentially bathe the tissue in sterile beakers containing approximately 100 mL of: 10% Betadine (45 s); 70% Ethanol (20 s); 70% Ethanol (20 s); and finally PBS (30 s).
5. Place the skin in a new sterile 100 mm cell culture dish, with the dermal side facing up. Use a pair of forceps to hold the

tissue and a scalpel to scrape away the excess fat and blood vessels from the dermis (see Note 16).

6. Float the skin with the dermal side facing down (hair facing up) in 10 mL pre-warmed 37°C trypsin (0.25%) in a 100 mm cell culture dish. Skin should be completely unfolded.
7. Trypsinize the skin for 2 h 15 min at 37°C (or alternatively overnight at 4°C).
8. Hold the tissue at one corner using a pair of forceps and gently scrape the epidermal cells from the dermis using a sterile scalpel. The epidermis should be easy to remove and will come off in small pieces (see Note 17). In contrast, the dermis is very tough. At the end of this procedure the dermis is discarded.
9. Inactivate the trypsin by adding 10 mL of pre-warmed FAD low calcium complete media (see Note 18).
10. Completely mince the epidermis using two scalpels.
11. Use a 10 mL serological pipette to pipette the mixture up and down several times to isolate individual keratinocytes.
12. Filter the mixture through a 70 μm cell strainer into a 50 mL conical tube.
13. Move tissue from the cell strainer back to the 10 mm cell culture dish. Add 10 more mL of FAD low calcium complete media and repeat steps 10–12 on the residual tissue to maximize the number of keratinocytes isolated.
14. Cells may be stored on ice while preparing additional samples.
15. Spin tubes at 500 × *g* for 8 min at 4°C to pellet the cells.
16. Carefully aspirate the supernatant without disrupting the cell pellet. Resuspend the cells in 5 mL of FAD low calcium complete media.
17. Remove a 20 μl aliquot, dilute in an equal volume of trypan blue, and count cells using a hemocytometer (see Note 19).
18. Cells are now ready to be used for flow cytometry analyses (see Subheading 3.5 and 3.6). Please note that protocols for analysis of the self-renewal and differentiation potential of FACS-enriched cells in tissue culture and in vivo regeneration assays have been described elsewhere (28–31).

3.5. Immunofluorescence Identification and Cell Cycle Analysis of Epidermal Stem Cells

In this section, we describe the identification of hair follicle bulge cells by means of specific antibodies directed against the cell surface markers CD49f (α6-integrin), a marker of basal keratinocytes, and CD34, which is specifically expressed in follicular stem cells of the bulge (21, 22). However, this protocol could be adapted to identify and perform cell cycle analyses on other epidermal subpopulations, including putative progenitor and stem cells (see Subheading 1).

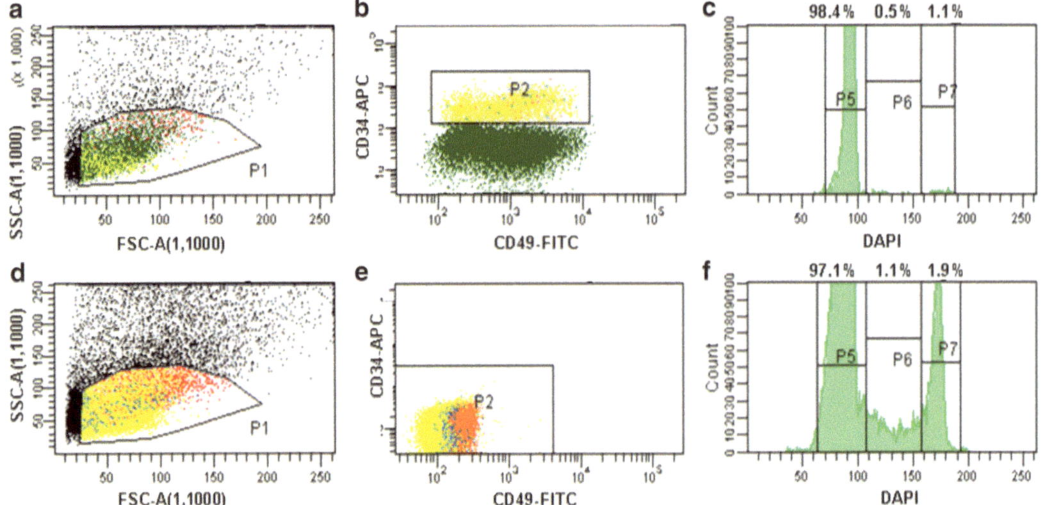

Fig. 2. FACS profile of cell cycle analysis. Freshly isolated keratinocytes from 7-week-old telogen stage mice were immunolabeled with anti-CD34 and anti-CD49 antibodies and DAPI was added to determine the DNA content of cells. (**a**) Forward scatter (FSC)/Side scatter (SSC) plot which allows the identification of epidermal progenitors based on size and granularity (P1). (**b**) Histogram plot. CD34-APC fluorescence intensity is shown on the *y*-axis and CD49-FITC on the *x*-axis. Gate P2 contains the CD34+ve basal population of the bulge. (**c**) DAPI fluorescence signal intensity defines the cell cycle stage. While 98.4% of cells are at G_0 stage of the cell cycle (P5), only 0.5% and 1.1% of cells are in the synthesis (S) phase or G_2 phase, respectively. (**d–f**) Profile of all sorted (unlabeled) cells. Both populations are mitotically quiescent at this stage of the hair cycle.

Flow cytometry based cell cycle analysis takes advantage of the fact that the DNA content of a cell changes in the course of the cell cycle, which can be monitored by a number of fluorescent DNA intercalating dyes that bind to DNA in a stoichiometric manner; thus the fluorescent signal is equal to the amount of DNA. Cells in the G_0 or G_1 phase of the cell cycle have a normal diploid set of chromosomes ($n=2$) and have half the fluorescent signal of cells in the G_2 phase, which, just before entering mitosis, are tetraploid ($n=4$). Cells in the S phase of the cell cycle are in the process of synthesizing DNA and are either $2n$ or $4n$. As a result, the fluorescence intensity appears in between G_0/G_1 and G_2. The percentage of the total population in each stage of the cell cycle can be determined in a histogram plot of fluorescence (DNA content) versus cell numbers (see Fig. 2).

1. To a 500 mL bottle of Minimal Essential Medium (MEM) add 10 mL fetal bovine serum (2% final concentration) (see Note 20).
2. Centrifuge cells (*from* Subheading 3.4, Step 17) at $300\times g$ for 5 min at 4°C. Remove supernatant and adjust cell density to 1×10^7 cells/mL with FBS-complemented MEM.
3. Transfer 100 μl per sample to a round bottom FACS tube. Include one sample for each isotype control and set aside some

unstained cells for adjusting the instrument settings for the flow cytometry analysis. Although fluorochromes with minimal spectral overlap have been chosen, it is nevertheless recommended to include single labeled cells or compensation beads for compensation of spectral overlap.

4. Add 1 μl FITC-conjugated anti-CD49f (α6-integrin) antibody (1:100 dilution in FBS-complemented MEM) and 1 μl Alexa Fluor® 647-conjugated anti CD34 antibody (1:100), which is equivalent to 2 μg/test. To separate tubes, add 20 μl of either Alexa Fluor® 647- or FITC-conjugated IgG2a isotype as a negative control (see Note 21).
5. Incubate cells at 4°C for 20 min inverting the tubes regularly. Alternatively, cells can be incubated on a shaker at 50–100 rpm.
6. Following incubation, centrifuge cells at 300 × *g* for 5 min, remove the supernatant, and resuspend the pellet in 3 mL FBS-complemented MEM.
7. Centrifuge again at 300 × *g* for 5 min, remove the supernatant, and resuspend in FBS-complemented MEM. Repeat the washing step two times.
8. Remove the supernatant, resuspend the pellet in ice-cold 70% ethanol, and vortex vigorously.
9. Incubate the cells for 30 min at 4°C and centrifuge at 1,900 × *g*.
10. Remove the supernatant and wash the cells in 3 mL FBS-complemented MEM followed by centrifugation at 1,900 × *g* at 4°C. Repeat the washing step two times.
11. Resuspend the pellet in 100 μl RNaseA in 1× PBS (100 μg/mL) and incubate for 15 min at 37°C.
12. Add 200 μl FACS buffer and 2–3 μl DAPI stock solution (to a final concentration of 1 μg/mL) (see Note 22).
13. Transfer cells though a 70 μm cell strainer into a new round bottom FACS tube.
14. To obtain meaningful results, the correct instrument configuration is crucially important. The flow cytometer needs to be equipped with Argon, Helium/Neon, and UV Lasers in order to achieve excitation of the FITC (515–545 nm BP filter), APC (650–670 nm BP filter), and DAPI (425–475 nm BP filter) fluorochromes, respectively (see Note 23).
15. Set the correct voltage for Forward Scatter (FSC) and Side Scatter (SSC) using unstained cells. FSC and SSC reflect the size and complexity of the cells, respectively, which allows selection of the desired population by setting a gate (P1) excluding differentiated cells, debris, and cell doublets (Fig. 2a).
16. Analyze the isotype controls. Any signal obtained in the APC or FITC laser channel is considered background. Take this into

consideration when analyzing the signal obtained in the CD49-FITC/CD34-APC-positive sample.

17. Analyze the CD49-FITC/CD34-APC-labeled sample. Epidermal bulge keratinocytes are $CD49^{+}$/CD34+ (Fig. 2b). In order to exclude other cell types from the plot, set a gate around the positive region.
18. Create a histogram by plotting cell count (*y*-axis) against DAPI fluorescent intensity (*x*-axis). Create a gate around the $DAPI^{low}(G_0/G_1)$, $DAPI^{medium}(S)$ and the $DAPI^{high}(G_2)$ population in order to determine the percentage of cells in each phase (Fig. 2c).

3.6. Analysis of Apoptosis in Epidermal Stem Cells by Annexin V Staining

Annexin V is a protein that specifically binds to phosphatidyl-serine residues in the membrane of cells that are about to undergo programmed cell death (apoptosis). Phosphatidyl-serine residues are normally localized on the inner cytoplasmic membrane facing the cytosol, but are transported to the cell surface when cell death is induced (32, 33). In consequence, apoptotic cells can be detected by means of fluorophore coupled Annexin V in FACS analysis. A typical FACS profile for Annexin V staining is shown in Fig. 3.

1. Immunolabel freshly isolated keratinocytes to identify basal cells and bulge stem cells (as described in Subheading 3.5) or other cellular subpopulations as required.
2. Wash cells 2× with ice-cold PBS and resuspend in 1× binding buffer at a concentration of 1×10^6/mL.
3. Transfer 100 μl (1×10^5 cells) to a 5 mL culture tube and add 5 μl of Annexin V-PE.
4. Gently vortex the cells and incubate for 15 min at room temperature in the dark.
5. Add 400 μl of 1× binding buffer to each tube and add DAPI to a final concentration of 250 ng/mL.

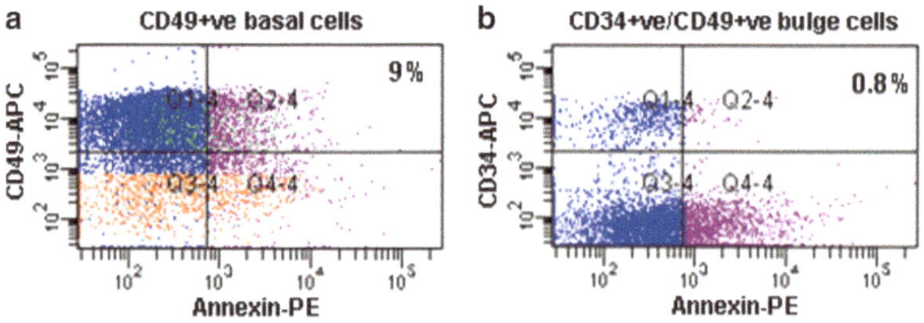

Fig. 3. FACS plot of Annexin V labeled follicular basal keratinocytes and stem cells. Freshly isolated keratinocytes from 7-week-old telogen stage mice were immunolabeled either with anti-CD49 antibody alone or a combination of anti-CD34/anti-CD49. PE labeled anti-Annexin V antibody was added to each preparation. (**a**, **b**) Annexin V expression of (**a**) $CD49^{+}/CD34^{-}$ or (**b**) $CD49^{+}/CD34^{+}$ populations are shown. Double-positive populations are located in the *top right* quadrant of the plot. DAPI was used to exclude dead cells from the assay.

4. Notes

1. All procedures using live rodents must be carried out in accordance with institutional and governmental regulations.
2. Sodium azide is very hazardous in case of skin or eye contact; therefore appropriate precautions should be taken.
3. The first two hair growth cycles occur in a synchronized fashion moving in waves from the anterior toward the posterior of the mouse. After the second hair cycle, hair growth becomes largely asynchronous (5, 6). Typically, mice are assessed at time points when hair growth is synchronous, so follicles in wild-type mice will be a predictable phase of the hair cycle (e.g., 5-week-old mice in anagen; 7-week-old mice in telogen).
4. There are several published methods for generating label-retaining cells (to identify putative stem cells) in mice (10). In our laboratory, 10-day-old mice are injected with BrdU (50 mg/kg body weight) every 12 h for a total of four injections.
5. To achieve short-term labeling of cells undergoing DNA synthesis, postnatal mice are injected with BrdU (100 mg BrdU/kg body weight) 1 h prior to sacrifice.
6. Most of the hair should separate with the epidermis. This can be monitored by the naked eye or by inspecting the epidermis and dermis under a dissecting microscope. If a significant amount of hair remains in the dermis, incubate the skin in EDTA/PBS solution at 37°C for a longer period of time.
7. This protocol is particularly useful for visualizing expression patterns in the basal layer of the epidermis, sebaceous gland, and hair follicle. Due to the limitations of antibody penetrance and confocal imaging, the method is not very effective for visualizing changes within the suprabasal layers of the epidermis.
8. For immunodetection of BrdU incorporation (Fig. 1a–b), insert the following additional steps after the Blocking Buffer blocking. Aspirate Blocking Buffer. To denature DNA, incubate epidermal sheets in 2 M HCL pre-warmed to 37°C for 20–30 min. Rinse 4× with PBS. Proceed to primary antibody step as usual.
9. Apoptotic cells can be detected in epidermal sheets by using the DeadEnd™ Fluorometric TUNEL System (Promega) (Fig. 1c) (24).
10. In some cases, primary antibodies directly conjugated to fluorescent dyes may be used; however generally it is better to use dye-conjugated secondary antibodies to amplify the fluorescence signal. For double-immunolabeling, antibodies

generated in two different species may be added simultaneously or the protocol may be repeated sequentially.

11. Definition of key terms used in describing confocal microscopy.
    - *Intensity*—The brightness of the pixels in the image. This is directly correlated to the number of photons from the fluorescent dye hitting the detector. The more photons detected, the higher the intensity.
    - *Detector*—On point scanning confocals images are typically collected on photomultiplier tubes.
    - *Pinhole*—Adjustable point in the conjugate plane to the focal plane; its purpose is to remove out of focus light, so only in focus light from the specimen reaches the detectors.
    - *Optical Slice*—The thickness (in *Z* axis) of a confocal slice. Determined by the magnification and numeric aperture of the objective lens and the wavelength of the light traveling through the objective.
12. It is possible to increase laser power if the sample is very dim.
13. This procedure reduces background noise from the detector. Together, the net result of reducing the scan speed and increasing scan averaging is a marked improvement in the signal-to-noise ratio in the image; however, this process causes an increase in data acquisition time.
14. Sixteen bit images allow recording of 65,535 intensity levels between 0 (black) and 65,536 (white) and are currently the optimal choice for saving data from a high contrast image.
15. Therefore, the 20× 0.5NA objective lens used in Fig. 1 (panels c, d, f, g) used a *Z* step interval of 2.02 μm if Alexa Fluor® 568 is the longest wavelength dye used or 1.82 μm if Alexa Fluor® 488 is the longest wavelength dye.
16. It is important to remove all the fat to permit optimal trypsinization of the skin. This is the key step to ensure both a high quantity and a pure preparation of epidermal cells.
17. If separation of the dermis is difficult after incubation, the skin can be incubated in trypsin for another 30 min at 37°C prior to epidermal isolation.
18. This protocol has been written using FAD low Calcium complete medium, as this is the media that we use to culture mouse keratinocytes. If the keratinocytes are only to be used for flow cytometry analyses, then alternative low Calcium media may be used (e.g., MEM, see Subheading 2.5). However, it is important that the media is supplemented with FBS to inactive trypsin.
19. As a guide, the expected yield of viable cells from adult dorsal epidermis isolated during the telogen (resting) phase of the

hair cycle is $15 \pm 5 \times 10^6$. Cell yield is significantly lower during the anagen (growth) phase of the hair cycle.

20. MEM for the staining may be replaced with FACS buffer or other Calcium-free medium. Calcium-containing medium should be avoided since it has been reported to stabilize cell–cell contacts, which makes it difficult to obtain a single-cell suspension.
21. Primary keratinocytes are very sensitive. Keep cold and minimize pipetting steps. If other antibodies are used, it is desirable to use directly conjugated primary antibodies to minimize washing and incubation steps. It is recommended to optimize antibody concentrations before use.
22. When using a different combination of fluorochromes, make sure they are compatible (i.e., their emission wavelength are different). Instead of DAPI, other nucleic acid probes can be used such as propidium iodine (1 μg/mL, Sigma-Aldrich) or 7-AAD (25 μg/mL, Sigma-Aldrich) (final concentrations).
23. FACS technology was initially developed for the analysis of immune cells. Since keratinocytes are larger in size, the use of a 100 μm nozzle is recommended to avoid blockage. In this protocol, a LSRII (BD Biosciences) is used; however individual settings may vary between instruments and experiments. Please refer to your FACS core facility manager for advice.

Acknowledgements

We thank Professor Fiona Watt, for supporting development of the human and murine epidermal whole-mount techniques in her laboratory. We also thank Dr. Danielle Lavery for helping to develop epidermal flow cytometry techniques and useful discussions. We are grateful for support from the Medical Research Council (KB), Barts and The London Charity (KB, KH), and Wellcome Trust (KH).

References

1. Levy V, Lindon C, Harfe BD, Morgan BA (2005) Distinct stem cell populations regenerate the follicle and interfollicular epidermis. Dev Cell 9:855–61
2. Oshima H, Rochat A, Kedzia C, Kobayashi K, Barrandon Y (2001) Morphogenesis and renewal of hair follicles from adult multipotent stem cells. Cell 104:233–45
3. Ghazizadeh S, Taichman LB (2001) Multiple classes of stem cells in cutaneous epithelium: a lineage analysis of adult mouse skin. EMBO J 20:1215–22
4. Taylor G, Lehrer MS, Jensen PJ, Sun TT, Lavker RM (2000) Involvement of follicular stem cells in forming not only the follicle but also the epidermis. Cell 102:451–61
5. Stenn KS, Paus R (2001) Controls of hair follicle cycling. Physiol Rev 81:449–494
6. Hardy MH (1992) The secret life of the hair follicle. Trends Genet 8:55–61

7. Cotsarelis G, Sun TT, Lavker RM (1990) Label-retaining cells reside in the bulge area of pilosebaceous unit: implications for follicular stem cells, hair cycle, and skin carcinogenesis. Cell 61:1329–37
8. Morris RJ, Potten CS (1999) Highly persistent label-retaining cells in the hair follicles of mice and their fate following induction of anagen. J Invest Dermatol 112:470–5
9. Tumbar T, Guasch G, Greco V, Blanpain C, Lowry WE, Rendl M, Fuchs E (2004) Defining the epithelial stem cell niche in skin. Science 303:359–63
10. Braun KM, Watt FM (2004) Epidermal label-retaining cells: background and recent applications. J Investig Dermatol Symp Proc 9: 196–201
11. Fuchs E (2009) The tortoise and the hair: slow-cycling cells in the stem cell race. Cell 137: 811–9
12. Cotsarelis G (2006) Epithelial stem cells: a folliculocentric view. J Invest Dermatol 126: 1459–68
13. Ito M, Liu Y, Yang Z, Nguyen J, Liang F, Morris RJ, Cotsarelis G (2005) Stem cells in the hair follicle bulge contribute to wound repair but not to homeostasis of the epidermis. Nat Med 11:1351–4
14. Morris RJ, Liu Y, Marles L, Yang Z, Trempus C, Li S, Lin JS, Sawicki JA, Cotsarelis G (2004) Capturing and profiling adult hair follicle stem cells. Nat Biotechnol 22:411–7
15. Lavker RM, Miller S, Wilson C, Cotsarelis G, Wei ZG, Yang JS, Sun TT (1993) Hair follicle stem cells: their location, role in hair cycle, and involvement in skin tumor formation. J Invest Dermatol 101:16S–26S
16. Snippert HJ, Haegebarth A, Kasper M, Jaks V, van Es JH, Barker N, van de Wetering M, van den Born M, Begthel H, Vries RG, Stange DE, Toftgard R, Clevers H (2010) Lgr6 marks stem cells in the hair follicle that generate all cell lineages of the skin. Science 327:1385–9
17. Jaks V, Barker N, Kasper M, van Es JH, Snippert HJ, Clevers H, Toftgard R (2008) Lgr5 marks cycling, yet long-lived, hair follicle stem cells. Nat Genet 40:1291–9
18. Jensen KB, Collins CA, Nascimento E, Tan DW, Frye M, Itami S, Watt FM (2009) Lrig1 expression defines a distinct multipotent stem cell population in mammalian epidermis. Cell Stem Cell 4:427–39
19. Horsley V, O'Carroll D, Tooze R, Ohinata Y, Saitou M, Obukhanych T, Nussenzweig M, Tarakhovsky A, Fuchs E (2006) Blimp1 defines a progenitor population that governs cellular input to the sebaceous gland. Cell 126:597–609
20. Nijhof JG, Braun KM, Giangreco A, van Pelt C, Kawamoto H, Boyd RL, Willemze R, Mullenders LH, Watt FM, de Gruijl FR, van Ewijk W (2006) The cell-surface marker MTS24 identifies a novel population of follicular keratinocytes with characteristics of progenitor cells. Development 133:3027–37
21. Blanpain C, Lowry WE, Geoghegan A, Polak L, Fuchs E (2004) Self-renewal, multipotency, and the existence of two cell populations within an epithelial stem cell niche. Cell 118:635–48
22. Trempus CS, Morris RJ, Bortner CD, Cotsarelis G, Faircloth RS, Reece JM, Tennant RW (2003) Enrichment for living murine keratinocytes from the hair follicle bulge with the cell surface marker CD34. J Invest Dermatol 120:501–11
23. Jensen UB, Lowell S, Watt FM (1999) The spatial relationship between stem cells and their progeny in the basal layer of human epidermis: a new view based on whole-mount labelling and lineage analysis. Development 126: 2409–18
24. Braun KM, Niemann C, Jensen UB, Sundberg JP, Silva-Vargas V, Watt FM (2003) Manipulation of stem cell proliferation and lineage commitment: visualisation of label-retaining cells in wholemounts of mouse epidermis. Development 130:5241–55
25. Silva-Vargas V, Lo Celso C, Giangreco A, Ofstad T, Prowse DM, Braun KM, Watt FM (2005) Beta-catenin and Hedgehog signal strength can specify number and location of hair follicles in adult epidermis without recruitment of bulge stem cells. Dev Cell 9:121–31
26. Schweizer J, Marks F (1977) A developmental study of the distribution and frequency of Langerhans cells in relation to formation of patterning in mouse tail epidermis. J Invest Dermatol 69:198–204
27. Schweizer J, Marks F (1977) Induction of the formation of new hair follicles in mouse tail epidermis by the tumor promoter 12-O-tetradecanoylphorbol-13-acetate. Cancer Res 37:4195–201
28. Jensen KB, Driskell RR, Watt FM (2010) Assaying proliferation and differentiation capacity of stem cells using disaggregated adult mouse epidermis. Nat Protoc 5:898–911
29. Nowak JA, Fuchs E (2009) Isolation and culture of epithelial stem cells. Methods Mol Biol 482:215–32
30. Redvers RP, Kaur P (2005) Serial cultivation of primary adult murine keratinocytes. Methods Mol Biol 289:15–22
31. Wu WY, Morris RJ (2005) Method for the harvest and assay of in vitro clonogenic keratino-

cytes stem cells from mice. Methods Mol Biol 289:79–86

32. Fadok VA, Voelker DR, Campbell PA, Cohen JJ, Bratton DL, Henson PM (1992) Exposure of phosphatidylserine on the surface of apoptotic lymphocytes triggers specific recognition and removal by macrophages. J Immunol 148:2207–16
33. Martin SJ, Reutelingsperger CP, McGahon AJ, Rader JA, van Schie RC, LaFace DM, Green DR (1995) Early redistribution of plasma membrane phosphatidylserine is a general feature of apoptosis regardless of the initiating stimulus: inhibition by overexpression of Bcl-2 and Abl. J Exp Med 182: 1545–56

Chapter 20

Murine Aggregation Chimeras and Wholemount Imaging in Airway Stem Cell Biology

Ian R. Rosewell and Adam Giangreco

Abstract

Local tissue stem cells are known to exist in mammalian lungs but their role in epithelial maintenance remains unclear. We therefore developed murine aggregation chimera and wholemount imaging techniques to assess the contribution of these cells to lung homeostasis and repair. In this chapter we provide further details regarding the generation of murine aggregation chimera mice and their subsequent use in wholemount lung imaging. We also describe methods related to the interpretation of this data that allows for quantitative assessment of airway stem cell activation versus quiescence. Using these techniques, it is possible to compare the growth and differentiation capacity of various lung epithelial cells in normal, repairing, and diseased states.

Key words: Lung, airway, stem cell, chimera, wholemount, confocal, immunofluorescence, progenitor cell.

1. Introduction

Previous chemical and transgenic lung injury models have determined that stem cells associated with intrapulmonary neuroepithelial bodies (NEBs) and bronchio-alveolar duct junctions (BADJs) are necessary for airway repair (1–3). NEB- and BADJ-associated stem cells are characterized by expression of Clara cell secretory protein (CCSP), robust proliferation capacity and multipotent differentiation potential, and mitotic DNA label retention (2, 4). Despite these observations, it has remained unclear whether endogenous stem cells maintain airway homeostasis. Intrapulmonary airways exhibit low cellular turnover, significant regional and functional heterogeneity, and airway epithelial proliferation is less than 1% per day (5–8).

It has been suggested that organs that exhibit low proliferation and turnover may not require stem cell populations to maintain

Kimberly A. Mace and Kristin M. Braun (eds.), *Progenitor Cells: Methods and Protocols*, Methods in Molecular Biology, vol. 916, DOI 10.1007/978-1-61779-980-8_20, © Springer Science+Business Media, LLC 2012

homeostasis (9, 10). Classically, this has been addressed through the use of embryo aggregation chimera models in which progenitor cell activity is determined by comparing chimeric patch size versus frequency (11). This model supports the existence of active stem cells in organs exhibiting high cellular turnover but not within more quiescent tissue types (11–14). More recently, in vivo transgenic lineage tagging approaches have been used to address this question. Transgenic models have several advantages, including the use of cell-specific promoters, the ability to introduce lineage tags at clonal frequency, and the potential to temporally regulate tag initiation (15, 16). Unfortunately, the use of both aggregation chimera and transgenic techniques for lung stem cell analysis has remained difficult due to the complex branching pattern of mammalian airways.

We recently addressed this problem by combining embryo aggregation chimerism with three-dimensional wholemount imaging (17). In this chapter we describe how lung aggregation chimeras were generated and visualized using three-dimensional wholemount imaging. Using these approaches, it was possible to compare the contribution of different chimeric patches of adult lung epithelial cells during homeostasis and repair of murine conducting airways. We determined that in the absence of injury, single, randomly distributed progenitor cells maintain normal epithelial homeostasis. In contrast we found that severe lung injury resulted in the generation of large chimeric cell patches that were associated with previously identified stem cells. We concluded that endogenous stem cells were dispensable for normal homeostasis but serve to repopulate airways following severe lung injury.

2. Materials

2.1. Animal Husbandry, Embryo Fertilization, Harvesting, Preparation

1. Mouse strains: C57BL/6-Tg(CAG-EGFP) 1Osb/J and FVB/NCrl (WT) (Jackson Labs, Bar Harbor, Maine USA).
2. Caging (Techniplast, Italy).
3. Flushing holding medium (FHM) (Millipore, Watford UK).
4. Potassium simplex optimized medium (KSOM) (Millipore, Watford UK).
5. Mineral oil (embryo tested).
6. Pregnant mare gonadotrophin (PMSG) (Intervet, Milton Keynes, UK).
7. Human chorionic gonadotropin (HCG) (Intervet, Milton Keynes, UK).
8. CO_2 incubator.
9. Petri dishes (6 cm).

2.2. Embryo Aggregation, Implantation, Offspring Identification

1. Acid Tyrode's (Millipore, Watford UK).
2. Aggregation needles (BLS Ltd. Hungary).
3. Nonsurgical embryo transfer tips (Paratechs Corporation, Lexington, KY, US).
4. FHS/F-01 GFP Goggles (BLS Ltd. Hungary).

2.3. Tissue Fixation and Microdissection

1. 10% Neutral buffered formalin (NBF).
2. Dissection board and instruments.
3. 4/0 suture silk.
4. 21 G cannula.
5. Flexible tubing, stop valve, 50 ml syringe.
6. 1X Phosphate buffered saline (PBS).
7. Microdissecting scissors and forceps.
8. Stereo dissecting microscope.

2.4. Immunostaining, Imaging, and Image Analysis

1. 1× PBS (+10% fetal bovine serum, 0.2% Tween-20, 0.2% fish skin gelatin).
2. Rabbit-anti-calcitonin gene related peptide (CGRP) (C8198, Sigma).
3. Goat-anti-Clara cell secretory protein (CCSP) (goat #899, gift of Prof. Barry R. Stripp, Duke University, USA).
4. Roller mixer.
5. Donkey-anti-rabbit Alexafluor 555 (Invitrogen).
6. Donkey-anti-goat Alexafluor 633 (Invitrogen).
7. Parafilm.
8. Clear nail varnish.
9. Microscope slides.
10. Leica TCS-SP5, Resonance scan head confocal with motorized *XYZ* stage (Leica Microsystems).
11. Volocity 5.4 Imaging Software (Perkin Elmer, Waltham, MA, USA).
12. Microsoft Excel.

3. Methods

The first chimeric mice were produced through the aggregation of early stage embryos by Professor Andrzej K. Tarkowski at Bangor University in 1961 (18). Aggregation chimera mice can be created from multiple strains or combinations of ES cells and embryos, any of which might be mutant, tagged, or nontransgenic. This technique

therefore allows cellular interactions to be evaluated within a unique shared tissue environment. While the importance of aggregation chimeras is today most appreciated as a prelude to the generation of transgenic offspring from pluripotent cells, at the time it formed a cornerstone for the study of mammalian embryology.
More recently, murine aggregation chimera models have been used to assess progenitor cell dynamics in various tissues. These include the intestinal epithelium, the aortic endothelium, the retinal pigment epithelium, thymus, and pancreas (12–14, 19, 20). To obtain reliable results, careful mathematical and statistical interpretation of resultant chimeric clones is needed (11, 13). Success is also aided by the use of tissue wholemounts that provide a much greater overview of tissue structure and progenitor cell organization (12).

Previous attempts to assess conducting airway progenitor cell dynamics using aggregation chimera mice have proven unsuccessful due to the highly complex branching structure of mammalian lungs (B. Ponder, personal communication). We herein describe methods for preparation and imaging of chimeric whole lung conducting airways. We also provide simple analysis methods that facilitate resolution of stochastic versus stem cell-dependent lung homeostasis on the basis of patch size and frequency. Although designed specifically for lung airways, these methods should prove useful for a wide variety of epithelial tissues.

3.1. Preparation of Fertilized Embryos

1. Sexually mature male mice and 3–5-week-old female mice of each strain are maintained under a 12 h light–dark cycle within individually ventilated cages and allowed access to food and water ad libitum. Male mice are individually housed, while female mice are initially group housed.
2. Superovulations are initiated with the administration of 5 IU, PMSG, 0.1 ml saline via intraperitoneal injection (IP). This is followed 48 h later with a further IP injection of HCG. Each injection is given at approximately the midpoint of the light cycle.
3. Immediately following HCG administration, female mice are placed into the cages containing the male mice of the same strain (WT with WT; CAG-EGFP with CAG-EGFP).
4. The following morning female mice are checked for the presence of a copulatory plug.
5. At 2.5 days post coitum (dpc, approximately 48 h after discovery of the plug) the plugged female mice are culled by cervical dislocation and the oviducts are removed intact and placed into FHM.
6. Preimplantation staged embryos are then collected (separately) from each strain by locating the infundibulum at the distal end of the oviduct, holding this loosely with superfine forceps while carefully introducing a 30 G needle and flushing with FHM medium.

7. Once all oviducts are flushed the embryos are collected, pooled, washed free of debris, and selected based on morphology. Embryos are transferred into pre-equilibrated KSOM (5% CO_2, 37°C) drops under mineral oil with minimal FHM carryover (see Note 1).

3.2. Embryo Aggregation and Implantation

1. At least 1–2 h before the aggregation, aggregation drops are made as described in (21). Briefly, 5 μl drops of KSOM are pipetted onto the surface of a 6 cm diameter Petri dish and overlaid with mineral oil.
2. Wells are then formed in the plastic surface of these dishes by gently pressing and rotating an aggregation needle so that a small depression results in the plastic surface without overt cracks. 5–6 wells are formed within each 5 μl KSOM drop and culture dishes are then placed into an incubator (5% CO_2, 37°C; see Note 2).
3. To enable embryos to aggregate the *zona pellucida* is first removed from each embryo. This is achieved through the use of acid Tyrode's as previously described (21) (see Note 3).
4. As soon as all embryos are judged to be *zona* free, the group should be quickly moved to a second FHM drop and from there to the third drop.
5. Once all embryos have been denuded, they are placed back into equilibrated KSOM drops under oil with minimal carry over.
6. Small embryo groups are moved into the aggregation drops such that each well contains one embryo of each of the two strains (see Note 4).
7. Once all embryos are placed into the aggregation drops the dishes are placed back into incubator for overnight culture.
8. The following morning embryos will either have formed large compacted morula, or in other cases formed early or advanced stage blastocysts. A minority of embryos sometime fail to develop or development occurs without aggregation (see Note 5).
9. Blastocysts are routinely transferred into 2.5 dpc pseudopregnant foster mice. This is best achieved using a Nonsurgical Embryo Transfer (NSET) technique (22). Alternatively, 15–20 embryos are reimplanted of into either one or both sides of the uterus by conventional surgical techniques (21). Where 2.5 dpc mice are unavailable 0.5 dpc foster mice and oviduct transfer can be used with similar success (birth rates).

3.3. Identification of Chimeric Offspring

1. In order to evaluate the extent of chimerism first perform visual coat color inspection. Skin appears albino if derived from FVB/n strain embryos and black if derived from C57BL/6 (CAG-EGFP) strain mice. However, this assessment of chimerism does not necessarily reflect true GFP chimerism and therefore all mice are additionally assessed for GFP expression.

2. Chimeric offspring are evaluated for GFP expression both by conventional ear or tail DNA genotyping (via PCR) as well as through the use of specialist GFP-sensitive goggles. All fur free regions, eyes (retina), toes, ears, tail, and nose exhibit robust, patchy GFP expression where chimerism is present and serves as a rapid, sensitive, and effective readout of overall animal chimerism.

3.4. Tissue Preparation and Microdissection

1. Animals are killed by intraperitoneal injection of sodium pentobarbital (Euthatal, 1.2 mg/g body weight). The peritoneal cavity is opened using scissors and forceps and the animal exsanguinated by severing the aorta. The diaphragm is punctured using sharp scissors, the ribcage removed using blunt-ended scissors, and the trachea exposed by careful dissection.
2. Insert the cannula into the opened trachea, tighten the knot over the trachea, and connect to a 50 ml syringe with tubing and a stop valve containing 10% neutral buffered formalin at 10 cm H_2O pressure (filled to 10 cm above the open chest cavity). Lungs are insufflated by opening the stop valve and allowed to fix in situ for 5 min (Fig. 1a–c).
3. After fixation individual lung lobes are isolated and washed 5 × 5 min each in 1 × PBS (see Note 6).
4. Microdissection is performed using a stereo dissecting microscope and microdissecting instruments. Briefly, the internal lobar bronchus is identified anatomically, and the mainstem bronchi exposed by careful dissection. The upper third of this airway is then removed, revealing the airway branching pattern. Subsequent bronchi and bronchioles are further dissected using this technique to the level of airway terminal bronchioles (Fig. 1d–f) (see Note 7).
5. Microdissected lung lobes can again be stored up to 1 month in 1× PBS/0.2% azide at 4°C or directly immunostained.

3.5. Immunofluorescent Antibody Staining

1. Block microdissected lung lobes in 10% fetal bovine serum/0.2% Tween 20/0.2% fish skin gelatin made in 1× PBS for a minimum of 2 h at room temperature. Blocking should be performed in a minimum 5 ml solution/lung on a roller mixer.
2. Primary antibodies should be applied at appropriate concentrations (1:4,000 rabbit CGRP, 1:5,000 goat CCSP) in blocking solution. Primary antibodies are applied at 1 ml volume/lung lobe incubated for 24 h at 4°C with rolling.
3. Wash each lung lobe 4 × 15 min each in 1× PBS/0.2% Tween-20 rotating at room temperature.
4. Secondary antibodies are diluted 1:300 in 1 ml blocking solution/lung lobe and incubated for 24 h at room temperature.

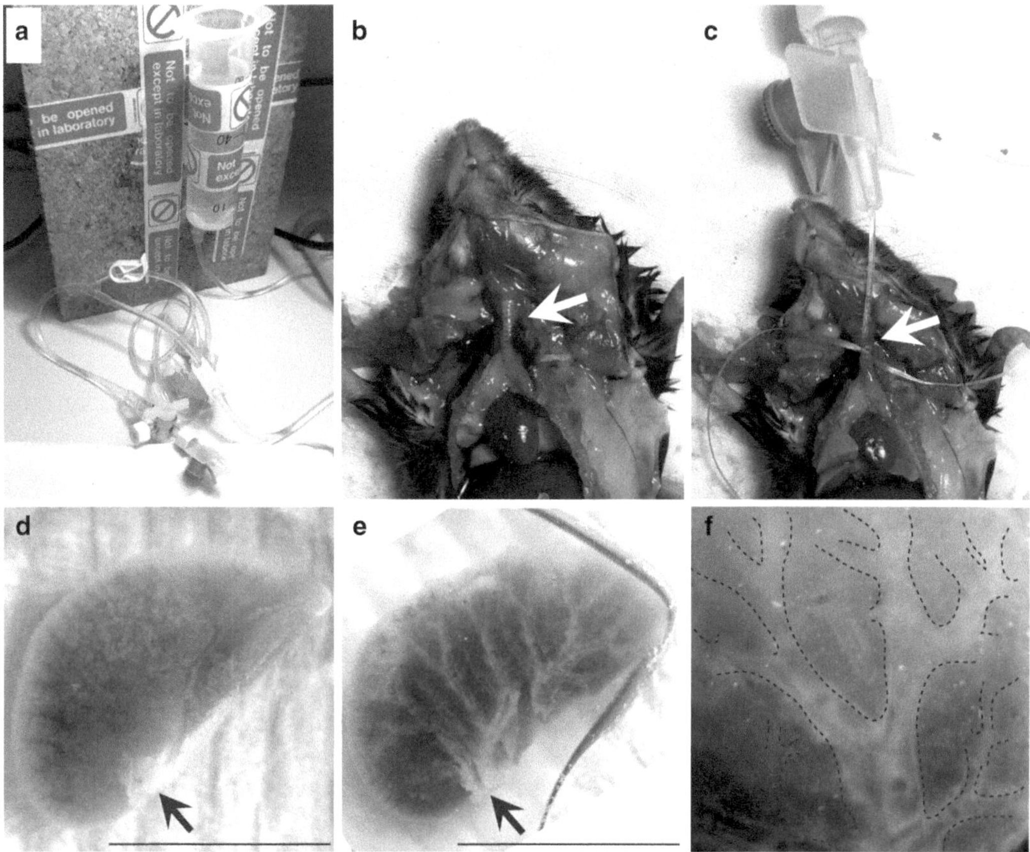

Fig. 1. Lung tissue preparation and microdissection. (**a**) Representative insufflation apparatus and cannula with stop valve. (**b**, **c**) Gross dissection of trachea and mainstem bronchi before (**b**) and after tracheal cannulation (**c**). Note partially inflated lungs in (**c**). (**d**, **e**) Left lung lobe before (**d**) and after conducting airway microdissection (**e**). *Arrow* (**d**, **e**) denotes internal lobar bronchus. (**f**) Detailed image of fully microdissected conducting airway. *Scale bars* (**d**, **e**) are 1 cm.

Fluorescent secondary antibodies are donkey anti-rabbit (Alexafluor 555) and donkey anti-goat (Alexafluor 633) (see Note 8).

5. After 24 h, remove excess secondary antibody from each lung lobe by washing 4×15 min in 1× PBS/0.2% Tween-20 at room temperature.
6. Counterstain for 10 min using 2 μg/ml DAPI resuspended in 1 ml 1× PBS, wash twice in 1× PBS to remove excess DAPI.
7. Stained, microdissected lungs can be stored up to 1 month in 1× PBS/0.2% sodium azide or imaged directly as below (see Note 9).

3.6. Confocal Imaging

1. Turn on the Leica TCS-SP5 microscope, lasers, and acquisition software approximately 30 min prior to image acquisition to allow stabilization of laser signals. Activate and initiate the Leica resonance scan head, the motorized *XYZ* stage, and choose the 10× objective lens for whole-lung imaging.

2. Stained lungs are placed flat and gently compressed between two microscope slides. Three edges of the slide are sealed using either Parafilm or clear nail varnish. This is then backfilled using 1× PBS and the remaining edge sealed using parafilm or clear nail varnish.
3. The slide-lung unit is mounted on the Leica TCS-SP5 microscope stage such that the microdissected airway faces the objective lens. Open the pinhole to >100 μm as it will not be possible to achieve true confocal images at this magnification (see Note 10).
4. Using resonance scanning, adjust the detection wavelength, laser power, gain, and offset to maximally excite the brightest portion of the sample while maintaining minimal background signal intensity and bleed through to other detection channels (see Note 11).
5. For whole lung lobe imaging, adjust the min and max *Z*-plane to image through the entire lung at its thickest point. Set the *Z*-step size to 2.5–3.5 μm to ensure adequate cellular resolution of all lung optical sections. We also used ten frame averages/frame/*Z*-slice.
6. To image the lung, move to the upper right portion of the sample and set this *XY* position using "tiled" acquisition parameters. Next find the lower left *XY* position and set this using the same parameters. A large *XY* field should now be selected in the acquisition software. Ensure there is no rotation indicated for images; image tiling is not currently compatible with image rotation in the Leica software (see Note 12).
7. To begin acquisition press the "start" button in the Leica software. Whole lung scans typically take 2–3 h for acquisition plus 1–2 h for digital post-processing. If successful a single, 2–5 GB image of the entire lung should be produced that can be subsequently saved and analyzed in Volocity.

3.7. Image Processing and Data Analysis of Chimeric Patches

1. Individual data files are opened using Volocity Image Analysis software (see Note 13).
2. For analysis, three-dimensional images are digitally flattened using the "merge planes" feature in Volocity.
3. A "region of interest (ROI)" is drawn using the "Freehand ROI tool" to encompass the entire contiguous conducting airway. Under the "measurements" menu the total surface area of this ROI can then be measured.
4. For individual GFP chimeric patch analysis and overall GFP chimerism assessments, select "find objects by ROI" in the drop down menu to determine the total surface area of the conducting airway then select "find objects by intensity" within the "Measurements tab" and adjust settings based on GFP brightness (green channel) intensity to discriminate GFP (+) and GFP (-) lung cells.

5. To identify individual GFP-expressing cells and cell patches within the conducting airway ROI select "clip objects to ROI" from within the measurements toolbar (left side of image).
6. To eliminate single cells and subcellular debris from chimeric patch size analysis, we selected the "Exclude objects by size" parameter (see Note 14).
7. At this point there should be a large list of individual ROI objects with a minimum size equivalent to two or more cells. This list is exported into Microsoft Excel (see Note 15).
8. In Excel, convert the average patch size (in μm^2) of each object to cell number by dividing each ROI area by the previously determined average single cell size.
9. Sort all cell clusters according to cell number (from lowest to highest). This is the "Data array". In a separate column, make a "Bins array" corresponding to cell number doublings (starting with two cells).
10. Use the "Frequency" command to determine the relative cell cluster abundance within each cell "Bin". This will calculate how often values occur within a range of established values (Bins).
11. Graph this data as Number of cells/patch (Patch size, *X*-axis) versus Patch frequency (*Y* axis). It is also possible to determine the average patch size and patch size deviation between multiple lungs using standard Microsoft Excel "Average" and "Standard deviation" commands (see Note 16).

4. Notes

1. Around 15–20 viable embryos result from each plugged female mouse. Plugging rates are strain variable but for inbred strains a plug rate of 30–50% is to be expected. Most (90%+) embryos will successfully aggregate. 15–20 aggregated embryos are transferred into each pseudo-pregnant foster mouse. To achieve 3–4 transfers would therefore require the use of 10–15 male and female mice from each strain.
2. The aggregation wells allow the embryos to fall into contact allowing them to adhere as a prelude to aggregation and allowing for the formation of a single embryo.
3. A group of 50–100 embryos were placed into one FHM drop and 10–20 embryos in turn were placed into acid Tyrode's. The *zona* can at first be seen to thin and then finally will be seen to be absent. This can happen rapidly and exposure to the acid Tyrode's, once the *zona* is absent, should be minimized, to prevent decompaction of the embryos.

4. The embryos can be pushed or sucked to enable contact to be made. Once established and with careful handling of the dish, the embryos should remain in contact and merge.
5. All aggregated embryos irrespective of their developmental stage are transferred. The nonaggregated embryos are discarded or may be used to supplement transfer numbers where numbers of aggregated embryos are low.
6. Lungs may then be stored for up to 1 month in 1× PBS/0.2% sodium azide.
7. Great care should be taken to avoid dissecting pulmonary vessels as these can closely resemble airways.
8. It is important to protect secondary antibodies from light; wrap incubation vials in aluminum foil during this time.
9. Storage should be at 4°C and samples protected from light.
10. It is important to use digital magnification as 10× objective tiled images can produce uneven *XY* field illumination at low magnification, resulting in poor stitched image quality (a very obvious "tiling" defect). A digital magnification of 2.35× was used for all our image acquisitions.
11. Simultaneous 4-color detection of DAPI, eGFP (endogenous), Alexafluor 555, and Alexafluor 633 is possible using proper detection wavelength, laser power, gain, and offset settings. This is most easily achieved using the "glowover" function for each individual channel.
12. Using these settings it is sometimes necessary to acquire a high number of image tiles that contain no tissue. At this time, if there is an error message indicating that the image exceeds the maximum possible image size, reduce the tile resolution (we used between 250 × 250 to 400 × 400 (*XY*) pixels/tile).
13. It is not possible to open more than one image at a time due to image file size.
14. In separate experiments, we drew individual ROIs (>50) around single GFP-expressing cells and determined that the average surface area of a single conducting airway cell in uninjured lung was approximately 30 μm^2. We therefore set a cutoff to eliminate any objects <45 μm^2 based on this analysis. A similar analysis should be performed for each experiment to ensure that the average cell size does not change significantly.
15. Adding the size of all selected ROI objects together allows one to determine the total surface area encompassed by GFP expressing cells. Divide this by the total ROI surface area to obtain the overall percent airway GFP chimerism.
16. Due to the wide distribution of patch sizes observed in our study, we plotted the *X*-axis using a logarithmic scale.

Acknowledgments

We gratefully acknowledge Doug Winton and John Stingl for suggestions regarding interpretation of epithelial chimerism, Jessica Gruninger for embryo aggregations, and Virgilio Failla for assistance with Volocity Image analysis. The original study upon which this chapter is based was supported by funding from the US National Institutes of Health (to A.G.) and Cancer Research UK (to A.G. and I.R.). A.G. is currently supported by funding from the European Research Council.

References

1. Reynolds SD, Giangreco A, Power JH, Stripp BR (2000) Neuroepithelial bodies of pulmonary airways serve as a reservoir of progenitor cells capable of epithelial regeneration. Am J Pathol 156:269–278
2. Giangreco A, Reynolds SD, Stripp BR (2002) Terminal bronchioles harbor a unique airway stem cell population that localizes to the bronchoalveolar duct junction. Am J Pathol 161:173–182
3. Kim CF, Jackson EL, Woolfenden AE, Lawrence S, Babar I, Vogel S, Crowley D, Bronson RT, Jacks T (2005) Identification of bronchioalveolar stem cells in normal lung and lung cancer. Cell 121:823–835
4. Hong KU, Reynolds SD, Giangreco A, Hurley CM, Stripp BR (2001) Clara cell secretory protein-expressing cells of the airway neuroepithelial body microenvironment include a label-retaining subset and are critical for epithelial renewal after progenitor cell depletion. Am J Respir Cell Mol Biol 24:671–681
5. Rawlins EL, Hogan BL (2008) Ciliated epithelial cell lifespan in the mouse trachea and lung. Am J Physiol 295:L231–234
6. Rawlins EL, Ostrowski LE, Randell SH, Hogan BL (2007) Lung development and repair: contribution of the ciliated lineage. Proc Natl Acad Sci USA 104:410–417
7. Stripp BR, Reynolds SD (2008) Maintenance and repair of the bronchiolar epithelium. Proc Am Thorac Soc 5:328–333
8. Evans MJ, Cabral-Anderson LJ, Freeman G (1978) Role of the Clara cell in renewal of the bronchiolar epithelium. Laboratory Investigation; A Journal of Technical Methods and Pathology 38:648–653
9. Dor Y, Brown J, Martinez OI, Melton DA (2004) Adult pancreatic beta-cells are formed by self-duplication rather than stem-cell differentiation. Nature 429:41–46
10. Dor Y, Melton DA (2004) How important are adult stem cells for tissue maintenance? Cell Cycle (Georgetown Tex) 3:1104–1106
11. Mead R, Schmidt GH, Ponder BA (1987) Calculating numbers of tissue progenitor cells using chimaeric animals. Dev Biol 121: 273–276
12. Ponder BA, Schmidt GH, Wilkinson MM, Wood MJ, Monk M, Reid A (1985) Derivation of mouse intestinal crypts from single progenitor cells. Nature 313:689–691
13. Schmidt GH, Ponder BA (1987) From patterns to clones in chimaeric tissues. Bioessays 6:104–108
14. Schmidt GH, Wilkinson MM, Ponder BA (1986) Clonal analysis of chimaeric patterns in aortic endothelium. J Embryol Exp Morphol 93:267–280
15. Clayton E, Doupe DP, Klein AM, Winton DJ, Simons BD, Jones PH (2007) A single type of progenitor cell maintains normal epidermis. Nature 446:185–189
16. Barker N, van Es JH, Kuipers J, Kujala P, van den Born M, Cozijnsen M, Haegebarth A, Korving J, Begthel H, Peters PJ, Clevers H (2007) Identification of stem cells in small intestine and colon by marker gene Lgr5. Nature 449:1003–1007
17. Giangreco A, Arwert EN, Rosewell IR, Snyder J, Watt FM, Stripp BR (2009) Stem cells are dispensable for lung homeostasis but restore airways after injury. Proc Natl Acad Sci USA 106:9286–9291
18. Tarkowski AK (1961) Mouse chimaeras developed from fused eggs. Nature 190: 857–860

19. Swenson ES, Xanthopoulos J, Nottoli T, McGrath J, Theise ND, Krause DS (2009) Chimeric mice reveal clonal development of pancreatic acini, but not islets. Biochem Biophys Res Commun 379:526–531
20. Jenkinson WE, Bacon A, White AJ, Anderson G, Jenkinson EJ (2008) An epithelial progenitor pool regulates thymus growth. J Immunol 181:6101–6108
21. Nagy A (2003) Manipulating the mouse embryo: a laboratory manual, 3rd edn. Cold Spring Harbor Laboratory Press, Cold Spring Harbor, NY
22. Green M, Bass S, Spear B (2009) A device for the simple and rapid transcervical transfer of mouse embryos eliminates the need for surgery and potential post-operative complications. Biotechniques 47:919–924

Chapter 21

Live Imaging of Primitive Endoderm Precursors in the Mouse Blastocyst

Joanna B. Grabarek and Berenika Plusa

Abstract

The separation of two populations of cells—primitive endoderm and epiblast—within the inner cell mass (ICM) of the mammalian blastocyst is a crucial event during preimplantation development. However, many aspects of this process are still not very well understood. Recently, the identification of platelet derived growth factor receptor alpha (Pdgfrα) as an early-expressed protein that is also a marker of the later primitive endoderm lineage, together with the availability of the $Pdgfr\alpha^{H2B\text{-}GFP}$ mouse strain (Hamilton et al. Mol Cell Biol 23:4013–4025, 2003), has made in vivo imaging of primitive endoderm formation possible. In this chapter we present two different approaches that can be used to follow the behavior of primitive endoderm cells within the mouse blastocyst in real time.

Key words: Real time imaging, Mouse embryo, Chimaerae, Primitive endoderm, Epiblast, Lineage formation, PDGFRα

1. Introduction

Preimplantation development in the mouse is characterized by the successive segregation of three cell lineages: the trophoblast (TB), the primitive endoderm (PrE) and a pluripotent lineage, the epiblast (EPI). The trophoblast is specified by positional signals between inner and outer cells of the morula that lead to epithelialization in the outer layer of cells. In the newly formed blastocyst, TB surrounds the remaining inner cells, which form the inner cell mass (ICM), and the blastocyst cavity.

The mechanisms directing the specification of the next two lineages to emerge—the EPI and the PrE—within the ICM are still not well understood. In accordance with the mode of endoderm formation observed on the outside of embryonic stem (ES) cell-derived embryoid bodies (reviewed in (1)), it was proposed that

Kimberly A. Mace and Kristin M. Braun (eds.), *Progenitor Cells: Methods and Protocols*, Methods in Molecular Biology, vol. 916, DOI 10.1007/978-1-61779-980-8_21, © Springer Science+Business Media, LLC 2012

positional signals induce the ICM cells lining the cavity roof to form the PrE, while the remaining inner cells form EPI. However, this model was challenged by the more recent observation that markers of EPI and PrE—Nanog and Gata6—show a salt-and-pepper expression pattern within the ICM (2, 3). This finding suggested that EPI and PrE precursors are predetermined before sorting into their respective layers.

A transgenic mouse line that expresses a histone H2B–GFP fusion reporter from the *Pdgfrα* locus (*Pdgfrα*$^{H2B\text{-}GFP}$) (4) and functions as a marker of PrE precursors (5) proved recently to be an invaluable tool for understanding the process of sorting of ICM lineages. Localization of H2B–GFP to active chromatin allows identification and tracking of individual reporter-expressing (putative PrE) cells in vivo (6, 7) providing direct evidence not only for cell sorting but also for the influence of positional information on PrE cell behavior. Live imaging experiments combined with an immunohistochemical analysis of lineage-specific factors revealed a series of sequential and distinct phases in the process of PrE formation (5) leading to development of a revised model for PrE formation.

Thus, live imaging of embryos can provide invaluable information concerning the crucial processes that take part during preimplantation stages of development.

2. Materials

2.1. Mouse Strains

1. *CAG::mRFP1*: A strain constitutively expressing a transgene reporter in all cells from the beginning of development (8). Expression of cytoplasmic mRFP is readily visualized using wide field epifluorescence and laser scanning confocal microscopy.
2. *PdgfrαH2B–GFP*: A strain expressing a knock-in H2B–GFP transgene from the mouse platelet-derived growth factor receptor alpha PDGFRα locus (4). This provides an early marker of PrE in heterozygous mice. Nuclear localization of GFP is readily visualized using wide field epifluorescence and laser scanning confocal microscopy.
3. *CD1*: A wild-type albino strain, available from commercial suppliers (e.g., Charles River).

2.2. Embryo Recovery (Or Collection) and Culture

1. Plastic culture dishes, 35 mm diameter (Nunclon or Corning).
2. Glass Pasteur pipettes or other glass pipettes to prepare embryo handling pipettes.
3. Plastic tubing (Fisher) for use with glass embryo handling pipettes.
4. Precision forceps, No 3.0 and 4.0 (Scientific Laboratory Supplies Ltd or other).

5. Small scissors (John Weiss or other).
6. Small gas burner.
7. Heating stage (Semic, Poland).
8. Light paraffin oil, embryo tested (Sigma).

2.3. Media for Embryo Collection and Culture

All the media described below are produced fresh in the lab from the concentrated stocks (Tables 1 and 2, see Note 1). All the stocks were modified from recipes from (9). Once prepared, these stocks can then be stored at −20°C for 3–6 months. All the chemicals used to produce stocks and media are embryo or cell culture tested. Unless otherwise stated, they are purchased from Sigma. Embryo culture tested AnaLar water (VWR) is used to prepare all stocks and working solutions.

1. M2 medium: This handling medium is a variation of M16 culture medium (10) with added HEPES that allows embryo handling and short culture without need to supply CO_2. Add components M-A to M-F from Table 1 for a final concentration of 1×. Upon preparation, M2 medium is further supplemented with BSA (fraction V) and the pH adjusted to 7.2–7.4. This medium can be stored at 4°C for 4–6 weeks and requires pre-warming before each use.
2. M2 medium without Ca^{2+} and Mg^{2+}: This handling medium is a variation of M2 culture medium that lacks calcium and magnesium ions. The medium promotes decompaction of embryos. Prepare as above, replacing M-d and M-F components with the same volume of water. Upon preparation, the medium is

Table 1
M2 stocks

M2 medium stock name	Ingredients	For 100 ml (g)
M-A 10×	NaCl	5.534
	KCl	0.356
	KH_2PO_4	0.162
	Na-Lactate 60% syrup	3.2 ml
	Glucose	1.0
	Penicillin	0.06
	Streptomycin	0.05
M-B 10×	$NaHCO_3$	2.101
	Phenol red	0.01
M-C 100×	Na pyruvate	0.36
M-D 100×	$CaCl_2 \times 2H_2O$	2.52
M-E 10×	HEPES	5.958
M-F 100×	$MgSO_4 \times 7H_2O$	2.93

Table 2
KSOM stocks

KSOM medium stock name	Ingredients	For 100 ml (g)
K-A 10×	NaCl	5.5527
	KCl	0.1864
	KH_2PO_4	0.0476
	$MgSO_4 \times 7H_2O$	0.0493
	Na-Lactate 60% syrup	1.42 ml
	Glucose	0.036
K-B 10×	$NaHCO_3$	2.101
K-C 100×	Na Pyruvate	0.22
K-D 100×	$CaCl_2 \times 2H_2O$	2.517
K-E 10×	HEPES	5.206
K-F 10,000×	$NaEDTA \times 2H_2O$	3.72
K-G 100×	L-Glutamine (Gibco)	1.46
K-H 100×	Glucose	9.656

further supplemented with BSA (fraction V) and the pH adjusted to 7.2–7.4. The medium can be stored at 4°C for 4–6 weeks and requires pre-warming before each use.

3. KSOM^{AA}: This culture medium (11) requires buffering in an atmosphere of 5% CO_2 in air to reach the required pH of 7.2–7.4. Prepare medium by adding K-A through K-H components, except K-E, from Table 2 to a final concentration of 1×. Upon preparation, KSOM^{AA} is supplemented with nonessential and essential amino acids (Gibco), and BSA. This medium does not contain antibiotics, although they can be added if necessary. KSOM^{AA} can be stored at 4°C for 1–2 weeks maximum.
4. HEPES-KSOM: This culture medium includes HEPES, which allows embryo culture without the need to supply CO_2. The medium does not contain antibiotics, although they can be added if necessary. Prepare medium by adding K-A through K-H components, including K-E, from Table 2 to a final concentration of 1×. Upon preparation, HEPES-KSOM is supplemented with BSA, and HEPES and the pH adjusted to 7.2–7.4. This medium can be stored at 4°C for 1–2 weeks maximum and requires pre-warming prior to each use.

2.4. Production of Chimaerae

1. M2 medium.
2. M2 medium without Ca^{2+} and Mg^{2+}.
3. KSOM^{AA}/HEPES-KSOM.

4. BSA fraction V (Sigma).
5. Acidic Tyrode solution (Sigma).
6. Anti-mouse serum produced in rabbit (Sigma).
7. Complement serum from guinea pig (Sigma).
8. 0.5% Trypsin (10×) with EDTA (Gibco Invitrogen).
9. Agar (Sigma).
10. NaCl (Sigma).
11. TransferTip ES ready-made micromanipulation micropipettes (Eppendorf).
12. Glass capillaries without filament (GC100T-10, Harvard Apparatus Ltd).
13. Glass Pasteur pipettes for embryo handling.
14. Plastic culture dishes, 35 mm (Nunclon or Corning).
15. Glass-bottom dishes, 35 mm (14 mm diameter of glass) (MatTek).
16. Cavity microscope slides (Fisher Scientific or other).
17. Needle with blunted end/pin.
18. Binocular microscope with transmitted light (Leica, Nikon, or other).
19. AirTram Oil (Eppendorf).
20. AirTram Air (Eppendorf).
21. Leica DMI 6000B inverted microscope (see Note 2).
22. Leica Mechanical Manipulators.

3. Methods

3.1. Real-Time Imaging of PrE Formation

1. Embryos are collected from spontaneously ovulating female mice mated with male studs on the day of oestrus. Noon of the day of plug detection is marked as 0.5 days post coitum (dpc) or at stage E0.5.
2. The strain of both females and males is chosen depending on the aim of the experiment. For example, Pdgfr$\alpha^{H2B\text{-}GFP/+}$ females can be mated with wild-type CD1 males (or vice versa) to observe the formation of PrE in real time (see Note 3).
3. Prior to the collection of embryos, 20–30-μl drops of M2 medium are placed on the bottom of a plastic culture dish and covered with paraffin oil. The dish is then placed on the heating stage calibrated to maintain a temperature of 37–37.5°C in the dish.

4. For real-time imaging (RTI) of early stages of PrE specification, morulae (8–16 cell stage) are recovered by flushing the oviducts of pregnant females in M2 medium at 2.5–2.75 dpc. Embryos are then collected, washed in fresh medium, and placed in the drops of M2 medium until further use.
5. For RTI of late stages of PrE specification, blastocysts (≥32 cells, cavity present) are recovered by flushing the uterine horns of pregnant females in M2 medium. Prior to RTI, *zonae pellucidae* should be removed from E3.5 embryos to avoid imaging difficulties due to the hatching process they undergo during maturation at about stage E3.75–E4.0 (or equivalent of such in the in vitro culture). Hatching is an abrupt process during which the whole volume of embryo is squeezed through a very narrow slit in the *zona pellucida*. During this process spatial relations between cells become very distorted, making it impossible to follow the position of every cell within the embryo.
6. The *zona pellucida* is removed by a brief exposure to Acidic Tyrode's solution, followed by thorough washing of embryos in fresh M2 medium. Embryos are then left in drops of M2 medium in a culture dish to recover for a minimum of 15 min before further use.
7. Glass bottom dishes are used for imaging. A single 40–50-μl drop of medium is placed in the center of the glass part of the dish and covered with paraffin oil. Embryos are placed in such prepared drops and imaged using a laser scanning confocal or spinning disk (see Notes 4 and 5).
8. Example of parameters for laser scanning confocal on a Zeiss LSM 510 META are as follows: 10–20 *XY* planes (depending on the size of the embryo), separated by 3–4 μm. Time intervals between *Z*-stacks: 7–15 min (Fig. 1).

3.2. Collection of Embryos for Monitoring PrE Precursors in Embryonic Chimaerae

1. Embryos are collected from spontaneously ovulating female mice mated with male studs on the day of oestrus. Noon of the day of plug detection is marked as 0.5 dpc or stage E0.5. The Pdgfrα$^{H2B-GFP/+}$, CAG:mRFP1$^{Tg/+}$, and CD1 strains are used for this experiment.
2. Morulae (8–16 cell stage) are collected from the oviducts of pregnant CD1 or Pdgfrα$^{H2B-GFP/+}$ females mated with Pdgfrα$^{H2B-GFP/+}$ males. Blastocysts are collected at 3.25, 3.6, and 4.5 dpc from the uteri of pregnant Pdgfrα$^{H2B-GFP/+}$ and CAG:mRFP1$^{Tg/+}$ females mated with CAG:mRFP1$^{Tg/+}$ and Pdgfrα$^{H2B-GFP/+}$ males, respectively.
3. Preparation of donor cells: Blastocysts used for the assay are preselected using a confocal microscope. Only blastocysts positive for both mRFP and H2B-GFP fluorescence are used for further experiments. To prepare donor cells, the *zona pellucida*

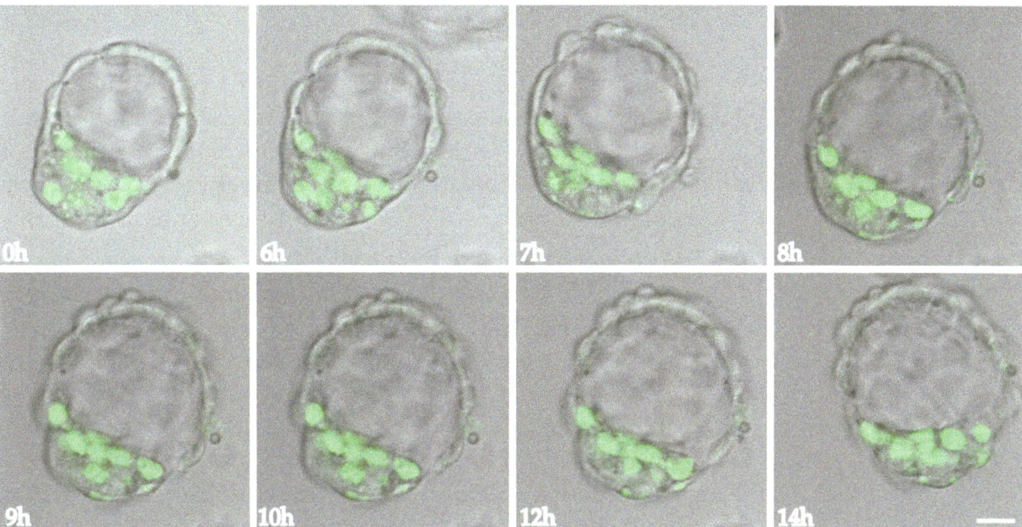

Fig. 1. Dynamic expression of GFP in Pdgfr$\alpha^{H2B-GFP/+}$ embryos visualized by 3D time-lapse microscopy. GFP expression in a subpopulation of cells in the inner cell mass (ICM) is shown in a single blastocyst over time. This display shows examples of images that were taken in real time every 15 min during 14 h culture. Primitive endoderm forms as a layer of cells lining blastocyst cavity at about 8 h of culture. Single optical sections of GFP fluorescence (*green*) are merged with bright-field images. *Scale bar* is 20 μm.

is removed from blastocysts by brief treatment with acid Tyrode's solution. Blastocysts are then left to recover in M2 medium for 15 min. To isolate ICMs, blastocysts are subjected to immunosurgery (12, 13) using anti-mouse serum antibody at 1:3 dilution in M2 and guinea pig complement serum at 1:4 dilution in M2. ICMs then should be thoroughly washed and cleaned of remaining trophoblast by vigorous pipetting in M2 using a narrow glass pipette. ICMs are then disaggregated as described previously (14), by incubation in 1× trypsin-EDTA in M2 without Ca^{2+} and Mg^{2+}. Single cells are immediately collected from the trypsin solution, washed thoroughly in M2, left for a minimum of 15 min in fresh drops of M2 medium under paraffin oil to recover and then sorted using a confocal microscope according to their level of nuclear H2B-GFP (Pdgfr$\alpha^{H2B-GFP}$) expression. GFP-negative cells represent putative EPI precursors. GFP-positive cells can be further separated into two groups based on the intensity of GFP fluorescence, as quantified using the confocal software. The intensity measurements are usually normalized against background (taken as 0%) and the highest intensity measure in the group (taken as 100%). Cells with GFP intensity of greater than 70% represent putative PrE precursors. Cells with GFP intensity of less than 70% represent a putative transient population between EPI and PrE precursors.

3.3. Aggregation

1. Aggregation chamber preparation: Prior to the procedure "aggregation chambers" need to be prepared. Small shallow indentations are made in the bottom of a plastic culture dish using a blunt-ended needle or pin. The indentations are then carefully covered with drops of M2 medium and the whole dish is covered with paraffin oil. The dish is placed on the heating stage, brought to 37°C, and kept there until further use.
2. To produce chimaerae by means of aggregation, the *zona pellucida* needs to be removed from recipient morulae by brief treatment with Acidic Tyrode's solution. Morulae are then left to recover in drops of M2 medium under paraffin oil for 15 min They are then placed in M2 medium without Ca^{2+} and Mg^{2+} for 10–15 min to induce decompaction. Single donor cells and single decompacted recipient morula pairs are then placed in an aggregation chamber. Each donor and recipient is then carefully brought together using a glass pipette. They are then left in the aggregation chamber till further use.

3.4. Micromanipulation

Prior to this stage of the experiment, several preparations are needed.

1. "Micromanipulation chamber"—a cavity microscope slide, holding a single 30–40-μl drop of M2 medium without Ca^{2+} and Mg^{2+} covered with paraffin oil.
2. Holding pipette—prepared by pulling a glass capillary over a gas burner, then trimming the end so that it is even, and then polishing the end by careful and brief exposure to the flame heat to reduce sharp surfaces at the end of the pipette. The holding pipette needs to be filled to at least two-thirds of its length with M2 medium without Ca^{2+} and Mg^{2+} and mounted in the holder connected to the AirTram Air. The holder needs to be then firmly secured on the micromanipulator prior to use.
3. Micromanipulation micropipette—these ready-made pipettes need to be fully filled with paraffin oil and mounted in the holder connected to the AirTram Oil. The holder needs to be firmly secured on the micromanipulator.
4. To produce chimaerae by means of micromanipulation, recipient morulae are first placed in M2 medium without Ca^{2+} and Mg^{2+} for 10–15 min to induce decompaction. Both donor cells and recipient embryos are then placed in the micromanipulation chamber in M2 medium without Ca^{2+} and Mg^{2+} and a single donor cell is introduced between cells of the decompacted morula using a micromanipulation pipette.
5. Chimaerae should be produced in small groups of 2–5 embryos at a time to minimize exposure to M2 medium without Ca^{2+} and Mg^{2+}. After the procedure they are thoroughly washed in M2 medium and left in drops of fresh M2 for at least 30 min to recover before further use.

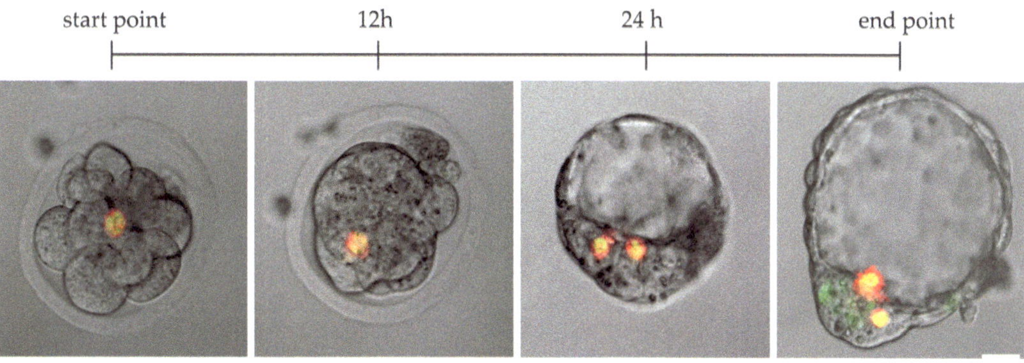

Fig. 2. Development of chimaerae during in vitro culture: time-point imaging. High GFP-expressing (presumptive PrE) progeny contribute to primitive endoderm. Primitive endoderm donor precursors maintain GFP expression and contribute to PrE layer at the end of the culture. Donor cells and their progeny are visualized as RFP-expressing (*red*) cells. *Green* represents expression of Pdgfr$\alpha^{H2B\text{-}GFP}$ in cells. Start point refers to 1–2 h after obtaining a chimera; end point represents the end of culture after 36 h. *Scale bar* is 20 μm.

6. The behavior of PrE precursors in the chimaerae can be monitored in real time in a similar way as in the intact Pdgfr$\alpha^{H2B\text{-}GFP/+}$ embryos. Chimaerae can also be monitored every 6–12 h (time-points) during the culture period using similar equipment as for RTI. Since introduced (donor) cells differ from recipient cells in that they express red fluorescent protein, it is possible to track them even if the time intervals are very spaced (Fig. 2). Imaging at selected time points allows the effect of phototoxicity to be kept to a minimum and allows many more sections to be taken at the time of imaging than for RTI (see Note 6).

4. Notes

1. Commercial versus "in-house" culture media: It is important to prepare the culture media freshly in the lab if possible. Although KSOM^{AA} is now commercially available, its quality diminishes rapidly which makes it less reliable in the long term if the lab has no access to fresh supplies every week. Some of the ingredients (e.g., sodium pyruvate and glutamine) are known to be unstable, especially if they are in the mixture with other chemicals such as salts and glucose. Other ingredients (such as amino acids and BSA) should be added freshly and not stored below 0°C. Freezing a working solution of KSOM^{AA} also proves to be unreliable as far as the quality is concerned. Therefore, separate stocks of various chemicals are prepared which can be mixed to form a working solution prior to use. This allows fresh medium to be prepared every time it is needed for RTI and used only for up to 1 week when stored at 4°C.

2. Examples of inverted microscope equipped with sufficiently sensitive imaging system to obtain high quality images with minimal phototoxicity:
   - Olympus inverted confocal microscope (Fluoview FV1000).
   - Zeiss LSM 510 META confocal microscope.
   - Perkin-Elmer RS3 spinning disk inverted microscope running Volocity software.
3. Use of the Pdgfr$\alpha^{H2B-GFP/+}$ strain for imaging: Pdgfr$\alpha^{H2B-GFP/+}$ is a mouse strain expressing a knock-in H2B–GFP transgene from the mouse platelet-dervived growth factor receptor alpha (*Pdgfrα*) locus to image PrE formation. The recent study by Kurimoto et al. (15) identified Pdgfrα as PrE-specific gene, which makes it a plausible marker to study PrE formation in vivo. The localization of H2B–GFP to active chromatin allows the identification and tracking of individual reporter-expressing cells in vivo (7, 8). Because of the histone turnover it is possible to observe dynamic changes in *Pdgfrα* expression as this system minimizes the problem of prolonged stability of GFP-tagged proteins—a significant drawback in conventional GFP reporter lines.
4. For short-term RTI (up to 16 h), M2 medium can be used provided it is freshly prepared (i.e., not older than 2 weeks). For RTI longer than 16 h, KSOM^{AA} is recommended. If the microscope system is supplied with the CO_2 chamber, it is best to choose KSOM^{AA}, as it provides the finest environment for maturing embryos. In this case it is important to make sure that the KSOM^{AA} drop on the imaging dish has equilibrated for a minimum 20 min to ensure it acquired a pH of 7.2–7.4. However, if a CO_2 chamber is not available, freshly prepared HEPES-KSOM should be used, which does not require constant buffering in CO_2. In both cases, it is important to keep the embryos at a constant temperature of 37–37.5°C. Keeping embryos below that temperature will inevitably slow and eventually stop their development.
5. RTI must be performed using an inverted microscope equipped with a sufficiently sensitive imaging system to obtain high-quality images with minimal phototoxicity caused by either laser or fluorescence lamp exposure. Usually the best results can be obtained using laser scanning confocal or spinning disk. The parameters used for RTI can vary depending on the equipment used for experiments. However, in order to ensure that all cells are traceable during the whole period of RTI, the time interval between series of images should not be longer than 15 min and successive focal planes should not be separated by more than 4–5 μm.

6. Viability test for RTI conditions: It is important to ensure that live imaging does not compromise either the viability of embryos or the process of lineage segregation. In order to test this, a group of embryos subjected to a full round of RTI should be subsequently transferred to pseudopregnant females and allowed to develop to term (9). If the newborn pups are viable and exhibit normal fertility in adult life, similar to that observed in stage-matched freshly flushed embryos, then it may be assumed that the RTI procedure does not compromise embryo development.

References

1. Yamanaka Y, Ralston A, Stephenson RO, Rossant J (2006) Cell and molecular regulation of the mouse blastocyst. Dev Dyn 235: 2301–14
2. Rossant J, Chazaud C, Yamanaka Y (2003) Lineage allocation and asymmetries in the early mouse embryo. Philos Trans R Soc Lond B Biol Sci 358:1341–8
3. Chazaud C, Yamanaka Y, Pawson T, Rossant J (2006) Early lineage segregation between epiblast and primitive endoderm in mouse blastocysts through the Grb2-MAPK pathway. Dev Cell 10:615–24
4. Hamilton TG, Klinghoffer RA, Corrin PD, Soriano P (2003) Evolutionary divergence of platelet-derived growth factor alpha receptor signaling mechanisms. Mol Cell Biol 23: 4013–25
5. Plusa B, Piliszek A, Frankenberg S, Artus J, Hadjantonakis AK (2008) Distinct sequential cell behaviours direct primitive endoderm formation in the mouse blastocyst. Development 135:3081–91
6. Hadjantonakis AK, Papaioannou VE (2004) Dynamic in vivo imaging and cell tracking using a histone fluorescent protein fusion in mice. BMC Biotechnol 4:33
7. Kanda T, Sullivan KF, Wahl GM (1998) Histone-GFP fusion protein enables sensitive analysis of chromosome dynamics in living mammalian cells. Curr Biol 8:377–85
8. Long JZ, Lackan CS, Hadjantonakis AK (2005) Genetic and spectrally distinct in vivo imaging: embryonic stem cells and mice with widespread expression of a monomeric red fluorescent protein. BMC Biotechnol 5:20
9. Nagy A, Gertsenstein M, Vintersten K, Behringer R (2003). Manipulating the mouse embryo, a laboratory manual (3rd Edn)
10. Whittingham DG (1971) Culture of mouse ova. J Reprod Fertil Suppl 14:7–21
11. Lawitts JA, Biggers JD (1993) Culture of preimplantation embryos. Methods Enzymol 255:153–64
12. Solter D, Knowles BB (1975) Immunosurgery of mouse blastocyst. Proc Natl Acad Sci USA 72:5099–102
13. Hogan B, Tilly R (1978) In vitro development of inner cell masses isolated immunosurgically from mouse blastocysts. II. Inner cell masses from 3.5- to 4.0-day p.c. blastocysts. J Embryol Exp Morphol 45:107–21
14. Gardner RL, Rossant J (1979) Investigation of the fate of 4-5 day post-coitum mouse inner cell mass cells by blastocyst injection. J Embryol Exp Morphol 52:141–52
15. Kurimoto K, Yabuta Y, Ohinata Y, Ono Y, Uno KD, Yamada RG, Ueda HR, Saitou M (2006) An improved single-cell cDNA amplification method for efficient high-density oligonucleotide microarray analysis. Nucleic Acids Res 34:e42

Part IV

Human Systems

Chapter 22

Utilizing DNA Mutations to Trace Epithelial Cell Lineages in Human Tissues

Sebastian Zeki, Trevor A. Graham, and Stuart A.C. McDonald

Abstract

Epithelial stem cells are typically multipotential and are likely the cell of origin of epithelial cancers. Tracing the expansion of a single stem cell's progeny, identifying and characterizing these cells in human tissue has proven difficult. Invasive labeling studies, which have led to much success in model organisms, are impracticable in humans. Instead, human studies must rely upon naturally occurring *clonal markers*: typically somatic DNA alterations that uniquely identify a population of cells with the same ancestry. In normal epithelium, nonpathogenic mitochondrial DNA mutations have proven useful. In premalignant and malignant disease, genomic DNA mutations within tumor suppressor genes or oncogenes can be used to trace the spread of mutant clones.

Key words: Mitochondrial DNA mutations, Genomic mutations, PCR, Laser capture microdissection, Stem cell

1. Introduction

That epithelial cells are derived from a common, multipotent stem cell has been established in several human tissues (1–4). The understanding that a cell is derived from another cell ultimately reveals the dynamics of tissue maintenance, repair, and pathology such as cancer. One technique to examine clonal relationships within tissues is known as *lineage tracing*. Because the parental relationship between cells cannot be morphologically distinguished, the researcher must rely on surrogate markers of clonality, such as genetic point mutations, to trace lineages. Furthermore, the accurate isolation of a population of cells is necessary to trace lineage to remove confounding genetic markers from contaminating cells. A clonal relationship between cells can be inferred in all the cells that share a somatic DNA point mutation or other rare genetic abnormality. The usefulness of genetic alterations as clonal markers is determined primarily

Kimberly A. Mace and Kristin M. Braun (eds.), *Progenitor Cells: Methods and Protocols*, Methods in Molecular Biology, vol. 916, DOI 10.1007/978-1-61779-980-8_22, © Springer Science+Business Media, LLC 2012

by two factors: the rarity of the alteration and the stochastic nature of the mutation. Rare mutations are unlikely to occur independently in neighboring cells, so offer a powerful clonal marker. However, scarcity can also be the Achilles' heal of a marker too: markers that occur very rarely will be uninformative in most samples. Thus, at a practical level, the most useful genetic clonal markers in human tissues are those which occur frequently enough to be readily findable, but infrequently enough to be reliable clonal markers. Furthermore, if there is a large segment of DNA that mutations occur randomly within, there is a high probability that neighboring cells with the same mutation are derived from a founder cell.

A further issue in clonality studies on human tissues is the specificity at which an alteration can be localized within the tissue. Interpretation of genomic studies can be confounded if, for example, whole biopsy samples are used as DNA from both epithelium and stroma will be analyzed. We therefore use laser capture routinely to isolate individual cells and small tissue structures, such as colonic crypts, of interest. In the remainder of this chapter, we discuss the relative merits of three types of genetic alterations as clonal markers and provide a methodology for their use:

1.1. Mitochondrial DNA Analysis

Normal human epithelium, by definition, has few if any genomic defects. Therefore finding a nuclear DNA mutation to act as a marker of clonal expansion is difficult. However, mitochondrial DNA (mtDNA) is subject to frequent mutation, even in normal tissue, as a result of its poor DNA repair mechanisms and lack of protective histones. Somatic mtDNA mutations are rarely pathological and cause no or very minor morphological or proliferative changes to the cell (5). Each cell contains many mitochondria each with multiple copies of the mtDNA genome, and so it is very likely that a mutant mtDNA takes many years to drift to a detectable level within the mtDNA pool (6). Therefore the only cell type that lives long enough for near-homoplasmy (presence of a mutation that affects all mitochondria in a cell, as opposed to heteroplasmy where a fraction of the mitochondria are affected) to occur is the stem cell. This method of lineage tracing is therefore best placed for the study of nonpathological tissue.

Cytochrome *c* oxidase (CCO) is a primarily mitochondrially encoded enzyme. Patches of cells deficient in CCO are frequently observed in aging tissues. CCO deficiency is frequently associated with mtDNA mutations, often, but not exclusively in the CCO gene. Staining for CCO activity, and using enzyme histochemistry for succinate dehydrogenase (SDH) activity as a counterstain (SDH is a genomically encoded, mitochondrially active enzyme), identifies CCO-deficient cells by staining them blue and CCO-proficient cells as brown. Subsequent polymerase chain reaction (PCR) amplification of the entire mitochondrial genome from cells within wholly CCO-deficient (blue) areas, for example, in individually laser captured cells from the adjacent CCO-deficient crypts in the colon, can prove that these cells are clonal—they have the same mutation and are therefore derived from a common precursor cell. The protocol for this technique is described below and has successfully applied to human fresh frozen colon (7), small intestine (8), pancreas (7), stomach (9), liver (10), and skin and is an adaption of a protocol first published by Taylor et al. (11).

1.2. Genomic Point Mutations

Possibly the strongest marker of clonal expansion in premalignant disease and cancers is the point mutation of genes specifically involved in tumor progression, for example the *TP53* tumor suppressor gene is commonly mutated in human cancers. Analysis of a single mutation indicates where a (tumorigenic) clone has spread through a lesion. The analysis of multiple genes simultaneously can reveal the relative order in which the mutations occurred. The *founder* mutation will be present throughout a lesion, whereas later mutations will be present only as a subclone of the founder mutation. Thus if clone A has one mutation, and clone B the same mutation as well as an additional one, then clone A can be said to have arisen before clone B. This multi-gene analysis is termed *genetic dependency analysis.* Genetic analysis of individually laser captured epithelial units from pre-neoplastic or cancerous lesions can reveal the histological and genetic pathways of tumor development.

1.3. Microsatellite Analysis

Microsatellites are hypervariable regions composed of short, repetitive DNA sequences and are ubiquitous throughout the genome. This hypervariability means that frequently maternal and paternal alleles can contain different numbers of motif repeats at any particular microsatellite. Thus, microsatellites offer a serendipitous means to distinguish between maternal and paternal alleles. PCR amplifying a heterozygous microsatellite locus produces PCR products of two different sizes: the two products can be distinguished by a sensitive electrophoresis system. In this manner, microsatellites can be used to demonstrate allelic loss, typically loss of heterozygosity (LOH). LOH is a frequent causative event in neoplastic and pre-neoplastic disease, and consequently microsatellite-LOH analysis has been used widely for clonality assessment of a number of different human premalignant disease and cancers (3).

It must be stressed that LOH analysis is difficult. More common than pure loss of an allelic peak is the reduction in size of one peak relative to another and the thresholds for what is called “significant” have not been standardized. Furthermore “no loss” and homozygous deletions cannot be distinguished as the analysis will detect the contaminating DNA; this can happen despite the attempts to minimize contamination through the use of microdissection. Conceptual problems also exist. If cancer is being studied, polyploidy, if it occurs before an LOH event, may confound LOH detection because the extra ploidy will give either a false negative (in tetraploidy two LOH events will still result in a “no LOH” result) or a false positive due to chromosome segregation.

2. Material

2.1. Mitochondrial DNA Analysis: Histochemistry

20 μm frozen sections are cut on to membrane slides for laser microdissection as above; for general tissue histochemistry sections can be cut 6–12 μm (depending on concentration of mitochondria) on normal charged glass sides and are useful as a preliminary screening of tissue samples.

1. Phosphate Buffer: 0.2 M dihydrogen orthoposphate/sodium phosphate dibasic, 0.2 M dihydrogen orthophosphate/ sodium phosphate monobasic, add 19.5 ml dibasic to 30.5 ml monobasic, pH to 7.0 and make up to 100 ml with ddH_2O.
2. CCO stock solutions: Diaminobenzidine (5 mM DAB in phosphate buffer pH 7.0) stored in 800 μl aliquots; Cytochrome *c* (500 μM cytochrome *c* in phosphate buffer pH 7.0 stored in 250 μl aliquots); Catalase 20 microgrammes/ml (Sigma-Aldrich, UK).
3. SDH Stock solutions: Sodium succinate (1.3 M sodium succinate in phosphate buffer pH 7.0 300 μl aliquots); NitroBlue Tetrazolium, (1.875 mM NBT in phosphate buffer pH 7.0 800 μl aliquots); Phenazine methosulphate, (2 mM PMS in phosphate buffer pH 7.0 150 μl aliquots); Sodium azide (100 mM sodium azide in phosphate buffer pH 7.0 50 μl aliquots).

2.2. Mitochondrial DNA Analysis: Mitochondrial DNA PCR

1. To the primers add Tris EDTA buffer or water to make a stock concentration of 200 mM. Then dilute 1:10 in ddH_2O to make working concentration of 20 mM.
2. Primers (Sigma Genosys, UK) see reference (12) for sequences.
3. Ultraviolet treated ddH_2O.
4. 10× buffer (Comes with ABI Amplitaq Gold, Applied Biosystems, Carlsbad, California).
5. 10× dNTPs (Roche, Welwyn Garden City, UK).
6. 25 mM Mg^{2+} solution (ABI $MgCl_2$ 25 mM).
7. AmpliTaq Gold (Applied Biosystems kit).
8. Extracted DNA.
9. 96-well plate and adhesive PCR film (ABGene, Epsom, UK).
10. 1.5 ml eppendorf tubes.
11. Big dye terminator® (BDT-Applied Biosystems).
12. ExoSAPIT (USB Corporation, Cleveland, Ohio, US) PCR product to increase the volume present 20 μL ddH_2O can be added to each well at this stage.
13. DyeEx 96 Kit (Qiagen). This contains spin columns in which there is a gel filtration resin to remove all dye terminators.
14. 96-well thermocyler.

2.3. Genomic DNA PCR

1. To the primers add Tris EDTA buffer or water to make a stock concentration of 200 mM. Then dilute 1:10 in ddH_2O to make working concentration of 20 mM.
2. Primers (Sigma Genosys, UK).
3. Ultraviolet treated ddH_2O.
4. 10× PCR buffer.
5. 10× dNTPs (Roche, Welwyn Garden City, UK).
6. 25 mM Mg^{2+} solution (Applied Biosystems $MgCl_2$ 25 mM).
7. Simple Red Taq (Sigma-Aldrich, UK).
8. Extracted DNA.
9. 96-well plate and adhesive PCR film (ABGene, Epsom,UK).
10. 1.5 ml eppendorf tubes.
11. Big dye terminator® (BDT-Applied Biosystems, Carlsbad, California, US).
12. ExoSAPIT (USB Corporation, Cleveland, OH, USA)
13. DyeEx 96 Kit (Qiagen).
14. 96-well thermocyler.

2.4. Microsatellite Analysis

1. 0.85 ml Multiplex Mix contains Taq polymerase, Buffer solution, Magnesium chloride (Qiagen multiplex PCR kits).
2. ddH_2O.
3. Primer mix for multiplex PCR. Each stock primer is 200 mM; the working solution is 20 mM.
4. 96-well plate.
5. DNA isolated as above.

The multiplex solution, Q and ddH_2O should be placed on ice and all PCR done in an ultraviolet cabinet.

3. Methods

3.1. Samples

This protocol assumes the availability of a laser capture microdissection system (LMD). Most models will be suitable; however, our specific model is P.A.L.M Combisystem (Zeiss, Germany). Tissue sections are cut on to membrane slides P.A.L.M (Zeiss) to enable laser microdissection. Formalin-fixed paraffin-embedded (FFPE) or fresh frozen specimens are suitable; however, FFPE sections should be dewaxed and all sections completely dried before laser capture.

For microsatellite and genomic point mutation analysis, use 6 μm formalin-fixed, paraffin-embedded sections. Pretreating slides with ultra violet (UV) light improves section adhesion. For mtDNA mutation analysis, a single 20 μm frozen section is used that had been stained for CCO and SDH activity and cut single cells (or as near as possible) in areas of CCO-deficiency, with nearby CCO-normal cells as a wild-type control. Staining for CCO activity does not adversely affect subsequent genetic analysis. Laser capture is performed on 5–6 serial sections 4 μm thick, cutting out the same epithelial unit through all sections (in our case, the colonic crypt).

3.2. Extraction of DNA from Cells

1. All reaction tubes and ddH_2O are pretreated with 30 min of UV-light exposure.
2. Digestion buffers are made up in an ultraviolet PCR flow cabinet to minimize contamination (see Note 1).
3. Following laser capture, the unit or cell will be present on the cap of the adhesive cap 500 opaque (Zeiss) tube. PicoPure (Arcturus Bioscience, Mt View, California, USA) is placed directly on the cap after microdissection (see Note 2).
4. For optimal yields, frozen tissues should be incubated for 3 h at 65°C and FFPE tissue should be incubated at 65°C overnight.
5. Following digestion, tubes are briefly centrifuged, then denatured at 95°C for 10 min and again briefly centrifuged. Store supernatant at −20°C until use (see Note 3).

3.3. Mitochondrial DNA Analysis: Histochemistry

1. Thaw CCO media and DAB at 55°C/under hot tap.
2. Prepare CCO incubation medium as follows:
   - Add 200 μl Cytochrome *c* solution to 800 μl DAB solution.
   - Add a small amount (roughly the size of a match head—not necessary to be absolutely precise) of catalase powder, vortex, and filter with a 0.2 μm syringe filter.
   - Add 50–100 μl medium to the tissue section, ensuring the entire section is covered.
3. Incubate for 50 min at 37°C in a humid chamber. Incubation time is dependent on cellular mitochondria numbers. Wash section twice in PBS.
4. Prepare SDH incubation medium. To 800 μl NBT add 100 μl sodium succinate solution, 100 μl PMS solution (see Note 4), and 10 μl sodium azide solution. Filter with a 0.2 μm syringe filter.
5. Wipe any excess PBS from the slide with a tissue being careful not to touch the section.
6. Add 50–100 μl medium to the tissue section, ensuring the entire section is covered.

7. Incubate for 45 min at 37°C in a humid chamber. Wash sections twice in PBS.
8. Dehydrate in increasing ethanol concentrations (70%, 95%, 100%, and 100%). Incubate sections in the final concentration of 100% ethanol for 10 min.
9. Leave to air dry for 1 h. Dry slides can be stored at −40°C or used immediately for laser capture.
10. For normal glass slide tissue processing, dehydrate through ethanol gradients (final 100% only requires 2–5 min depending on tissue thickness) and clear with Histoclear, mount in Permount.

3.4. Mitochondrial DNA Analysis: Mitochondrial DNA PCR

A nested PCR protocol is necessary. In this protocol, the entire mitochondrial genome is split up into nine first round primer sets. These are designed as overlapping primer pairs giving fragments of ~2 kb. A second round of PCR is then performed using first round PCR product as the DNA template. To facilitate direct sequencing of PCR product, the second round primers are tagged with the M13 universal primer sequence (see (12) for nucleotide sequence). A nested PCR protocol is necessary so that the very small amounts of starting DNA are maximally amplified at the same time as reducing the signal-to-noise ratio of the sequencing result (see Note 5).

3.4.1. First Round PCR

1. For each primer pair, aliquot 1.5 μl of forward and reverse primer into each tube, then aliquot 1 μl lysate into the appropriate tubes. Use one primer pair and set up an additional two tubes for the lysis buffer controls, together with one tube of positive control DNA and one tube for mastermix negative control.
2. Make up a mastermix for the number of tubes + 1 as below (volumes are given per tube):

33.65 μl Ultraviolet treated ddH_2O
5.0 μl 10× buffer
5.0 μl 10× dNTPs
25 mM Mg^{2+} solution −2.0 μl
AmpliTaq Gold −0.35 μl

3. Run on the thermocycler as follows: 10 min at 95°C, then cycle 38 times through 94°C for 45 s, 58°C for 45 s, 72°C for 2 min. After this round of cycles, incubate for 8 min at 72°C and thereafter at 4°C.

3.4.2. Second Round PCR

1. Aliquot 1 μl of each forward and reverse primer into each tube.
2. Add 1 μl of each first round PCR product to the appropriate tubes. Set up two additional tubes for a positive and negative

control. Make up a mastermix as follows (volumes are given per tube):

16.87 μl Ultraviolet treated ddH_2O
2.5 μl 10× buffer
2.5 μl 10× dNTPs
0.13 μl AmpliTaq Gold
22.0 μl Total volume

3. Run on the thermocycler as follows: 10 min at 95°C, then cycle 30 times through 94°C for 45 s, 58°C for 45 s, 72°C for 1 min, and 72°C for 8 min. After this round of cycles, incubate for 8 min at 72°C and thereafter at 4°C. Run out 5 μl each PCR product on 1.5% agarose and use 5 μl 100 base pair ladder as standard.
4. Assuming that PCR product is detected at the gel electrophoresis stage, add 5 μl of each PCR product to the appropriate well of a separate 96-well sequencing plate. With the plate on ice, add 2 μl ExoSAP-IT (USB Corporation, Cleveland, Ohio, US) to each well (the enzyme must be kept on ice when out of freezer). Place a rubber cover mat onto the plate, mix briefly and pulse spin down, and incubate at 37°C for 15 min then 85°C for 15 min.

3.4.3. Sequencing Preparation

1. Add concentrated BDT to ddH_2O in the ratio 10 μL BDT to 5 μL ddH_2O in a 1.5 mL eppendorf. Each sequencing reaction will need 15 μL of this mix so make up 10% more volume than is required for the number of reactions.
2. To each well of a 96-well plate, add 15 μL of BDT and ddH_2O mix and 1 μL primer. The forward primer is added to one well, the reverse to a different well. Do not mix primers at this stage. Add 4 μL PCR product to each well to make total of 20 μL.
3. Run on the thermocycler as follows: 25 cycles of 5 min at 96°C, then 96°C for 10 s, 50°C for 5 s, 60°C for 4 min, and 72°C for 8 min. Thereafter incubate at 8°C.
4. Run out 5 μl each PCR product on 1.5% agarose and use 5 μl 100 base pair ladder as standard.

3.4.4. Removal of Big Dye Terminator and Sequencing Preparation

1. Vortex the spin column. This will resuspend the resin.
2. Remove the bottom of the spin column and once this is placed in a 2 mL collection tube centrifuge for 3 min. Transfer the spin column to a further centrifuge tube and apply 15 μl of the sequencing reaction to the center of the gel without touching the side. Centrifuge for 3 min and remove the spin column. The eluate contains the purified DNA. The extension products are then eluted in 20 μL ddH_2O and placed on the 3730XL Sequencer (Applied Biosystems, Carlsbad, California, US).

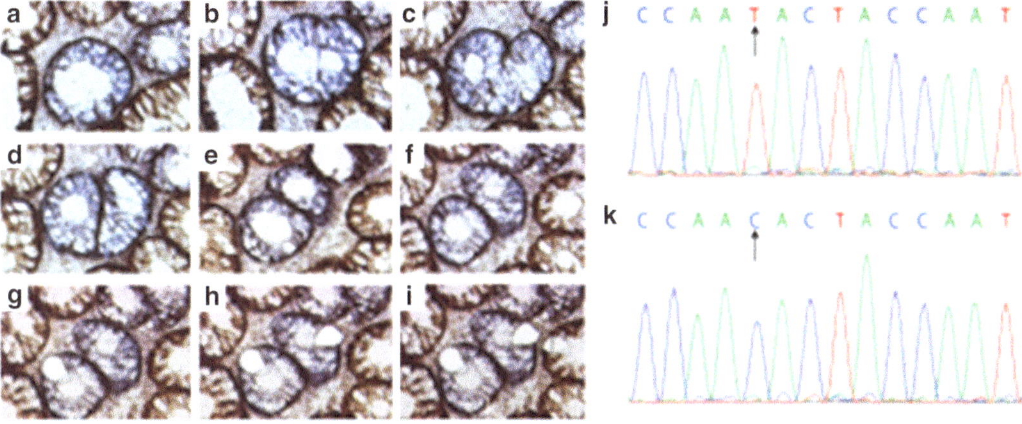

Fig. 1. **(a–j)** A colonic crypt in fission. The cytochrome *c* oxidase mitochondria have caused the cells to be stained blue. This crypt in fission has a 4,733 t > C transition in both dividing arms (**j**, **k**) confirming their common origin. Photo courtesy of publishers name from reference (8). Reprinted with permission. Copyright (2006) National Academy of Sciences, USA.

3.4.5. Results of Mitochondrial Staining and Analysis

The results should be returned as standard sequencing files. A mitochondrial mutation will either be shown by the presence of homoplasmy, in which case the reference nucleotide peak will have been replaced by another, or heteroplasmy in which case there are two peaks at one nucleotide locus. Determination of mutations is done by comparing the returned sequence with a reference sequence which can be found at: http://www.mitomap.org/MITOMAP. Automated software tools will identify mutations when compared to the reference sequence. These mutations should be checked to ensure they do not represent known polymorphisms (a list of these can be found in the Mitomap database). The sequence should also be checked visually to ensure the mutation is real and does not represent misreading by the software. Of the two peaks present in heteroplasmy, the dominant peak usually represents the mutation although this is not always the case, again the visual trace should be checked. Figure 1 below demonstrates the results of CCO staining and the homoplasmic mitochondrial sequence result that might be expected when tissue is laser captured from the blue colonic epithelium.

3.5. Genomic DNA PCR

The conditions used for genomic DNA PCR will vary depending on the gene being studied. These conditions should be ascertained during primer design and optimization.

3.5.1. First and Second Round PCR Methods

1. For each primer, make up a mastermix in a 1.5 mL Eppendorf tube as per the conditions established during optimization. This mastermix contains Taq polymerase, PCR buffer, Q solution, dNTPs, ddH_2O, and $MgCl_2$. Multiply the volume of mastermix needed per well by the number of PCR reactions needed.
2. Add 2 μl of DNA template to each well in a 96-well plate.

3. To each well add 23 μl of mastermix to establish a total volume of 25 μl.
4. Add the plate to the thermocycler. The cycle will again depend on the gene being amplified and should be established during primer design and optimization.
5. The First round PCR product is used as the DNA template in the second round of PCR. Again the constituents of the mastermix will depend on the conditions established during primer optimization. All other steps, including the total well volumes, are the same as in the first round.
6. Once the PCR reaction has been completed in the thermocycler, run out 5 μl each PCR on 1.5% agarose and use 5 μl 100 base pair ladder. As per the mitochondrial protocol, the second round PCR product will then need to undergo cleanup, labeling, and sequencing which can be run on the 3730XL Sequencer.

3.5.2. Analysis of Genomic Mutations

Once the genomic sequence has been returned, the search for mutations can begin. Two traces will be returned: one for the forward primer and one for the reverse. The forward primer trace should be visually inspected for areas in which there are two peaks at a nucleotide location (usually one peak under another). The normal DNA will often appear as the dominant peak, so simple automated comparison of the study sequence against a reference is often insufficient. Identification of double peaks requires a clean sequence trace so care should be taken at all steps from DNA digestion to sequencing to ensure this is the case, otherwise false positive identifications can arise.

Once a presumed mutation has been found, this should be confirmed by comparison firstly with a database of known mutations (e.g., COSMIC at http://www.sanger.ac.uk/genetics/CGP/cosmic) to confirm whether previous mutations have been found in this area. Secondly, polymorphisms should be excluded by consultation with databases such as Ensembl (http://www.ensembl.org). In the instance that a mutation is listed already and is also found on the polymorphism database, constitutional DNA should be analyzed. If the presumed mutation is also present in the constitutional DNA, then the mutation is in fact a polymorphism.

3.6. Microsatellite Analysis

This protocol assumes that primers for specific microsatellites have already been designed and optimized to determine whether Q solution will be needed or not, and to determine the optimal annealing temperature. We use multiplex reactions for microsatellite analysis using FAM- and HEX- labeled primers which give varying PCR product size and/or a different fluorescent-labeled product. The challenge here is to design complementary primer sets which produce distinguishable PCR products and which can be co-amplified in a single PCR reaction. It is assumed that such primer sets have been designed. All materials used are sourced from

Qiagen. Constitutional DNA (*i.e.*, not from an area containing pathology) should be obtained at the same time as capturing tissue from the area under study.

3.6.1. PCR

1. If primer optimization requires Q solution, add 5 μl Q solution, 25 μl of the multiplex solution, 5 μl primer mix, and 13 μl ddH_2O to a 1.5 mL eppendorf. If Q is not necessary, add 18 μl ddH_2O. Vortex the mixture and store at −20°C.
2. Add 2 μl from the extracted DNA sample to each well on the plate and to this add 48 μl of the master mix to each well containing DNA template to make a total volume per well of 50 μl.
3. Run the multiplex PCR reaction on the thermocycler at 57°C as follows: 95°C for 15 min, 94°C for 15 min, 57°C for 1.5 min, and 72°C for 1 min. Cycle to step 2 for 34 more times, and then incubate at 60°C for 30 min and at 8°C thereafter.
4. Run out 5 μl each PCR on 1.5% agarose and use 5 μl Hyperladder IV as standard.

3.6.2. Genescan Preparation

The constitutional DNA should now be prepared to assess whether the patient is informative at the studied allele locus.

1. Firstly, prepare dilutions of the PCR product using ultraviolet-treated ddH_2O from the normal DNA at each of: 4 μl of undiluted DNA, one sample at 1:20, one sample at 1:50, and one sample at 1:100 dilution. These dilutions should all be placed in separate wells in a Genescan specific 96-well plate.
2. In a separate container mix the Genescan loading buffer in the ratio: RoxGS500 size standard—0.5 μl. Sodium hydroxide—0.5 μl, and HLDI Formamide—10 μl.
3. Add 8 μl of the loading buffer to the volume in each well on the Genescan plate to create a total volume of 12 μl per well.
4. The product is run on a 3100 Genetic Analyzer.

Having established that the patient is informative at a locus and having established the optimal dilution for this, dilute the PCR product of the rest of the extracted DNA. There should be 4 μl of PCR product at the appropriate dilution in each well. The plate can then be submitted for Genescan analysis using the same preparation as discussed above.

3.6.3. Analysis of Results

The interpretation of LOH can be difficult. It consists of two steps. Firstly the constitutive DNA needs to be analyzed using software such as Gene Mapper (Applied Biosystems, Carlsbad, California) to assess whether it is informative or not. This means whether the alleles being studied have been detected by the microsatellite primers. If the DNA is not informative, no conclusions can be reached about the allele from the pathological specimen being examined (Fig. 2, top panel).

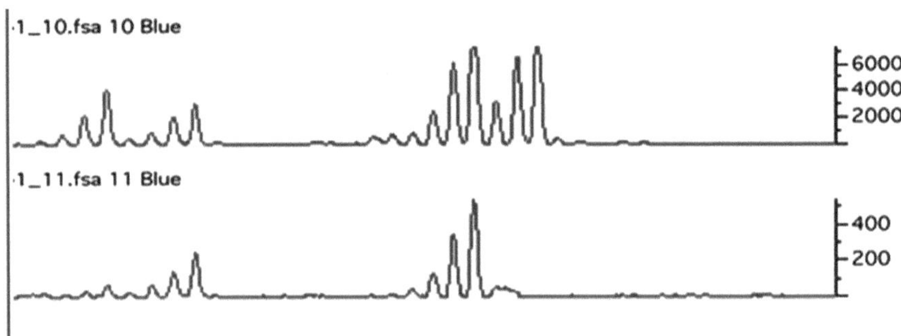

Fig. 2. Examples of Uninformative DNA and LOH. The top trace demonstrates an informative locus as both alleles of two genes are seen. The bottom trace is of the same patient taken from an oesophageal tumor. Here the patient can be said to have LOH of both genes as one trace of each pair is not present.

If informative, the pathological DNA can be studied and the sample is then analyzed as above. Possible results are shown (Fig. 2, bottom panel). Definite LOH can be concluded if only one peak is shown at a locus where the constitutive DNA had two peaks. If there is a difference in height in the two peaks, then this difference should be analyzed statistically. If the area under one allelic peak in the affected crypt is less than 0.5 times or greater than two times that of the other allele, LOH is considered present, after correcting for the relative areas using constitutional DNA.

4. Notes

1. Normal tissue culture cabinets would suffice as long as they have an ultraviolet light. Our lab uses an Astec Microflow system (Bioquell, Hants, UK). We use PicoPure (Arcturus, CA) proteinase K for DNA extraction.
2. 14 μl PicoPure is sufficient for single cell applications. For other applications this must be titrated.
3. A negative control of the PicoPure buffer should always be included. It should be noted that mtDNA sequence analysis cannot be performed on FFPE tissue using this protocol.
4. It is light sensitive so always keep covered when not in use.
5. DNA extraction and first round PCR must be done in the ultraviolet cabinet and filter tips used throughout first round PCR setup.

References

1. Alonso L, Fuchs E (2003) Stem cells of the skin epithelium. Proc Natl Acad Sci USA 100(Suppl 1):11830–11835
2. Bearzi C, Rota M, Hosoda T, Tillmanns J, Nascimbene A, De Angelis A, Yasuzawa-Amano S, Trofimova I, Siggins RW, Lecapitaine N, Cascapera S, Beltrami AP, D'Alessandro DA, Zias E, Quaini F, Urbanek K, Michler RE, Bolli R, Kajstura J, Leri A, Anversa P (2007) Human cardiac stem cells. Proc Natl Acad Sci USA 104:14068–14073
3. Ng IO, Guan XY, Poon RT, Fan ST, Lee JM (2003) Determination of the molecular relationship between multiple tumour nodules in hepatocellular carcinoma differentiates multicentric origin from intrahepatic metastasis. J Pathol 199:345–353
4. Stemple DL, Anderson DJ (1993) Lineage diversification of the neural crest: in vitro investigations. Dev Biol 159:12–23
5. McDonald SA, Preston SL, Greaves LC, Leedham SJ, Lovell MA, Jankowski JA, Turnbull DM, Wright NA (2006) Clonal expansion in the human gut: mitochondrial DNA mutations show us the way. Cell Cycle 5:808–811
6. Elson JL, Turnbull DM, Howell N (2004) Comparative genomics and the evolution of human mitochondrial DNA: assessing the effects of selection. Am J Hum Genet 74:229–238
7. Taylor RW, Taylor GA, Durham SE, Turnbull DM (2001) The determination of complete human mitochondrial DNA sequences in single cells: implications for the study of somatic mitochondrial DNA point mutations. Nucleic Acids Res 29:E74–74
8. Greaves LC, Preston SL, Tadrous PJ, Taylor RW, Barron MJ, Oukrif D, Leedham SJ, Deheragoda M, Sasieni P, Novelli MR, Jankowski JA, Turnbull DM, Wright NA, McDonald SA (2006) Mitochondrial DNA mutations are established in human colonic stem cells, and mutated clones expand by crypt fission. Proc Natl Acad Sci USA 103:714–719

Chapter 23

NF-Ya Protein Delivery as a Tool for Hematopoietic Progenitor Cell Expansion

Alevtina D. Domashenko, Susan Wiener, and Stephen G. Emerson

Abstract

The clinical potential of therapeutic quantities of primary hematopoietic cells, either unmodified or altered via genetic modification, has stimulated the search for techniques that allow the production of large numbers of hematopoietic precursors, more primitive progenitors, and perhaps hematopoietic stem cells (HSC) themselves. Modifications of in vitro culture conditions to promote progenitor cell expansion have included combinations of polypeptide cytokines, small molecules, and transcription factors. Here we describe the methods for use of the transcription factor linked to a TAT-based protein transcription domain, in combination with cytokines and serum-free culture condition to stimulate the proliferation of primary cells. Human peripheral blood (PB) $CD34^+$ cells treated with TAT-NF-Ya fusion protein and grown in vitro for 1 month proliferate four times more than did cells in cultures that contained only cytokines, including increased production of hematopoietic cells of all maturities. These results and techniques should be suitable for multiple applications of ex vivo generation of hematopoietic cells using protein transduction.

Key words: Hematopoietic progenitors, Human CD34+ cells, NF-Ya transcription factor, TAT-NF-Ya fusion protein, Protein transduction

1. Introduction

Modification of HSC by retroviral gene transduction delivery is efficient and powerful, but concerns about irreversible biological effects and insertional mutagenesis very likely preclude its application to clinical settings. Early clinical trials employing retroviral gene delivery reported the development of a T-cell leukemia-like syndrome in two young patients treated for X-linked severe combined immunodeficiency (SCID-X1) with γc gene transfer into bone marrow $CD34^+$ cells (1). Dogs and macaques transplanted with primary $CD34^+$ cells transduced by a retroviral vector overexpressing *HoxB4* (2) suffered from myeloid leukemia 2 years later

Kimberly A. Mace and Kristin M. Braun (eds.), *Progenitor Cells: Methods and Protocols*, Methods in Molecular Biology, vol. 916, DOI 10.1007/978-1-61779-980-8_23, © Springer Science+Business Media, LLC 2012

(2 of 2 dogs and 1 of 2 macaques). Given these theoretically and empirically demonstrated complications of techniques that modify genomes, particularly in nonhomologous fashion, it is essential that nonpermanent, non-retroviral techniques for manipulating gene expression be developed.

Unfortunately, many of the nonviral methods of gene transfer that have been employed to date, including microinjection, electroporation, and liposomes, exhibit significantly lower efficiencies than viral delivery and may themselves introduce significant toxicities (3). Cell-penetrating peptides (CPPs), on the other hand, possess the ability to overcome these limitations, and the use of CPPs appears to be safe and effective (4). CPPs are short peptides, often consisting of about 30 amino acids, that are capable of penetrating the cell and nuclear membranes while carrying with them functional cargo, such as proteins, nucleic acids, liposomes, or nanoparticles. CPPs were first discovered by Frankel and Pabo (5) and Green and Loewenstein (6) who independently determined that the transactivator of transcription (TAT) protein of the HIV-1 virus was able to enter cells and translocate to the nucleus. The TAT protein contains a sequence of 11 amino acids, the protein transduction domain, that is necessary and sufficient for cellular uptake (7). Schwarze et al. (8) demonstrated a practical application of CPP by showing that fusion of a protein (β-galactosidase) to the 11 amino acid TAT protein transduction domain enabled the delivery of the biologically active fusion protein to all tissues in mice. Since then, the family of CPPs has grown to dozens (reviewed by Heitz et al. (4)) and the technique has advanced to the development of automated delivery systems tested with cell cultures and suitable for clinical applications (9).

In earlier studies, our laboratory had found that Nuclear Factor Y (NF-Y), the trimeric transcription factor, is involved in the regulation of HSC self-renewal and differentiation through the activation of several genes, including the Hox4 paralogs *HoxB4*, *HoxC4*, and *HoxD4* as well as *Hes-1*, *LEF1*, *Notch1*, *p27*, and telomerase (10). NF-Y, ubiquitously expressed in mammalian tissues, consists of three subunits, NF-Ya, NF-Yb, and NF-Yc. Constitutively expressed NF-Yb and NF-Yc forms a heterodimer via a histone-fold motif, which then interacts with NF-Ya to form a complete heterotrimeric complex that recognizes the CCAAT consensus binding site (11) present in human promoters (12–14). All three subunits are necessary for high affinity sequence specific binding to promoters. The cellular level of available trimeric NF-Y is controlled by the level of expression of its regulatory subunit NF-Ya (15).

NF-Y regulates the expression of many genes implicated in both essentially cellular renewal processes and in specific developmental programs: *cyclin A2*, *cyclin B1*, *cyclin B2* (16), several erythroid-specific genes (17, 18), multidrug resistance gene *MDR1* (19), tumor suppressor *GADD45* γ (20) are some important examples. Given its many targets, a unique role for NF-Y in early mouse development is not surprising, and indeed genetic deletion of NF-Ya results in early embryonic lethality (21).

We have previously shown that NF-Y activates one of its targets, *HOXB4*, in hematopoietic cells by recognizing CCAAT promoter sequence (22). We found that the *HOXB4* core promoter contains two critical DNA-binding sites: HOX response element 1 (HxRE-1), a CCAAT box, and HxRE-2, an E-Box.

NF-Y, via NF-Ya subunit, binds to HxRE-1 and interacts with one of two closely related molecules, USF 1 and USF 2 (Upstream stimulating factors), that bind to HxRE-2. The formation of a NF-Y-USF complex increases the DNA-binding affinity of both the NF-Y and USF proteins to HxRE-1 and HxRE-2, and is necessary for the full activity of the *HOXB4* promoter. NF-Ya, the regulatory subunit of the NF-Y complex, is preferentially expressed within hematopoiesis in primitive HSC and expression declines with differentiation (10). Most importantly, overexpression of NF-Ya in murine HSCs prevents terminal differentiation to granulocytes and biases early stem cell divisions towards the preservation and expansion of HSCs and progenitor cells. Thus taken together, these data indicate that NF-Ya, the potential regulator of HSC renewal, might be considered as a therapeutic tool for HSC/ progenitors maintenance and ex vivo expansion.

Based on these findings, we used protein transduction methods to deliver a TAT-NF-Ya fusion protein (see Subheading 3.1) to both hematopoietic cell lines (K562) and primary human hematopoietic cells ($CD34^+$ peripheral blood cells).

HOXB4 mRNA expression measured by quantitative PCR was used as a readout for the evaluation of NF-Ya nuclear translocation and activity. Employing in vitro cell growth and clonogenic methylcellulose assays, we demonstrated that treatment of human PB $CD34^+$ cells with TAT-NF-Ya promotes cell proliferation, resulting in fourfold expansion of hematopoietic progenitor cell, and an increase of their multilineage colony-forming potential. These results confirm that protein transduction is a rapid and efficient technique for delivery of genetic material to cells and can be taken into consideration as an alternative to viral delivery.

In this report, we provide the complete details for our procedures, including cloning and purification of the fusion protein, its introduction into hematopoietic cell cultures, and the assaying of its efficacy via in vitro and in vivo assays.

This research was originally published in *Blood* (Domashenko AD, Danet-Desnoyers G, Aron A, Carroll MP, Emerson SG (2010) TAT-mediated transduction of NF-Ya peptide induces the ex vivo proliferation and engraftment potential of human hematopoietic progenitor cells. *Blood* 116(15):2676–2683. © the American Society of Hematology).

2. Materials

2.1. Expression and Purification of TAT-NF-Ya Proteins

2.1.1. Bacterial Culture

1. LB Broth Miller (Fisher).
2. Chloramphenicol (Calbiochem).
3. Glucose.
4. Isopropyl-B-D-thiogalactopyranoside (IPTG).
5. Protease inhibitor cocktail P8465 (Sigma).
6. Rosetta(DE3)pLysS (Novagen).

2.1.2. Protein Purification

1. 1,4-Dithiothreitol (DTT) (Roche Diagnostics).
2. *N*-Lauroylsarcosine sodium salt solution, 20% (Sigma).
3. Triton X-100 (Roche Diagnositcs).
4. Sonifier Cell Disruptor (Ultrasonics, Inc.).
5. Glutathione sepharose™ 4B (GE Healthcare).
6. L-Glutathione reduced, minimum 99% (Sigma).
7. Slide-A-Lyser dialysis cassettes (Fisher).
8. Amicon Ultra-4 10 K centrifugal filter device (Millipore).

2.2. Western Blotting

1. Ready Gel 10% Tris–Glycine precast gel (Bio-Rad).
2. 10× Tris/Glycine/SDS electrophoresis buffer (Bio-Rad) following dilution to 1× with water.
3. 2× Laemmli sample buffer (BioRad).
4. 10× Tris/Glycine transfer buffer (BioRad).
5. SDS-PAGE standards, broad range (Bio-Rad).
6. Pre-stained SDS-PAGE standards, broad range (Bio-Rad).
7. Simply Blue (Invitrogen).
8. Ponceau S (Sigma).
9. Nitrocellulose membrane (Bio-Rad).
10. Transfer buffer: 1× Tris–glycine buffer (Bio-Rad), 20% methanol.
11. TWEEN-20 (Sigma).
12. Mouse anti-GST-HRP monoclonal antibody (Santa Cruz Biotechnology).
13. ECL Plus (GE Healthcare).
14. Amersham Hyperfilm ECL (GE Healthcare).
15. Iscove modified Dulbecco medium (IMDM) (Gibco-BRL).
16. Densitometer (Molecular Dynamics).

2.3. Cell Culture and Protein Delivery

1. K562 human leukemia cell line (ATCC).
2. Human $CD34^+$ cells from peripheral blood (Fred Hutchinson Cancer Center).

3. Flt3/Flk-2 ligand, stem cell factor (SCF), and thrombopoietin (TPO) 100 ng/mL each (BD Biosciences Pharmingen).
4. IL-3 and IL-6, 20 ng/mL each (BD Biosciences Pharmingen).
5. StemSpan H3000, serum-free medium (Stem Cell Technologies).

2.4. RNA Purification and Real Time PCR

1. RNeasy mini kit (QIAGEN).
2. TaqMan Reverse Transcription Reagents kit (Applied Biosystems).
3. 2× TaqMan Universal PCR Master Mix (Applied Biosystems).
4. TaqMan primers/probe set for human HOXB4 (Hs00256884_m1) and GAPDH (Hs99999905_m1) (Applied Biosystems).
5. 7500 Real-Time PCR System (Applied Biosystems).

2.5. Methylcellulose

1. MethoCult SF H4436 (Stem Cell Technologies).
2. 35 mm culture dishes (Falcon catalog # 351008 or equivalent).

3. Methods

3.1. TAT-NF-Ya Fusion Protein Expression and Purification

Our purification strategy is aimed at optimization of protein preparation under native conditions, which is essential for protein activity in functional applications. To achieve this, the following parameters required for successful recombinant protein expression and purification should be considered: type of vector, bacterial host strain, its growth conditions, induction parameters of protein expression, protein solubility, stability, and toxicity to the host cells.

A fusion protein construct (Fig. 1a) expressing the short form of human NF-Ya that is active in HSC (22) along with glutathione-S-transferase (GST) epitope tag for protein purification, TAT transduction domain sequence (YGRKKRRQRRR), and a hemagglutinin (HA) tag was created using the pGEX-6p1 expression plasmid, as described (23). The choice of the vector was influenced by its inclusion of a strong, non-leaky, IPTG-inducible *tac* promoter which provides high level of expression of GST-fusion proteins.

3.1.1. Bacterial Culture Growth Condition

In *E. coli* several eukaryotic codons such as Arg, Gly, Ile, Leu, and Pro are underrepresented, and bacterial systems do not have a sufficient amount of tRNAs corresponding to these rare codons. This can affect heterologous protein expression because of premature translation termination or frameshifting.

This problem can be solved by using bacterial strains that are designed to enhance the expression of eukaryotic proteins by supplementation of insufficient tRNAs (24). We found that about 10%

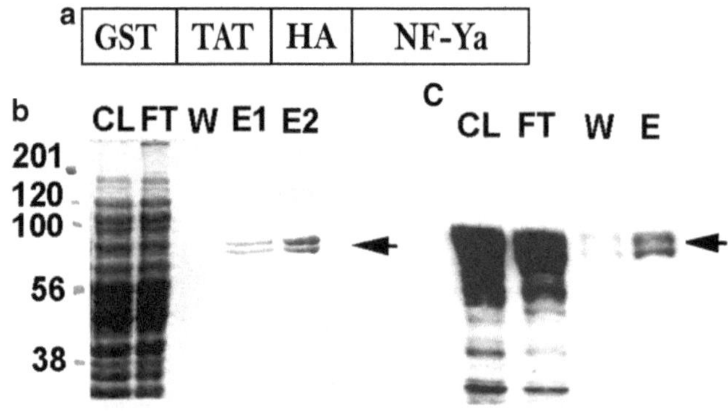

Fig. 1. *Purification of TAT-NF-Ya fusion protein.* (**a**) Human TAT-NF-Ya expression construct. A GST-TAT-HA-NF-Ya was designed by introducing SalI-TAT-HA-NF-Ya-NotI PCR fragment into pGEX-6p1 plasmid. The PCR product was verified by sequencing. The fusion protein was expressed using *E. coli* strain Rosetta(DE3)pLysS and induced for 3 h with 0.1 mM IPTG at 37°C. (**b**) Affinity purification of GST-TAT-HA-NF-Ya: Coomassie blue stained SDS gel and immunoblot (**c**) with anti-GST Ab. Arrow indicates fusion protein. *NI* noninduced culture, *CL* clear lysate, *FT* flow through, *W* wash, *E* eluate. (This figure is from part of Fig. 1 from our paper originally published in *Blood.* Domashenko AD, Danet-Desnoyers G, Aron A, Carroll MP, Emerson SG. TAT-mediated transduction of NF-Ya peptide induces the ex vivo proliferation and engraftment potential of human hematopoietic progenitor cells. *Blood.* 2010; 116(15):2676–2683. © the American Society of Hematology).

of the codons in the TAT-NF-Ya sequence are rare codons; therefore the bacterial strain from the Rosetta series, designed to supply tRNAs corresponding to rare codons, was used. Another major advantage of these strains is that they are deficient in the *ompT* and *lon* proteases, which are known to cause protein degradation during purification (25). The plasmid containing TAT-NF-Ya fusion protein construct was propagated in the bacterial host Rosetta(DE3) pLysS.

1. Inoculate 20 ml of LB broth supplemented with 50 μg/ml carbenicillin, 34 μg/ml chloramphenicol, and 0.5% glucose in a 200 ml flask with a single colony containing a recombinant plasmid. Grow the culture for 12–15 h at 25°C with vigorous shaking (see Note 1).
2. Inoculate 1 L of prewarmed LB medium supplemented with 50 μg/ml carbenicillin, 34 μg/ml chloramphenicol with 20 ml of the overnight culture and grow at 37°C with vigorous shaking to OD_{600} ~0.6–0.8.
3. Induce fusion protein expression by adding isopropyl-b-D-thiogalactoside (IPTG) to a final concentration of 0.1 mM and agitating culture for an additional 3 h at 37°C.
4. Centrifuge the culture at 6,000 × *g* for 15 min at 4°C.
5. Discard the supernatant, drain the pellet, and store at −80°C or proceed with protein purification.

3.1.2. Protein Purification

GST-TAT-NF-Ya protein is purified under native conditions to avoid a protein refolding step, which might reduce protein yield and physiological activity. GST binds to glutathione-based sepharose with high specificity, which allows rapid purification of GST-tagged proteins from cell lysates (see Note 2).

1. Resuspend cell pellet at 1:3 pellet/buffer ratio in ice-cold STE lysis buffer (150 mM NaCl; 1 mM EDTA; 10 mM Tris–HCl, pH 8.0) supplemented with protease inhibitor cocktail (1 ml of cocktail per 4 g of cell pellet).
2. Add to lysate: DTT to 5 mM, sarcosyl to 1.4%, and sonicate on ice using 15 s bursts at output power control 2 or 3 with a 15 s cooling between each burst until lysate is no longer viscous (see Note 3).
3. Centrifuge lysate at 20,000 × *g* for 25 min at 4°C to remove the cellular debris.
4. Transfer supernatant to a 50 ml conical tube, add 10% Triton X-100 and STE buffer to the final concentration of 2% for Triton X-100 and 0.7% for Sarcosyl. Rotate solution at room temperature for 30 min.
5. Use batch/column purification procedure. Add 2 ml of the 50% slurry of Glutathione Sepharose 4B to the lysate and incubate overnight at 4°C with rotation.
6. Sediment Glutathione Sepharose 4B matrix by centrifuging at 500 × *g* for 5 min. Carefully remove supernatant.
7. Wash matrix with ten bed volumes of cold PBS three times. Centrifuge at 500 × *g* for 5 min after each wash.
8. Using a pipette, transfer washed Glutathione Sepharose 4B with bound fusion protein to a disposable column.
9. Elute the fusion protein at 4°C and flow rate 0.5 ml/min by adding 1 ml of elution buffer (20 mM reduced glutathione; 150 mM NaCl; 50 mM Tris–HCl, pH 9.0) per 1 ml of beads volume. Monitor elution at 280 nm.
10. Dialyze collected protein fractions against PBS-20% glycerol at 4°C using Slide-A-Lyser dialysis cassettes.
11. Concentrate purified protein using Amicon Ultra-4 10K centrifugal filter device. Spin at 4000 × *g* in a swinging bucket rotor at 4°C until a desirable volume of protein is obtained (see Note 4).
12. Freeze protein in small aliquots and store at –80°C.
13. Monitor purification process by taking aliquots of samples at each step, including noninduced cell culture, cells after IPTG induction, cell lysate, flow-through fraction, wash samples, and protein eluate. Add to these aliquots Laemmli sample buffer and store at –20°C for SDS-PAGE.

3.2. Western Blotting

1. Separate samples collected during protein purification by SDS-PAGE (Fig. 1b): Mix 10 μl of each fraction or 10 μg of pure protein with equal volume of Laemmli buffer, boil for 5 min, cool on ice, and load on 10% Tris–Glycine polyacrylamide gel. Run two gels in the same apparatus (one for Coomassie staining, another one for western blotting) at 100–150 V for stacking and separating gel correspondingly. Use two protein markers: a pre-stained molecular weight marker to determine the endpoint of the electrophoresis, and non-stained marker for precise determination of molecular weight. Stain one gel with Simply Blue to determine protein purity.
2. Use the second gel for wet transfer onto nitrocellulose membrane. Run transfer at 100 V in 4°C cold room for 1 h with stirring. Stain the membrane with Ponceau S staining solution (0.1% Ponceau S, 5% acetic acid) to check for successful transfer and visualize protein markers. Mark the position of protein markers with Sharpie pen and destain membrane with water for 10 min.
3. Block nonspecific background binding of antibodies to the membrane by incubating the membrane in 5% nonfat dried milk, 0.1% Tween-20 in PBS (PBS-T) for 1 h at room temperature or overnight at 4°C with agitation. Rinse the membrane twice for 2 min each time in PBS-T.
4. Incubate the membrane with mouse anti-GST-HRP monoclonal antibody diluted 1:5,000 in PBS-T for 1 h at room temperature on a rocking platform.
5. Rinse the membrane with PBS-T and then wash for 15 min followed by 3 × 5 min washes at room temperature on an orbital shaker. Drain the excess wash buffer from the membrane, place on SaranWrap protein side up, load ECL Plus detection reagents to cover the surface of the membrane, and incubate for 5 min at room temperature.
6. Drain off detection reagent, wrap up the blot with SaranWrap, smooth out air bubbles.
7. In a dark room with safe light expose the membrane to autoradiography film for a suitable time. The results showing western blot analysis of the recombinant protein purification and stability are presented in Fig. 1c and Fig. 2a correspondingly.

3.3. TAT-NF-Ya Protein Stability

We found that TAT-NF-Ya is very sensitive to proteases present in culture serum, particularly in fetal bovine serum (FBS), and also in cells. To determine how protein stability is affected by culture conditions, we incubate cells (K562 or peripheral blood CD34$^+$ cells) with the fusion protein in a medium with or without serum up to 4 h followed by western blot analysis and densitometry (Fig. 2). After 15 min of incubation with the cells in the presence of FBS in

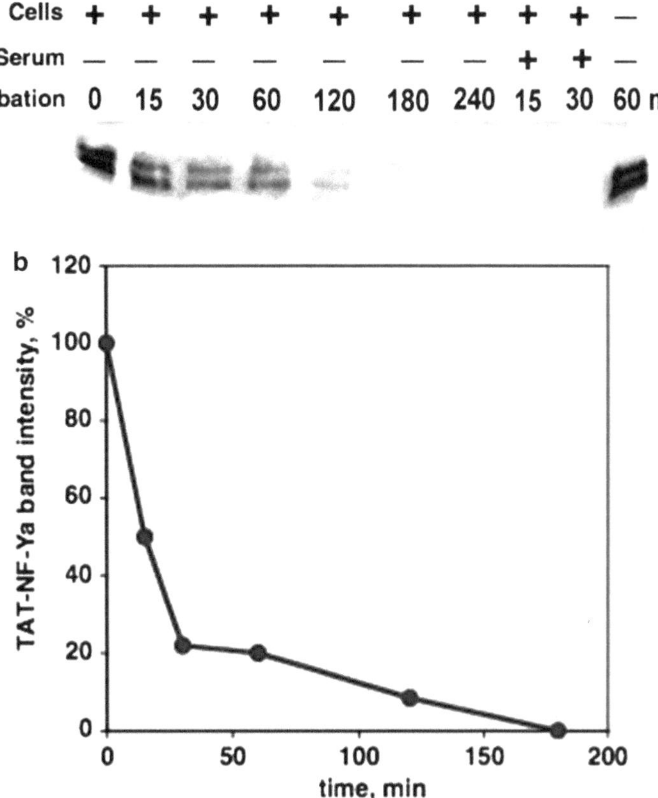

Fig. 2. *TAT-NF-Ya protein stability in cell culture.* (**a**) Western blot analysis with anti-GST antibody showing protein sensitivity to cell and serum proteases. (**b**) Densitometric quantitation of immunoreactive NF-Ya following exposure to K562 or CD34+ peripheral blood cells, in the presence or absence of fetal bovine serum. TAT-NF-Ya was stable in cell-free, serum-free medium, but the presence of serum in the culture causes complete disappearance of detectable NF-Ya within 15 min. Culture medium without serum inclusion of hematopoietic cells also leads to TAT-NF-Ya degradation, but at a slower rate. (This figure was originally published in *Blood.* Domashenko AD, Danet-Desnoyers G, Aron A, Carroll MP, Emerson SG. TAT-mediated transduction of NF-Ya peptide induces the ex vivo proliferation and engraftment potential of human hematopoietic progenitor cells. *Blood.* 2010;116(15):2676–2683. © the American Society of Hematology).

the medium, the TAT-NF-Ya fusion protein is completely degraded (Fig. 2a, b), suggesting the presence of proteases in serum. In culture medium without serum, the TAT-NF-Ya fusion protein is ~50% degraded in 15 min, and up to 75% degraded in 30 min when incubated with the cells. TAT-NF-ya fusion protein incubated in medium without cells at 37°C for 1 h does not show any sign of degradation (Fig. 2a, last right band), indicating that proteases produced by the cells also degrade the fusion protein.

1. Plate ~3×10^4 cells/well in a 96-well plate in Iscove modified Dulbecco medium (IMDM) with or without 10% FBS. Add 20 nM TAT-NF-Ya and incubate at 37°C in a CO_2 incubator for 15 min to 4 h.

2. Harvest cells at different time points by centrifugation at 800×*g*. Collect supernatants and separate proteins by 10% SDS-PAGE.
3. After transfer to nitrocellulose membrane, probe blot with anti-GST-HRP antibody (see Subheading 3.2).
4. Measure the protein bands' density by scanning densitometry of bands on Hyperfilm and quantify by Image Quant software.

3.4. Protein Delivery to the Cells Growing in Suspension

Analysis of TAT-NF-Ya protein stability in cell culture indicates that the optimal timing for fusion protein delivery is every 15 min for a total 1.5–2 h (see Note 5). TAT-NF-Ya intranuclear transport and functional activity can be assessed by measuring the changes in expression of its direct target *HOXB4* using quantitative RT-PCR. *HOXB4* mRNA levels increase up to 3.5-fold after treatment of peripheral blood CD34⁺ cells with TAT-NF-Ya fusion protein. Upregulation of *HOXB4* expression can be observed for 2 weeks following exposure to the protein (Fig. 3).

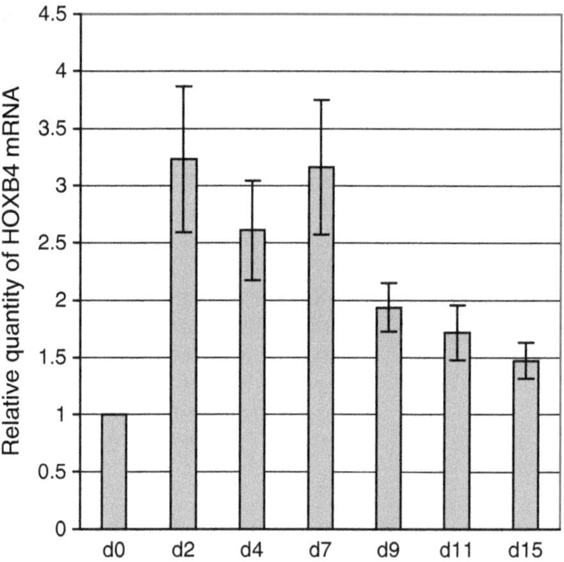

Fig. 3. *Induction of HoxB4 activity in peripheral blood (PB) CD34⁺ cells by TAT-NF-Ya transduction.* Time course of *HOXB4* mRNA induction following treatment with TAT-NF-Ya. Human PB CD34⁺ cells were treated with 60 nM TAT-NF-Ya plus 250 mM sucrose, then cultured with myeloid cytokines (see Subheading 3) from 2 to 15 days, and endogenous *HOXB4* mRNA was measured by quantitative PCR. Compared to CD34⁺ cells cultured in cytokines without TAT-NF-Ya, *HOXB4* mRNA increased 2.5–3-fold from day 2 to day 7 and remained elevated throughout the culture. (This figure is part of Fig. 5 from our paper originally published in *Blood.* Domashenko AD, Danet-Desnoyers G, Aron A, Carroll MP, Emerson SG. TAT-mediated transduction of NF-Ya peptide induces the ex vivo proliferation and engraftment potential of human hematopoietic progenitor cells. *Blood.* 2010;116(15): 2676–2683. © the American Society of Hematology).

1. Incubate 0.5–1 × 10^6 peripheral blood $CD34^+$ cells with TAT-NF-Ya in StemSpan H3000 serum-free medium supplemented with a cytokine cocktail of Flt3/Flk-2, SCF, TPO (100 ng/ml each), and IL-3, IL-6 (20 ng/ml each) in a 48-well tissue culture plate at 37°C in a 5% CO_2 incubator. Add aliquots of 60–70 nM TAT-NF-Ya fusion protein together with 250 mM sucrose every 15 min for a total 1.5 h (see Note 5).
2. Harvest cells, wash with PBS twice to remove non-internalized protein, and continue to grow cells in the medium described above (see step 1).
3. For validation of TAT-NF-Ya functional activity by quantitative RT-PCR, culture cells after protein application for at least 24 h (for time course cells were cultured for 2 week with the medium change every other day). Extract total RNA from cells with RNeasy mini kit and reverse transcribe 1 μg RNA from each sample using TaqMan Reverse Transcription Reagents kit according to manufacturer's protocol. Run quantitative PCR in 20 μL reaction volume containing reverse transcribed cDNA, TaqMan Universal PCR Master Mix (1× final concentration), and TaqMan primers/probe set (1× final concentration) for human *HOXB4* and *GAPDH* endogenous control gene with all samples in triplicate.
4. Calculate variations of gene expression using the ΔΔCt method.

3.5. Expansion of Hematopoietic Progenitors After TAT-NF-Ya Treatment

1. Use the same cell culture conditions as described above (see Subheading 3.4, step 1), applying TAT-NF-Ya in combination with the same set of cytokines and serum-free medium.
2. Grow human peripheral blood $CD34^+$ cells for 30 days, changing the medium every other day. Take aliquots of cell suspension, stain with trypan blue, and count viable cells (Fig. 4a).

3.6. Colony-Forming Cells Assay

Semisolid media, such as methylcellulose, supplemented with cytokines allows detection of the ability of cells to proliferate and differentiate into multipotential and/or lineage restricted progenitors.

1. Plate human peripheral blood $CD34^+$ cells in triplicate at 250 and 500 cells/35-mm dish with 1 ml of methylcellulose at day 4 and 9 after TAT-NF-Ya protein treatment and incubate at 37°C under 5% CO_2 in a humidified atmosphere.
2. Count individual colonies after 14 days for colony-forming unit erythroid (BFU-E), colony-forming unit granulocyte macrophage (CFU-GM), and after 21 days for colony-forming unit granulocyte, erythrocyte, macrophage, megacaryocyte (CFU-GEMM) progenitors (Fig. .4b).

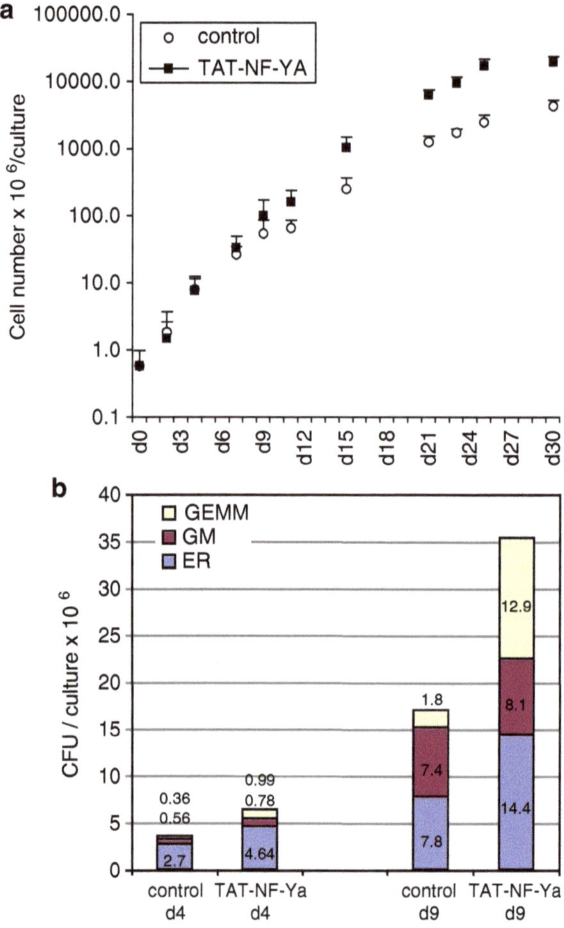

Fig. 4. Ex vivo expansion and colony forming potential of PB CD34+ cells after treatment with TAT-NF-Ya. Human PB CD34+ cells were treated with TAT-NF-Ya (or not treated) for 1.5 h and then cultured for 30 days in serum-free medium supplemented with Flt3/Flk-2, SCF, Tpo (100 ng/ml each), IL-3, and IL-6 (20 ng/ml each). Total mononuclear cells (**a**) were counted; the number of colony forming erythroid (BFU-E = Er), granulocyte-macrophage (CFU-GM), and erythroid/granulocyte/macrophage/megakaryocyte (CFU-GEMM)) were measured by methylcellulose assay (**b**). (This figure is part of Fig. 6 from our paper originally published in *Blood*. Domashenko AD, Danet-Desnoyers G, Aron A, Carroll MP, Emerson SG. TAT-mediated transduction of peptide induces the ex vivo proliferation and engraftment potential of human hematopoietic progenitor cells. *Blood*. 2010;116(15): 2676–2683. © the American Society of Hematology).

4. Notes

1. Bacterial culture growth temperature can influence recombinant protein expression level: lower temperature (20–25°C) and IPTG concentration (0.1 mM) reduces expression level of total protein which can accumulate as inclusion bodies, but increases the amount of soluble, native protein.

2. The GST tag is large (26 kD) and is more susceptible to degradation by proteases. Therefore purification of GST-proteins should be performed as quickly as possible with ice-cold reagents and protease inhibitors.
3. Sonicator should be equipped with a microtip. Immerse microtip up to 1/4 in. from the surface of solution. During sonication avoid foaming and oversonication which may cause denaturation of the fusion protein and or copurification of host proteins.
4. Do not overconcentrate protein as this may induce precipitation.
5. TAT-NF-Ya was found to be toxic to the cells: Incubation with human peripheral blood $CD34^{+}$ cells and cytokines in serum-free conditions at the optimal fusion protein concentrations 60–70 nM for 3 h or for 1.5 h, but at higher than optimal concentrations, showed significant cell death. The fact that the toxicity was from NF-Ya and not because of TAT peptide was confirmed by transduction of cells with TAT-ß-galactosidase (ß-Gal) fusion protein. Results showed that even twofold higher concentrations of TAT-ß-Gal for 4 h did not affect cell viability.

Acknowledgments

We are grateful to Steve Dowdy (University of California, San Diego) for kindly providing the pTAT-HA vector and Jiang Zhu (Shanghai Second Medical School, Shanghai, China) for the NF-Ya plasmid.

This study was supported by National Institute of Health (grant RO1-CA090833).

References

1. Hacein-Bey-Abina S, Von Kalle C, Schmidt M et al (2003) LMO2-associated clonal T cell proliferation in two patients after gene therapy for SCID-X1. Science 302:415–419
2. Zhang X, Beard B, Trobridge G et al (2008) High incidence of leukemia in large animals after stem cell gene therapy with a HOXB4-expressing retroviral vector. J Clin Invest 118:1502–1510
3. Gao X, Kim K, Liu D (2007) Nonviral gene delivery: what we know and what is next. AAPS J 9:E92–104
4. Heitz F, Morris M, Divita G (2009) Twenty years of cell-penetrating peptides: from molecular mechanisms to therapeutics. Br J Pharmacol 157:195–206
5. Frankel A, Pabo C (1988) Fingering too many proteins. Cell 53:675–675
6. Green M, Loewenstein P (1988) Autonomous functional domains of chemically synthesized human immunodeficiency virus tat trans-activator protein. Cell 55:1179–1188
7. Vivès E, Brodin P, Lebleu B (1997) A truncated HIV-1 Tat protein basic domain rapidly translocates through the plasma membrane and accumulates in the cell nucleus. J Biol Chem 272:16010–16017
8. Schwarze S, Ho A, Vocero-Akbani A, Dowdy S (1999) In vivo protein transduction: delivery of a biologically active protein into the mouse. Science 285:1569–1572

9. Csaszar E, Gavigan G, Ungrin M et al (2009) An automated system for delivery of an unstable transcription factor to hematopoietic stem cell cultures. Biotechnol Bioeng 103:402–412
10. Zhu J, Zhang Y, Joe G, Pompetti R, Emerson SG (2005) NF-Ya activates multiple hematopoietic stem cell (HSC) regulatory genes and promotes HSC self-renewal. Proc Natl Acad Sci USA 102:11728–11733
11. Mantovani R (1999) The molecular biology of the CCAAT-binding factor NF-Y. Gene 239:15–27
12. FitzGerald PC, Shlyakhtenko A, Mir AA, Vinson C (2004) Clustering of DNA sequences in human promoters. Genome Res 14:1562–1574
13. Marino-Ramirez L, Spouge JL, Kanga GC, Landsman D (2004) Statistical analysis of over-represented words in human promoter sequences. Nucleic Acids Res 32:949–958
14. Suzuki YR, Yamashita M, Shirota Y et al (2004) Large-scale collection and characterization of promoters of human and mouse genes. In Silico Biol 4:429–444
15. Marziali G, Perrotti E, Ilari R et al (1999) The activity of the CCAAT-box binding factor NF-Y is modulated through the regulated expression of its A subunit during monocyte to macrophage differentiation: regulation of tissue specific genes through a ubiquitous transcription factor. Blood 93:519–526
16. Hu Q, Maity SN (2000) Stable expression of a dominant negative mutant of CCAAT binding factor/NF-Y in mouse fibroblast cells resulting in retardation of cell growth and inhibition of transcription of various cellular genes. J Biol Chem 275:4435–4444
17. Fang X, Han H, Stamatoyannopoulos G, Li Q (2004) Developmentally specific role of the CCAAT box in regulation of human gamma-globin gene expression. J Biol Chem 279: 5444–5449
18. Huang DY, Kuo YY, Lai JS, Suzuki Y, Sugano S, Chang ZF (2004) GATA-1 and NF-Y cooperate to mediate erythroid-specific transcription of Gfi-1B gene. Nucleic Acids Res 32: 3935–3946
19. Tabe Y, Konopleva M, Contractor R et al (2006) Upregulation of MDR1 and induction of doxorubicin resistance by histone deacetylase inhibitor depsipeptide (FK228) and ATRA in acute promyelocytic leukemia cells. Blood 107:1546–1554
20. Campanero MR, Herrero A, Calvo V (2008) The histone deacetylase inhibitor trichostatin A induces GADD45c expression via Oct and NF-Y binding sites. Oncogene 27:1263–1272
21. Bhattacharya A, Deng JM, Zhang Z, Behringer R, de Crombrugghe B, Maity SN (2003) The B subunit of the CCAAT box binding transcription factor complex CBF/NF-Y is essential for early mouse development and cell proliferation. Cancer Res 63:8167–8172
22. Zhu J, Giannola DM, Zhang Y, Rivera AJ, Emerson SG (2003) NF-Y cooperates with USF1/2 to induce the hematopoietic expression of HOXB4. Blood 102:2420–2427
23. Domashenko AD, Danet-Desnoyers G, Aron A, Carroll MP, Emerson SG (2010) TAT-mediatied transduction of NF-Ya peptide induces the ex vivo proliferation and engraftment potential of human hemaopoietic progenitor cells. Blood 116:2676–2683
24. Brinkmann U, Mattes RE, Buckel P (1989) High level expression of recombinant genes in Escherichia coli is dependent on the availability of the dnaY gene product. Gene 85: 109–114
25. Grodberg J, Dunn JJ (1988) ompT encodes the Escherichia coli outer membrane protease that cleaves T7 RNA polymerase during purification. J Bacteriol 170:1245–1253

Chapter 24

Exploring the Link Between Human Embryonic Stem Cell Organization and Fate Using Tension-Calibrated Extracellular Matrix Functionalized Polyacrylamide Gels

Johnathon N. Lakins, Andrew R. Chin, and Valerie M. Weaver

Abstract

Human embryonic stem cell (hESc) lines are likely the in vitro equivalent of the pluripotent epiblast. hESc express high levels of the extracellular matrix (ECM) laminin integrin receptor $\alpha6\beta1$ and consequently can adhere robustly and be propagated in an undifferentiated state on tissue culture plastic coated with the laminin rich basement membrane preparation, Matrigel, even in the absence of supporting fibroblasts. Such cultures represent a critical step in the development of more defined feeder free cultures of hESc; a goal deemed necessary for regenerative medical applications and have been used as the starting point in some differentiation protocols. However, on standard non-deformable tissue culture plastic hESc either fail or inadequately develop the structural/morphological organization of the epiblast in vivo. By contrast, growth of hESc on appropriately defined mechanically deformable polyacrylamide substrates permits recapitulation of many of these in vivo features. These likely herald differences in the precise nature of the integration of signal transduction pathways from soluble morphogens and represent an unexplored variable in hESc (fate) state space. In this chapter we describe how to establish viable hESc colonies on these functionalized polyacrylamide gels. We suggest this strategy as a prospective in vitro model of the genetics, biochemistry, and cell biology of pre- and early-gastrulation stage human embryos and the permissive and instructive roles that cellular and substrate mechanics might play in early embryonic cell fate decisions. Such knowledge should inform regenerative medical applications aimed at enabling or improving the differentiation of specific cell types from embryonic or induced embryonic stem cells.

Key words: Embryonic stem cell, Early embryonic differentiation, Epiblast, Epithelial organization, Apical constriction, Gastrulation, Rho kinase (ROCK), Polyacrylamide substrates, Extracellular matrix, Mechanics, Visco elasticity, Substrate stiffness

1. Introduction

Human embryonic stem cells (hESc) and their differentiated progeny hold enormous potential for the regenerative treatment of human diseases. Understanding how ES cells maintain pluripotency and

Kimberly A. Mace and Kristin M. Braun (eds.), *Progenitor Cells: Methods and Protocols*, Methods in Molecular Biology, vol. 916, DOI 10.1007/978-1-61779-980-8_24, © Springer Science+Business Media, LLC 2012

differentiate under defined conditions is essential to these applications. Signaling networks, involving morphogens and transcription factors, are essential players but these are widely understood only in strictly (bio)chemical terms, especially by researchers studying cells in two-dimensional culture. Recent findings suggest that mechanical and topological cues may also play pivotal roles in directing stem cell fate and guiding morphogenic processes (1–3).

At a higher level of organization, developmental processes, which largely guide in vitro experiments, take place in three dimensions (3D) and require cells to temporally rearrange their positions with respect to each other in order to coordinate developmental signaling (4, 5). The subcellular systems that execute these morphogenetic movements are spatiotemporally organized, and are dynamically rearranged, via cell–cell and cell–matrix adhesion receptors that are coupled to contractile actomyosin networks (4, 6). Gastrulation is one such dramatic rearrangement in which constriction of apical actomyosin networks in the single layered pseudostratified epiblastic epithelium facilitates internalization and formation of the endodermal and mesodermal germ layers (5, 7). Recent work in *Drosophila* suggests that actomyosin contractility may be more than just a "passive" effector but might also feed back to play a permissive or even instructive role in gastrulation (8–10). An in vitro model system capable of faithfully mimicking the in vivo organization of the actomyosin and cell adhesion systems in the epiblastic epithelium could facilitate experimental testing of these effector, permissive, and/or instructive roles. The availability of such an epithelial system could also serve as a starting point to enhance the efficiency of deriving medically useful differentiated cell types.

Transcriptional and epigenetic profiling suggests that, unlike mouse ESc, hES cells are the in vitro equivalent of the pluripotent epiblast (11). Morphologically, hES cells grow as single layered colonies and exhibit features of an apical–basal polarized epithelia (12). This includes the development of lateral cell–cell contacts including markers of adherens junctions (AJ; E-cadherin) and tight junctions (TJ; ZO1) and the expression of relatively high levels of $\alpha6\beta1$ integrins that enable robust adhesion to basal laminins in the extracellular matrix (ECM) (13, 14). However, this structural organization is well developed only at high cell densities, on standard tissue culture plastic. Instead, when not limited to a small substrate area, hESc grown on a 2D substrate typically propagate as a loose network of spread cells that exhibit a low profile (vertical projection; $z \sim 3$–$5\ \mu$m, Fig. 1a) with reduced or undetectable laterally localized E-cadherin (unpublished observations). Nuclei are discoidal and flattened against the substrate. In addition, although supra nuclear (apical) transverse actin filaments are evident, the apical circumferential networks associated with apical constriction and gastrulation movements are less well developed and basal actin stress fibers are abundant (Fig. 1a).

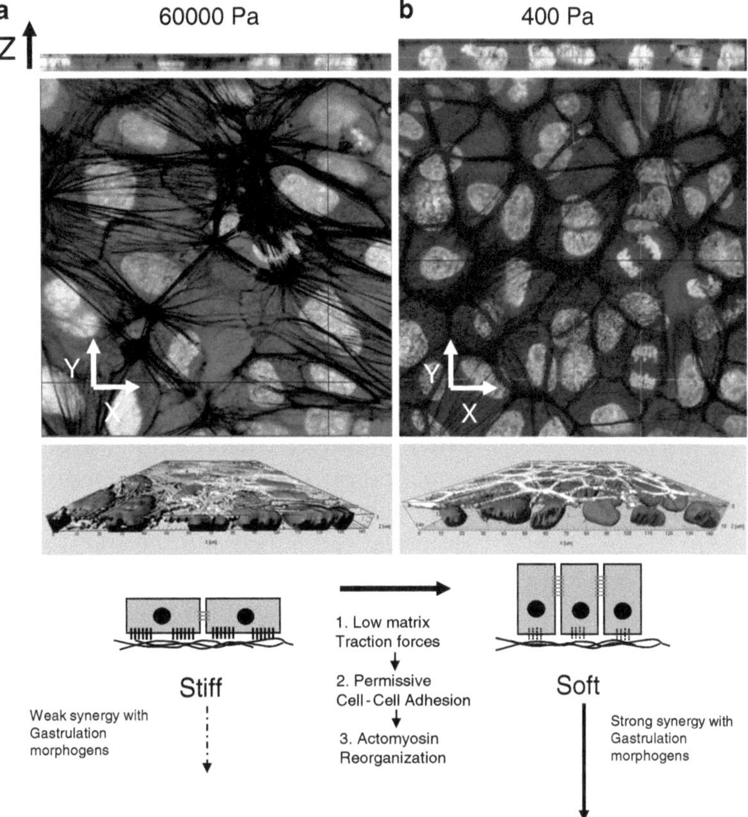

Fig. 1. Compliant substrates promote hESc apical–basal polarity. *Z*-stack confocal imaging of F-actin (Alexa Fluor 488 Phalloidin; *black*) and Nuclei (DAPI; *white*) in representative hESc colonies grown on stiff (**a**. *E*=60,000 Pa) and soft (**b**. *E*=400 Pa) Matrigel coupled polyacrylamide gels. *Top panel*; height (*Z*) profiles. *Upper middle panel*; *XY* plane overlaying *Z* slice at midpoint of nuclei (for nuclei) with *Z* slice near the apical surface (distal from gel) for F-actin. *Lower middle panel*; 3D rendering of confocal stacks using Imaris software with an end on top-down perspective (Nuclei; *Dark*, F-Actin, *Light*). Images show rounding and basal displacement of nuclei and apical (re)organization of F-actin in hESc grown on a soft substrate. Both the circumferential apical actin and transverse apical actin filaments characteristic of the epiblast in pre-gastrulation stage embryos are much better developed on the softer matrix. *Bottom panel*; Schema depicting the permissive influence of matrix compliance on the resulting hESc epithelial organization.

One major difference between tissues in vivo and standard 2D culture substrates is the rigidity of the substrate they grow on (15, 16). In contrast to the very rigid surfaces used to propagate hESc in standard cell culture, the epiblast develops on a much softer "substrate" (a basal lamina overlying another epithelial layer; the primitive endoderm (14). The primitive endoderm is orders of magnitude softer, with an Elastic modulus, E (in Pascals: Pa) likely to range between 10^2 and 10^4 versus $>10^9$ for standard tissue culture plastic (unpublished observations, (17)). Substrate rigidity has pleiotropic effects on adherent cells. At the single cell level, poorly deformable (stiff) substrates allow the development of high cell–matrix adhesion anchored actomyosin contractile tension promoting the assembly and maturation of integrin focal adhesions, stress fiber formation, and cell spreading through dynamic and reciprocal feedback mechanisms regulated by Rho-GTPase signaling and actin remodeling (3, 16, 18, 19). All of these effects are diminished, over a physiologically relevant range (for most soft tissues including muscle $E = 10^2–10^5$ Pa) as the substrate becomes increasingly deformable (16, 18). For groups of cells, such as epithelia, mechanically linked through cell–cell adhesions, substrate stiffness exerts permissive limits on the natural elaboration of these adhesions and thereby the overall tissue morphology (16, 20, 21). Figure 1b shows the result of growing hESc on a soft, Matrigel (reconstituted basement membrane) functionalized polyacrylamide (PA) gel. A more columnar epithelium develops naturally ($z \sim 10–12$ μm) when substrate deformation dependent mechanical feedback amplification of cell–matrix adhesions is sufficiently reduced. In consequence, nuclei are ovoid and basally displaced. E-cadherin AJ are better developed and the apical circumferential and transverse distribution of F-actin fibers more naturally resembles that of the epiblast in pre-gastrulation stage embryos. We and others have shown that morphological changes likely herald changes in the precise nature of signal integration from more classical soluble cytokines by a variety of mechanisms ranging from crosstalk and feedback with mechanoregulatory pathways to organizational states that affect signaling efficiency (3, 10, 16, 22–30). Ongoing work in our own lab suggests that morphological changes in hESc colonies are associated with biochemical and gene expression changes consistent with some form of permissive or instructive role in “gastrulation stage” germ layer differentiation (unpublished observations).

In this chapter we describe how to prepare adherent polyacrylamide surfaces of different elastic moduli that “tune” hESc cell–matrix mechanical interactions permitting varying degrees of cell–cell and polarized apical–basal epithelial organization. We describe a modification of an existing method for surface functionalization of PA gel with ECM molecules, which in our hands has proven to be versatile, consistent, scaleable, and inexpensive. And provide protocols by which to establish and study viable hESc colonies on such surfaces.

Our approach is to vary the ratio and amounts of acrylamide and bis-acrylamide to obtain PA gels with a range of physiologically relevant, reproducible elastic moduli. PA gels are first cast on glutaraldehyde modified circular coverglasses. We then modify the surface with an amine reactive functionality (acrylamido succinimidyl ester) linking to residual surface unsaturation using free radical photo-initiated polymerization. Finally proteins are coupled to the incorporated agent through side chain primary amines under mild conditions. In Subheading 3.1 we describe the synthesis of the acrylamido succinimidyl ester adapted by our lab for these gels. In Subheading 3.2 we describe glutaraldehyde modification of coverglasses, and in

Subheading 3.3 we describe preparation of protein coupled polyacrylamide gels on these glass supports. In Subheading 3.4 we provide protocols by which to establish viable hES cell colonies on ECM protein coupled PA gels. As described above hES cells express high levels of the laminin receptor α6/β1 integrin, adhere readily to tissue culture plates coated with the laminin rich basement membrane preparation, Matrigel, and onto which they can be maintained feeder free in a pluripotent state in the presence of fibroblast conditioned media (13). Accordingly, in our application the gels are functionalized with Matrigel and hESc are grown on them, feeder free in the presence of fibroblast conditioned medium. Alternately, more defined conditions can be provided by the use of purified laminin supplemented as above with fibroblast conditioned medium or with increasingly defined media.

2. Materials

2.1. Synthesis and Characterization of N6 (N-Succinimidyl Acrylamidohexanoic Acid)

1. 6-aminohexanoic acid (Sigma Cat# A2504). Store at +4°C.
2. Acryloyl chloride (Aldrich Cat #A24109). Handle with appropriate care. Very toxic (inhaled or ingested), corrosive, and flammable. Store at +4°C.
3. Calcium hydroxide: $CA(OH)_2$ (Sigma Cat# C7887). Store at room temperature.
4. Concentrated (12 N) HCl (Fisher Cat# A144S-500). Store at room temperature.
5. Cholorform (EMD Cat#3150). Store at room temperature.
6. Anhydrous 200 proof ethanol (Sigma Cat# 459836). Store at room temperature.
7. Anhydrous $Na_2(SO_4)$ (Fisher Cat# S421-3). Store at room temperature.
8. Activated charcoal, Norite A (USB Corp. Cat# 13365). Store at room temperature.
9. Diatomaceous earth (Sigma Cat# D3877). Store at room temperature.
10. *N*-hydroxysuccinimide (NHS) (Aldrich Cat# 130672). Store at +4°C.
11. 1-Ethyl-3-(3-dimethylaminopropyl)carbodiimide hydrochloride (EDAC) (Sigma Cat# E7750). Store at −20°C in a sealed secondary container with Dryerite.
12. Methylene chloride (Acros Organics Cat#AC32685).
13. Glacial acetic acid (Fisher Cat# A38S-500).
14. $NaHCO_3$ (Sigma Cat# S8875).

15. Anhydrous ethyl acetate (Aldrich Cat# 270989).
16. Drierite.
17. 4.7 and 9 cm circular glass fiber filters (GF/C; Whatmann, Cat#1822-025 and 1822-090).
18. 9 cm circular glass fiber filters (GF/F; Whatmann, Cat#1825-047).
19. 4.7 and 9 cm Büchner Funnels.
20. Side Arm Filter Flasks.
21. 1 l Separatory funnel.
22. Crystallization dish.
23. Rotary evaporator with several 500 ml rotary evaporation flasks.
24. Rubber policeman.
25. Nitrogen or argon gas cylinder.

2.2. Glutaraldehyde Activation of Coverglasses

1. 18 mm #1 coverglass circles (Fisher Cat# 12-545-100).
2. 50 mm #1 coverglass circles (ProSciTech, AUS Cat# G403).
3. 3-aminopropyltrimethoxysilane, 97% (Aldrich Cat #281778).
4. EM-grade 70% glutaraldehyde (Electron Microscopy Sciences Cat#16360) is purchased as 10 ml ampules and stored at –20°C (see Note 1).
5. Sonicating water bath (optional).

2.3. Preparation of Polyacrylamide Gels of Different Elastic Modulus Coupled to ECM Ligands

1. 40% (w/v) acrylamide in ultrapure water (BioRad Cat# 161-0140) (see Note 2). Acrylamide is toxic and should be handled with gloves.
2. 2% (w/v) bis-acrylamide in ultrapure water (BioRad Cat# 161-0142) (see Note 3). Bis-acrylamide is toxic and should be handled with gloves.
3. TEMED (BioRad Cat#161-0800). Store undiluted stock at +4°C (see Note 4).
4. 1% (w/v) ammonium persulfate or potassium persulfate (see Note 5).
5. 10× PBS. Autoclave to sterilize and store at room temperature.
6. 500 mM HEPES, pH 6. Sterilize through 0.22 μm filter. Store at +4°C.
7. 100 mM HEPES, 100 mM NaCl, pH 8. Sterilize through 0.22 μm filter. Store at +4°C.
8. 50 mM HEPES, 100 glycine, pH 8. Sterilize through 0.22 μm filter. Store at +4°C.
9. 0.9% (w/v) NaCl.
10. Absolute ethanol (200 proof).

11. 3% (w/v) Irgacure-2959 (Ciba-Geigy) solution in EtOH (see Note 6).
12. "N6" *N*-succinimidyl acrylamidohexanoic acid (see Note 7).
13. Rain•x (original) can be obtained from any automotive parts store. An alternative from scientific distributors is Enviro-Coat (EMD Cat# 4190).
14. Matrigel™ (8–10 mg/ml total protein) is obtained from BD Biosciences (Cat#354234) and is stored at –80° C (see Note 8).
15. Spectroline EN-180 (or equivalent) UV source with medium (306 nm, preferable) or long (365 nm) peak wavelength output.

2.4. Culture of Human Embryonic Stem Cells on Matrigel Coupled Polyacrylamide Gels

1. Fibroblast growth media is prepared on a weekly to bimonthly basis and consists of 18% M199 media (Gibco Cat# 12340-030), 72% high glucose DMEM (Gibco Cat# 11965-092), 10% fetal bovine serum (Hyclone Cat# SH30071.03), and supplemented to 1 mM with L-glutamine (Gibco Cat# 25030-081).
2. hESc media is prepared as needed on a weekly to bimonthly basis without bFGF and consists of 80% Knockout DMEM (Gibco Cat# 10829-018), 20% Knockout Serum Replacement (Gibco Cat# 10828-028), 1× MEM Non-Essential Amino Acids (100× Stock Gibco Cat#11140-050), 1 mM L-glutamine supplement (Gibco Cat# 25030-081), and 100 μM β-mercaptoethanol (6.9 μl of pure β-mercaptoethanol in 0.993 ml ultra-pure sterile water gives a 1,000× stock) (see Note 9).
3. Recombinant human bFGF is purchased from Millipore (Cat#GF003) (see Note 10).
4. Y-27632 hydrochloride is purchased from Cayman Chemical Company (Cat# 10005583) (see Note 11).
5. 0.1% (w/v) gelatin. Type A: porcine skin gelatin (300 Bloom, Sigma Cat#G-2500) (see Note 12).
6. 0.25% Trypsin EDTA (Gibco Cat# 25200-114).
7. Passage 3 CF1 mouse embryonic fibroblasts are obtained from Millipore (Cat# PMEF-CFL) expanded to passage 5 and then γ-irradiated to render them mitotically inactive (see Note 13).

3. Methods

3.1. Synthesis and Characterization of N6 (N-succinimidyl acrylamidohexanoic acid)

The two step synthesis of N6 is essentially as described by (31). In the first step a 6-acrylamidohexanoic acid intermediate is generated by condensation of acryloyl chloride and 6-aminohexanoic acid. After isolation this precursor undergoes dehydration with NHS ester via a carbodiimide generated O-acylisourea intermediate to yield the final product. N6 is then extracted and purified by crystallization. All steps should be performed in an appropriately vented fume hood by properly attired lab personnel, and all waste products should be collected and disposed of according to local guidelines.

3.1.1. Synthesis of 6-Acrylamidohexanoic Acid Intermediate

1. Place a clean 250 ml Erlenmeyer flask containing 150 ml of ultrapure water and a magnetic stir bar in an ice water slurry. With stirring allow the water to cool to 0°C before proceeding.
2. Add 10 g of 6-aminohexanoic acid with stirring and allow to completely dissolve before proceeding.
3. Add 10 g of $Ca(OH)_2$ while continuing to stir. Stir for 5 min. Some $Ca(OH)_2$ will not dissolve (see Note 14).
4. Add 7 ml of acryloyl chloride in 1 ml increments with 2 min intervals between additions while stirring.
5. 2 min following the last addition filter the reaction to remove suspended $Ca(OH)_2$ and return the clarified solution back to the ice water slurry. Filtration is easily accomplished into a filter flask under house vacuum using a 9 cm GF/C filter fitted in a Büchner funnel (see Note 15).
6. 6-Acrylamidohexanoate is neutralized and thereby precipitated from solution by acidification to pH 2.6 using concentrated hydrochloric acid. While stirring filtrate vigorously in an ice water slurry and monitoring with a pH electrode, add concentrated HCl slowly to lower pH to 2.6 (see Note 16).
7. Collect 6-acrylamidohexanoic acid intermediate by filtration using a 9 cm GF/C filter fitted in a Büchner Funnel (see Note 17).

3.1.2. Drying 6-Acrylamidohexanoic Acid

Residual water that can reduce the yield of N6 is removed from the 6-acrylamidohexanoic acid intermediate. From this point all work should be scrupulously anhydrous, using certified anhydrous solvents and thoroughly clean and dry glassware.

1. Prepare a diatomaceous earth filter in a Büchner Funnel with a 9 cm ultra fine GF/F glass fiber filter. Prepare a 50% slurry of diatomaceous earth in anhydrous ethanol and pour over the filter while drawing a vacuum, stopping when the plug reaches a depth of about 1 in.
2. Remove the filter, empty the collection flask, and rinse it out to clean out any fines that remain in the flask. Rinse the plug, under vacuum with some ethanol.
3. Repeat step 2 until no fines are evident in the collection flask.
4. Empty the collection flask and clean and dry thoroughly before proceeding or exchange for a clean and dry collection flask (see Note 18).

5. Transfer 6-acrylamidohexanoic acid precipitate to a clean dry flask and add 100 ml of a 1:1 (vol/vol) mix of chloroform and anhydrous ethanol and mix by swirling in an ice water slurry to facilitate dissolution.
6. Add 130 g of $Na_2(SO_4)$ and 6 g of activated charcoal and mix by swirling for 2–3 min to dehydrate (see Note 19).
7. Draw the slurry from step 6 through the diatomaceous earth filter under vacuum to remove $Na_2(SO_4)$ and activated charcoal and collect the filtrate (see Note 20).
8. Transfer the filtrate to a round bottom rotary evaporating flask.
9. Using a vacuum rotary evaporator, reduce volume to about 20–30 ml (see Note 21) and return the flask to ice water slurry.

3.1.3. Synthesis of N6

1. Dilute concentrated 6-acrylamidohexanoic acid into 200 ml cold anhydrous ethanol and mix thoroughly to ensure homogenous dispersion of the viscous concentrate before proceeding.
2. Place a clean dry stir bar into flask and with constant medium stirring add 8.7 g of NHS and 17 g of EDAC.
3. Stir continuously on ice water slurry for 2–4 h (see Note 22).
4. Using a vacuum rotary evaporator reduce volume to about 20–30 ml (see again Note 21).

3.1.4. Purification of N6 by Extraction and Crystallization

1. Prepare 500 ml of ice cold saturated Na_2HCO_3 in ultrapure water and have on hand 400 ml of ice cold ultrapure water (see Note 23).
2. Disperse product syrup from step 4 (Subheading 3.1.3) into 200 ml of methylene chloride and add 0.2 ml of glacial acetic acid.
3. Add a clean dry magnetic stir bar and while immersed in a ice water slurry stir for 10 min.
4. Transfer solution from step 3 into a 1 l separatory funnel.
5. Slowly and carefully add 200 ml of ice cold aqueous saturated Na_2HCO_3 to separatory funnel. Mix gently by slow back and forth inversion and then stand upright to allow phases to separate (see Note 24).
6. Open funnel stopcock and collect lower methylene chloride phase in a clean dry filter flask from step 2. Discard upper aqueous phase according to local guidelines.
7. Transfer methylene chloride solution back into separatory funnel and repeat extraction with saturated Na_2HCO_3.
8. Repeat steps 5–8 with two equal volumes of ice cold ultrapure water.
9. After the last extraction collect the methylene chloride phase and while immersed in a ice water slurry, add 65 g of $Na_2(SO_4)$

to dehydrate. Swirl to mix for 2–3 min and then filter under vacuum through a 9 cm GF/C filter in a 9 cm Büchner Funnel into a clean dry filter flask.

10. Transfer methylene chloride to a new clean and dry round bottomed rotary evaporator flask.
11. Using a vacuum rotary evaporator, reduce volume to about 20–30 ml (see again Note 21).
12. Add 200 ml of anhydrous ethyl acetate and mix well to disperse product from step 11.
13. Using a vacuum rotary evaporator reduce volume to about 100 ml (see Note 25).
14. Transfer to a clean dry crystallization dish; add some seed crystals if available and cover with Parafilm.
15. Place dish at +4°C for 1–2 h and then a freezer overnight. Slow cooling facilitates crystallization.
16. Recover crystals by vacuum filtration through a 4.7 cm GF/C filter in a Büchner Funnel and collect the filtrate in a clean dry filter flask (see Note 26)
17. Collect crystals by scraping, distribute small 0.5 g masses in separate small volume glass containers purged with an inert gas (nitrogen or argon) and store at –20°C in a sealed secondary container containing a small amount of Drierite (see Note 27).
18. We use low resolution electrospray ionization time-of-flight mass spectrometry to provide an analysis of the purity of the final product. In our hands strong peaks are observed at m/z 305 (one N6 plus sodium) and 587 (two N6 plus sodium) (see Note 28).

3.2. Glutaraldehyde Activation of Coverglasses

Coverglasses are washed to remove surface contaminants during manufacture, then derivatized with glutaraldehyde to provide a surface capable of covalently coupling to polyacrylamide during polymerization. This creates a bond between the gel and glass support durable enough to withstand the mechanical perturbations to which PA gels are subjected during normal workup and cell culture. The procedure is adapted from Damljanovi et al. (32).

1. Disperse coverglasses into a Petri dish (typically 100–150 mm depending on the number and size) containing 20 ml (100 mm) or 40 ml (150 mm) of 0.2 M HCl and incubate overnight at room temperature with gentle agitation on an orbital shaker (see Note 29).
2. The next day decant acid wash and wash coverglasses four or five times with 20 or 40 ml of ultrapure water using gentle agitation on an orbital shaker.
3. Replace last water wash with 20 or 40 ml of 0.1 N NaOH and incubate 1 h at room temperature with gentle agitation on an orbital shaker.

4. Decant base wash and wash coverglasses four or five times with 20 or 40 ml of ultrapure water using gentle agitation on an orbital shaker.
5. Replace last water wash with 20 or 40 ml as appropriate of 1:100 dilution of 3-aminopropyltrimethoxysilane in ultrapure water and incubate 1 h at room temperature with gentle agitation on an orbital shaker.
6. Decant diluted 3-aminopropyltrimethoxysilane waste and collect as required by local disposal ordinances. Wash coverglasses exhaustively in repeated 20 or 40 ml changes of ultrapure water with gentle agitation on an orbital shaker for 5–10 min each (see Note 30).
7. Replace last water wash with 20 or 40 ml, as appropriate of a 1:140 dilution of 70% glutaraldehyde in PBS and incubate 1 h at room temperature with gentle agitation on an orbital shaker.
8. Decant diluted glutaraldehyde waste and collect as required by local disposal ordinances. Wash coverglasses exhaustively in repeated 20 or 40 ml changes of best quality water with gentle agitation on an orbital shaker for 5–10 min each to remove the residuum.
9. Remove coverglasses individually using forceps and place on a lint free absorbent such as Kimwipe and allow to dry overnight. Transfer treated coverglasses to a clean dry sealed container for storage (see Note 31).

3.3. Preparation of Polyacrylamide Gels of Different Elastic Moduli Coupled to ECM Ligands

To maintain sterility all solutions are either filter sterilized or autoclaved, and all instruments that will handle gels are likewise sterilized by autoclaving or immersion in 70% ethanol. All operations are performed in a sterile flow hood. The method is divided into two parts: In the first part the polyacrylamide gel base that determines the compliance of the surface is prepared, and in the second part the surface is functionalized with Matrigel (or another protein). The latter is conceptually similar to (33) with points of surface unsaturation in the initial polymerization serving as points of attachment of an *N*-succidimidyl ester acrylamido derivative using UV photoinitiation. The first and second parts can be separated in time but once having started the functionalization, one must proceed as quickly and efficiently as possible to minimize hydrolysis of the succinimidyl ester. Two gel formats are commonly used: 18 mm circular coverglasses for single cell level experiments and 50 mm circular coverglasses for biochemical analysis (western blot etc.) (see Note 32).

3.3.1. Preparation of PA Gel Base

1. Gel formulations that yield polyacrylamide gels with defined Elastic Moduli, E, are given in Table 1, as well as the quantity of each stock required per ml of solution (see Note 33). Gel solutions are prepared by mixing all components except persulfate. For a circular 18 mm and 50 mm diameter coverglasses,

Table 1
Compositions of acrylamide and bis-acrylamide that yield polyacrylamide gels with a defined Elastic Modulus, E

Gel compositions/mechanical properties

Elastic modulus (Pa)	Final % acrylamide	Final % Bis-acrylamide	water (ml)	40% Acrylamide (ml)	2% Bis-acrylamide (ml)	10× PBS (ml)	1% TEMED (ml)	1% Potassium persulfate (ml)	Total Vol.(ml)
140	3	0.04	0.705	0.075	0.02	0.1	0.05	0.05	1
400	3	0.05	0.7	0.075	0.025	0.1	0.05	0.05	1
1,050	3	0.1	0.675	0.075	0.05	0.1	0.05	0.05	1
2,700	7.5	0.035	0.595	0.1875	0.0175	0.1	0.05	0.05	1
4,000	7.5	0.05	0.5875	0.1875	0.025	0.1	0.05	0.05	1
6,000	7.5	0.07	0.5775	0.1875	0.035	0.1	0.05	0.05	1
13,800	7.5	0.15	0.5375	0.1875	0.075	0.1	0.05	0.05	1
22,000	7.5	0.25	0.4875	0.1875	0.125	0.1	0.05	0.05	1
60,000	10	0.5	0.3	0.25	0.25	0.1	0.05	0.05	1

Elastic moduli were verified using an Atomic Force Microscope by fitting force curves versus surface indentation using a 5 μm spherical bead to a Hertz contact model. Physiological E is based on the assumption that hESc in vivo grow on the surface of another epithelial cell layer (the primitive endoderm) and that the cortical rheological properties of this layer will set this value. Although this value has not been measured, values from other cell types including undifferentiated hESc (unpublished observations) suggest that no cell is likely to deviate out of a range we have conservatively set between 10^2 and 10^4 Pa. hESc exhibit the highest rates of growth and lowest basal apoptosis for $1 \times 10^3 < E < 1 \times 10^4$ Pa (unpublished observations), consistent with optimal adaptation for pre-gastrulation expansion of the epiblast in vivo. For $E \leq 10^3$ cell spreading, growth and survival are progressively slower, and actomysosin driven colony compaction is most dramatic following removal of Y-27632. PA gels in this range support the most columnar colony organizations

25 and 195 μl, respectively, should yield a gel ~100 μm thick (see Note 34).

2. Mix solutions and degas in a vacuum flask or a bubble vacuum chamber under house vacuum for 30 min. The 1% persulfate stock solution should also be degassed simultaneously but separately in a flask/chamber (see Note 35).
3. Prepare some nonstick top coverglasses for gel casting. Apply Rain-x (or Enviro-coat) to 18-mm coverglasses with a dipped cotton swab or by smearing a small amount of solution (1–2 μl or 10 μl for 18 and 50 mm diameter respectively) over the surface with a pipet tip. Allow the Rain-x to dry (~5 min) and then buff coverglasses with a lint free absorbent such as Kimwipe or optical lens paper. Residual lint/dust can be blown off the coverglasses with compressed air.
4. Apply a small volume of water to a hard flat surface and seal a piece of Parafilm™ to the surface by gently sponging water out. Arrange the glutaraldehyde activated coverglasses on this piece of Parafilm (see Note 36).
5. After degassing gel solutions, working one gel solution at a time, add the appropriate volume of 1% persulfate, mix up and down gently by pipeting and immediately pipet appropriate volume of gel solution on to the glutaraldehyde activtated coverglass (25 and 195 μl for 18 and 50 mm coverglasses, respectively). Overlay top coverglass Rain-x side down on this bead of solution (see Note 37).
6. Gels are allowed to polymerize for about 60 min at room temperature (see Note 38).
7. Working one at a time and using a razor blade, insert the edge between the top and bottom coverglasses and lever off the top coverglass. Using a pair of tweezers take hold of the gel and bottom coverglass and dip briefly in 70% ethanol to surface sterilize the gel.
8. Transfer gels to a sterile Petri dish with sufficient sterile ultrapure water to keep gels hydrated. After a brief 10 min wash on an orbital shaker, gels can be processed further and functionalized with ECM ligand or can be stored indefinitely wrapped in Parafilm at +4°C.

3.3.2. Surface Functionalization with Matrigel

1. Table 2 gives the quantities of stock solutions of 0.2% bis-acrylamide, 3% Irgacure 2959, 0.5 M HEPES pH 6.0, 0.3% N6 (see Note 39), and ultrapure water per ml of solution for surface functionalization (see Note 40). 20–40 μl per 18 mm gel (200–400 μl per 50 mm) is typically used. Assemble water, 0.5 M HEPES pH 6.0 and 0.2% bis-acrylamide in a sterile 1.5 ml centrifuge tube and mix.
2. Degas in a vacuum flask or a bubble vacuum chamber under house vacuum for 30 min.

Table 2
Solution composition for photoinitiated incorporation of N6 on PA gels

Water (ml)	0.5 M HEPES NaOH pH = 6.0 (ml)	0.2% Bis-acrylamide (ml)	3% Irgacure 2,959 in ethanol (ml)	0.3% N6 in 50% ethanol (ml)	Total Vol.(ml)
0.48	0.1	0.05	0.03	0.34	1

3. While degassing weigh out the required amount of N6. Also prepare one Rain-x coated top coverglass per gel as described above. Just before degassing time is complete, dissolve N6 in 50% ethanol (in ultrapure water) (see Note 41).
4. After completion of degassing add required volume of 0.3% N6 and 3% Irgacure 2959 mix.
5. Working one at a time pick up gels with forceps and wick off as much surface fluid as possible using an absorbent paper towel (Kimwipe™). Apply 20–40 μl of N6 solution for 18 mm gel (200–400 μl for 50 mm) to the gel surface and using the flat edge of a 1 ml pipet tip move the fluid back and forth over the surface of the gel to ensure even coverage (see Note 42).
6. Overlay gel with a Rain-x coated top coverglass evenly spreading N6 and taking care not to trap any air (see Note 43) and place gel under a UV source to activate photoinitiator (see Note 44). Typical conditions are 1–2 in. from the UV source for 5–10 min with a medium wavelength UV source (Spectroline EN-180, 306 nm peak) or 15–20 min with a longer wavelength source (Spectroline EN-180, 365 nm peak) (see Note 45).
7. Using the edge of a razor blade lever off the top coverglass and immediately transfer gels into a sterile Petri dish on ice containing cold (0 to +4°C) sterile ultrapure water (see Note 46).
8. Wash gels gently for 5 min on ice using an orbital shaker.
9. Decant or aspirate water and replace with sufficient ice cold sterile 0.9% NaCl to cover gels.
10. Wash gels gently for 5 min on ice using an orbital shaker.
11. Repeat steps 9 and 10 one more time.
12. Transfer gels individually to a sterile dish or well on ice containing ice cold sterile 0.9% NaCl. The dish or well should be of adequate size to contain the coverglass with some but not a lot of additional room (see Note 47). 18 mm coverglasses can be placed gel side up in individual wells of a 12-well plate. 50 mm coverglasses can conveniently be placed into a 60 mm tissue culture dish.
13. Aspirate 0.9% NaCl.
14. Dilute Matrigel to 250 μg/ml in ice cold 100 mM HEPES, 100 mM NaCl, pH 8. Add 1 ml and 4 ml for 18 and 50 mm

diameter gels, respectively, and incubate with gentle agitation on ice in a covered container on an orbital shaker for 1–2 h. Transfer plates/dishes with gels to 0 to +4°C and leave overnight (see Note 48).

15. The next day aspirate Matrigel solution and replace with 0.5 ml and 2 ml for 18 and 50 mm diameter gels, respectively, of sterile 50 mM HEPES, 100 mM glycine, pH 8.0, and incubate for a minimum of 1 h at room temperature (see Note 49).
16. Aspirate glycine solution and replace with 1 and 4 ml for 18 and 50 mm diameter gels, respectively, of sterile PBS.
17. Agitate gels gently on an orbital shaker at room temperature for 10 min.
18. Aspirate PBS and repeat steps 16–18 four or more times (see Note 50).
19. Aspirate PBS and replace with 1 and 4 ml for 18 and 50 mm diameter gels, respectively, of some basal protein free media (we use DMEM/F12) containing antibiotics and transfer to a 37°C tissue culture incubator and incubate overnight (see Note 51).
20. Media is aspirated just prior to plating cells.

3.4. Culture of Human Embryonic Stem Cells on Matrigel Coupled Polyacrylamide Gels

Before starting, see Note 52.

3.4.1. Stock hESc Cultures

1. Primary stock cultures of hESc are maintained on γ-irradiated passage 5 CF-1 fibroblasts: 2×10^5 cells/35 mm gelatin coated dish (see Note 53) plated 1–2 days prior to passage (see Note 54) in 2.5 ml of hESc media with 10 ng/bFGF. Cultures are maintained at 37°C in a humid atmosphere in a 5% CO_2 and 95% air atmosphere. Selected colonies showing minimal morphological evidence of differentiation are collected and passaged by mechanically scraping (see Note 55). Stock cultures are fed daily by removing one-half of the spent media and replacing it with an equivalent volume of fresh media (see Note 56). Stock cultures are usually passaged every 4–5 days.
2. Secondary feeder free Matrigel stock cultures are prepared from feeder dependent hESc stock cultures and generated with each split of the feeder dependent stocks (see Note 57). Typically 50–60 medium sized colonies from the feeder dependent stock are mechanically picked and collected in the spent media of the culture, transferred to a sterile 15 ml polypropylene tube, titurated a few times with a 1 ml pipet to further break up colony fragments, and one-fourth of suspension transferred to a new sterile 15 ml polypropylene tube. Fresh hESc media with 10 ng/ml bFGF is added to a final volume of

2.5 ml in each tube and the 1:4 split is replated into a new 35 mm fibroblast stock plate while the 3:4 split is plated on to a 35 mm Matrigel coated dish (see Note 58). Secondary Matrigel stock cultures are fed daily by removing one-half of the spent media and replacing it with an equivalent volume of fibroblast conditioned medium (see Note 59) for 5–7 days prior to harvesting mechanically or enzymatically (trypsin, dispase, or collagenase) for plating on to PA gels. Typical yields of cells from a single 35 mm dish when passaged by trypsinization as described below are 2×10^5 to 5×10^5 cells.

3.4.2. Plating and Growing Single hES Cells Derived by Trypsinization on PA Gels

1. Aspirate medium from hESc growing on secondary 35 mm feeder free Matrigel coated stock dish and wash monolayer gently twice with 1 ml each wash of sterile PBS. Aspirate last wash.
2. Wash monolayer once with 0.4 ml of 0.25% trypsin-EDTA. Aspirate from plate (see Note 60). Add 0.4 ml of 0.25% trypsin. Quickly add 0.4 μl of 10 mM Y-27632 directly to the trypsin in the dish and incubate cells at 37°C for 10–15 min (see Note 61).
3. Titurate cells in trypsin solution relatively vigorously using a 200 μl pipettor to break up stubborn clumps of cells. Add 0.8 ml of fibroblast medium (see Note 62). Titurate a few times with a 1 ml pipettor and transfer cells to sterile 1.5 ml centrifuge tube.
4. Place tube in open top of a 15 ml polyproplene tube in swinging bucket low rpm centrifuge and spin $500\times g$ for 2 min at room temperature.
5. Aspirate supernatant to as low a residual volume as possible (see Note 63) taking care not to disturb cell pellet. Using a 10 μl pipet tip suction fitted on a Pasteur pipet, low vacuum aspirator obtains very fine control of aspiration.
6. To the cell pellet add 1 ml of hESc media with 10 ng/ml bFGF. Add 1 μl of 10 mM Y-27632 and resuspend cells well by titurating using a 1 ml pipettor.
7. Count 10 μl of cell suspension with hemacytometer. Expect between 20 and 50 cells per field corresponding to yields of 2×10^5 to 5×10^5 cells total depending on the final density when using a 35 mm secondary feeder free Matrigel stock culture prepared as described above.
8. Transfer an aliquot of your cell stock corresponding to the number of cells determined by the desired culture volume and density to fresh hESc media with a final concentration of 10 ng/ml bFGF and 20 μM Y-27632. Typically we plate about 2,500 cells/cm^2 of surface area in 0.25–0.3 ml media/cm^2 (see Notes 64 and 65).

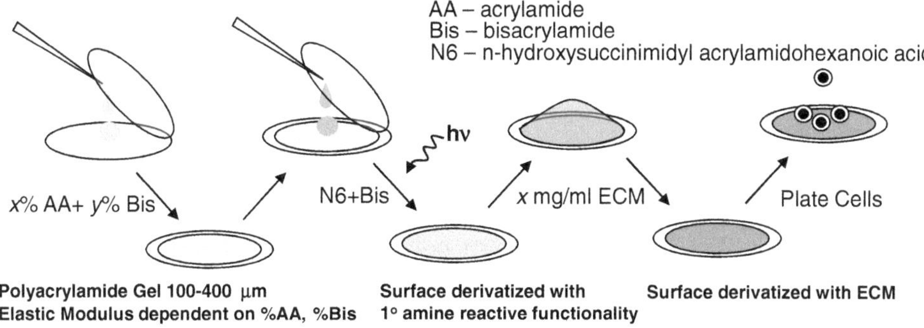

Fig. 2. Overview of protocol.

9. Return culture to tissue culture incubator. Cells should attach to PA gels rapidly and homogeneously within 30 min to 1 h (see Note 66).
10. Cells on PA gels are fed daily by carefully removing ~½ of the media and replacing with the same volume of fibroblast conditioned medium supplemented with 6 ng/ml bFGF (see Note 67). For the first feeding, the medium is supplemented with 20 μM Y-27632. For subsequent feedings, the medium is not supplemented with Y-27632 to enable stepwise removal of the inhibitor (see Note 68).
11. Depending on the compliance of the PA gel, 6–10 days are typically required for growth from single cells at these densities to reasonable sized colonies (see Note 69). An overview of the entire method is presented in Fig. 2.

4. Notes

1. As needed a single vial is opened and transferred to a small amber screw cap bottle for easier access and stored at –20°C. The reagent appears to be usable in this application for at least 6 months to 1 year.
2. Sterilize through 0.22 μm filter and store protected from light at +4 °C. Dilute an aliquot 1:4 into sterile ultrapure water to obtain a 10% (w/v) stock.
3. Sterilize through 0.22 μm filter and store protected from light at +4°C. Dilute an aliquot 1:10 into sterile ultrapure water to obtain a 0.2% (w/v) stock.
4. Prepare a fresh dilution 1 in 100 into sterile ultrapure water prior to each use as a working stock.
5. Dissolve 10 mg in 1 ml sterile ultrapure water and use soon after preparing. It is important to make this fresh each time as

even low temperature stored solutions become less effective in catalyzing polyacrylamide polymerization particularly when the gels are thin. Changes in polymerization translate into irreproducibility in the elastic modulus of the resulting gel. Note that the solid stocks stored at room temperature can also decompose over time.

6. Dissolution may be aided using a sonicating water bath. Store at 4°C protected from the light. This solution appears to be stable for at least a month or more.
7. N6 is sensitive to hydrolysis so it is ideally distributed in small bottles flushed with argon or other inert gas prior to sealing. Small bottles should be stored in a secondary sealed container with Drierite at −20°C. Warm secondary container to room temperature before opening to avoid condensation of atmospheric water. The alternative compound to N6 is acrylic acid NHS ester (Sigma Cat# A8060).
8. Matrigel is thawed overnight on ice in a sealed Styrofoam container at +4°C and stored in the same. Thawed Matrigel will gel if allowed to warm up. To maintain temperature at 0°C, the ice is periodically replenished as needed and during use is maintained on ice at all times. Accidentally gelled Matrigel (an expensive reagent) can be depolymerized by incubation overnight in a (table) salt–ice bath with enough salt to obtain a temperature of ~−5°C (by freezing point depression). Temperatures much lower than this will result in freezing without depolymerization.
9. When culturing hESc on PA gels, sterility may be maintained by addition of the following antibiotics: 100 units/ml Penicillin G, 100 μg/ml Streptomycin sulfate, 50 μg/ml Gentamycin, and 250 μg/ml Fungizone. 1 μl of 10 μg/ml bFGF and up to 2 μl 10 mM Y-27632 are added per ml of media as needed during routine feeding or plating.
10. 50 μg of lyophilized protein is dissolved on ice in the bottle in 1 ml of ice cold sterile filtered 0.2% (w/v) Bovine Serum Albumin (Cohn Fraction V, Sigma Cat#A-9647) in PBS for at least 1 h. This is then transferred to a sterile 15 ml polypropylene centrifuge tube containing 4 ml of 0.2% (w/v) BSA/PBS on ice to give a final bFGF concentration of 10 μg/ml. The solution is then sterile filtered through a 13 mm diameter 0.22 μm low protein binding PVDF syringe filter (Fisher Cat# 09-720-3) and aliquoted at 50 μl in each of up to 100 sterile 1.5 ml polypropylene centrifuge tubes, quick frozen on pulverized dry ice, and stored for 1 year or more at −80°C. Aliquots are thawed as needed (typically every week or so) and maintained at +4°C. To help maintain biostability, bFGF is always kept on ice when used and added to cultures directly as needed rather than making up large volumes of hESc medium containing bFGF.

11. 1 mg is dissolved in the bottle in 338 μl of absolute ethanol to give a stock concentration of 10 mM and transferred to an opaque (dark) 1.5 ml Eppendorf and stored at −20°C. In this form the compound appears to be stable for months. To help maintain activity, stock is always kept on ice when used and added to cultures only as needed rather than making up large volumes of medium.
12. Gelatin is simultaneously dissolved and sterilized by autoclaving at 0.1% (w/v) in ultrapure water and stored at +4°C.
13. Other means of mitotically inactivating fibroblasts (e.g., Mitomycin-C treatment) or of purchasing them already mitotically inactive are also possible, but if one has access to a γ-irradiator this is more economical. A single thawed vial containing $5–6 \times 10^6$ cells is plated in five 150 cm^2 gelatin coated tissue culture dishes in 20 ml each of fibroblast medium and allowed to grow in a tissue culture incubator at 37°C with 5% CO_2 for 3–4 days. This is considered passage 4. Cells are then trypsinized and replated in 20–25 150 cm^2 gelatin coated tissue culture dishes and allowed to continue to grow for 3 days. Media is then replaced and on the 4th day the cells (passage 5) are harvested by trypsinization, centrifuged and collected as a final suspension in 40–50 ml of ice cold fibroblast medium in a 50 polypropylene centrifuge tube (maintained on ice until introduced into the irradiator), and treated (at room temperature without warming up the medium) with a total of 3,200–4,000 rads (average dose of ~200–250 rads/min) in a γ-irradiator. Cells (typical yield of $\sim 50–80 \times 10^6$) are then divided into two equal aliquots for two different freezing densities and centrifuged at $\sim 500 \times g$ for 5 min to pellet (room temperature). One aliquot is resuspended at 1.2×10^6 cells/ml in ice cold freezing medium (10% DMSO/90% FBS) and 1 ml aliquoted per freezing vial on ice (20–25 vials). These are used for preparing fibroblast conditioned medium. The other aliquot is resuspended at 0.4×10^6 cells/ml in freezing media and 1 ml aliquoted per freezing vial (80–100 vials). These are used for propagating stock cultures of hESc on fibroblast feeder layers. Vials are slow frozen overnight at −80°C in a sealed Styrofoam container and then transferred within a day or two to liquid N_2 for long term storage. In the typical culture scales used in our lab, this stock of fibroblasts can last up to 1 year.
14. Solution is saturated in $Ca(OH)_2$ and some will remain undissolved, giving the solution a milky white appearance. $Ca(OH)_2$ is used to adjust the pH of the reaction to about 12 and to neutralize HCl generated as a byproduct of the condensation reaction thereby maintaining this high pH.
15. Filtered solution should have a translucent yellow color.

16. A dense slurry of white crystals results from this precipitation. A slow rate of addition is important to minimize rapid (catastrophic) precipitation which can generate a too-thick slurry that blocks the stirrer and/or the pH electrode and renders subsequent filtration difficult. HCl should be allowed to mix thoroughly between each addition. A rate of stirring sufficient to generate a significant vortex will facilitate this mixing. Approximately 9.5–10 ml of 12 N HCl is required to reach a pH of 2.6. Very little precipitation will occur until pH begins to drop below 5 (~2.5–3 ml of 12 N HCl) so up to that point the rate of addition may be somewhat more aggressive, however past this point the rate of addition should slow considerably (dropwise). The pH may plateau above 2.6 if the electrode becomes clogged or precipitate prevents adequate mixing.
17. If the slurry is very thick it may be difficult to pour. In this case it may be necessary to help transfer the solid to the filter using a spatula. Scraping the flask can maximize this yield or residual precipitate can be transferred by swirling the flask with a small amount of ice cold acidified (pH 2.6) ultrapure water. Moving the product around on the filter with a spatula for a few minutes while under vacuum can help remove excess water which will facilitate subsequent dehydration of the product. Wet weight yields are typically 20–30 g.
18. Prepare diatomaceous filter just before proceeding with step 5.
19. Small amounts of gas may be produced during this step. If the flask is stoppered be sure to relieve pressure build up by frequent venting during this process.
20. If particulates are evident in the filtrate these should be removed by filtration through a second diatomaceous earth filter before proceeding with solvent evaporation.
21. This evaporation removes most of the more volatile chloroform and reduces the volume of the remaining ethanol. The product is sensitive to excessive heating so a lower pressure, lower temperature evaporation is desirable. Under moderate vacuum a room temperature evaporation should suffice. After chloroform is removed the temperature may need to be raised to 40–45°C to facilitate further evaporation of the ethanol. At all stages of evaporation, aim for conditions of temperature and pressure that produce a gentle boil. A slightly viscous "syrup" is the desired endpoint. Marking the flask at 20–30 ml can help in determination of this point.
22. After a couple of hours color should change from yellow/brown and slightly opaque to a milder hue becoming more translucent.
23. Solubility of Na_2HCO_3 is 7 g per 100 g of water at 0°C. Add Na_2HCO_3 in excess and stir while immersed in an ice water

slurry until solubility limit is reached (~1 h) then let undissolved Na_2HCO_3 settle under gravity and carefully decant avoiding transfer of particulates during extractions.

24. Rapid mixing of methylene chloride during this and subsequent aqueous extractions can lead to formation of a stubborn emulsion in the aqueous phase. This emulsion contains the desired N6 product and can lead to substantial product losses.
25. Place a mark on the rotary evaporator flask at the 100 ml mark in order to more accurately gauge concentration to this point.
26. Run vacuum for 15–20 min in order to evaporate residual ethyl acetate away from crystals.
27. The highest yield of product is often from this first crystallization. The filtrate from step 16 can be reduced by one-half again by rotary evaporation and the crystallization procedure repeated to give a second "crop" and again for a third crop. It is a good idea to keep crops separate and analyze these separately until purity has been established before combining and aliquoting.
28. Extraction combined with a single crystallization usually obtains relatively pure preparations of N6 free from unreacted precursors and side products. However, recrystallization can on occasion rescue lower purity preparations. To recrystallize, the product is redissolved in 200 ml pre-warmed (60°C) anhydrous ethyl acetate. Stubborn redissolution can be facilitated using a sonicating water bath. After filtration of insoluble particulates, crystallization is achieved as described above.
29. Mechanically separate coverglasses that may have stuck together in order to ensure that all surfaces have efficient access to washes and chemical activation steps. A large number of coverglasses may be activated at once but when activating large numbers of 18 mm coverglasses or for large 50 mm coverglasses this is more effectively achieved in 150 mm petri dishes. As best as possible attempt to maintain the separateness of coverglasses in all subsequent steps by separating those that may have stuck together by surface tension. Cleaning during acid and base washes can be facilitated by putting coverglasses in a 100 mm diameter × 80 mm high flat bottomed glass petri dish and immersing the base in a sonicating water bath.
30. If excess residual 3-aminopropyltimethoxysilane is not removed, it will react subsequently with glutaraldehyde to yield a stubborn, translucent, milky white deposit on coverglasses that although without effect on the preparation of PA gels can compromise their optical clarity. Extensive water washes prior to glutaraldehyde treatment ensures efficient removal of this residuum.
31. Glutaraldehyde activated coverglasses stored at room temperature in this manner remain effective for this application for months. Some care should be taken to minimize touching coverglasses

or getting dust on them. However, oils from an accidental fingerprint can be wiped off with a KimWipe moistened with ethanol without apparent effect. One of us in particular favors handling coverglasses with ungloved hands through much of the workup for finer motor control.

32. Presented here is only the most easily applied format for preparation of PA gels. The final product is a gel attached to a coverglass moving relatively freely in the bottom of another plastic bottomed vessel. For some applications this is adequate. However, for some aspects of gel preparation and post-cell culture processing and for live cell microscopy in particular it is desirable to have the coverglass form the bottom of the culture well and for coverglass and gel to be fixed in position. A simple solution is to cast the gel in the center of a MatTek™ dish. Such dishes can be inexpensively prepared in the lab. We bore a 22 mm hole (using a regular 7/8″ wood spade bit using a drill press or with practice a hand drill) in the center of a 35 mm TC dish (or wells of a 6 well plate). After cleaning off burrs with sandpaper or file, a coverslip (25 mm square, glutaraldehyde activated) is glued to the outside of this using UV cured (15 min Spectroline EN-180, 365 nm peak long wavelength UV lamp) Norland Optical Adhesive #68 and then the gel (18 mm) is cast in the center of this. This allows imaging with high magnification/high NA objectives in inverted microscopes with full range of access to the gel surface.

33. If preparing small volumes of low percentage acrylamide and/or bis-acrylamide gels, greater accuracy can be obtained by using a stock of 10% acrylamide (1:4 dilution of 40% stock in ultrapure water) and 0.2% bis-acrylamide (1:10 dilution of 2% stock in ultrapure water).

34. We usually cast gels that are ~100 μm in thickness. According to Elasticity Theory, ~100 μm has been suggested to be a minimum theoretical thickness in order that the mechanical deformations exerted by surface bound cells are not propagated to the glass support (the cells "sense" the stiffness of the glass support). Engler et al. have tested this assumption experimentally and suggested that gels as thin as 20 μm can be used (34) and might under certain experimental circumstances be desirable (e.g., live cell microscopy). For other sizes of coverglasses, increase the volume of gel solution, scaling with the surface area, to obtain a 100 μm thick gel.

35. Removal of oxygen is required for efficient and reproducible polymerization.

36. Glutaraldehyde activated coverglasses can be sterilized either by autoclaving or by brief immersion in 70% ethanol and air drying in sterile flow hood.

37. No special care need be taken to ensure that top and bottom coverglasses are aligned. Surface tension will pull them into register. However, take care not to trap air bubbles as the gel will not polymerize up to ~0.5–1 mm around these bubbles leaving bare glass.
38. Polymerization is frequently accompanied by radial contraction and some dehydration around the edges (~1 mm from the edge). This often renders the edges of the gel unpredictable in terms of mechanical properties and cell plating. We typically make our observations only on cells at least this far from the edges.
39. As described the commercially available "N2": acrylic acid NHS ester (Sigma A8060) is a relatively inexpensive alternative to N6. The difference between N2 and N6 is a seven atom spacer which might ultimately affect presentation of some ligands for cell surface receptors but this has not been compared. We have done some experiments to show that N2 can be substituted for N6 in the conjugation of protein ligands to polyacrylamide gels but have done less to optimize the exact concentrations to use. A recommended starting point is to use it in the same molar ratio as N6. We suggest trying 0.06% N2 and 0.01% bisacrylamide (see Note 40).
40. Table 2 gives the most widely used conditions in our laboratory for the preparation of functionalized polyacrylamide gels for a variety of ligands and cell types. We have tried various concentrations/ratios of N6 and bisacrylamide in the surface conjugation (from 0.01–1% N6 and 0% bisacrylamide or 1:1–10:1 N6 to bisacrylamide by weight). In addition, we have used smaller and larger volumes per gel and varied the concentration of photoinitiator (from 0.1 to 0.2% final), and UV sources with long (365 nm) and medium (306 nm) wavelength peak outputs. All of these factors affect the nature of surface copolymerization in roughly predictable ways. However, recent feedback from other labs attempting to employ this method suggest there may be operator to operator differences that impact factors we may only partially understand. Some general guidelines (component ranges) and observations can be given but some lab to lab optimization of these factors may be necessary. In initial experiments using high concentrations of N6 and bisacrylamide, a thick copolymeric meshwork developed on the gels. The underlying gel may be partially screened if these are dense, and anisotropic discrete distributions of ligand along this fibrous mat can affect the development of cell–matrix adhesions in unpredictable ways. Lowering the bisacrylamide concentration especially reduced this significantly and at a ratio of 10:1 N6 to bisacrylamide by weight and total final concentration of 0.1% N6, gels with only sparse (at worst in our hands) focal ("dot like") copolymeric deposits developed (visible

under low power phase contrast microscopy). Using higher per gel volumes, higher concentrations of photoinitiator and/or shorter wavelength UV sources appears to increase the prevalence of these focal deposits. Therefore, variation in actual N6 amount (see below), purity, UV source power, residual oxygenation, etc., may account for some of the reported lab to lab variability. At present anecdotal observations suggest that somewhat reduced hESc spreading, growth, and survival may be related to the presence of these focal deposits. A working hypothesis is that high densities of ECM ligand in foci may prevent dynamic rearrangements of the integrins and cytoskeleton required for efficient cell spreading (effectively locking them in) and/or possibly activate weak cytotoxic responses in a manner analogous to the neurotoxic effects of contacted focal β-amyloid plaques on neurons. Recently (as of this submission), as per gel volumes of 50 μl (for 18 mm diameter gels), and using 0.2% final Irgacure 2959, with a 365 nm UV source, total final concentrations of 0.01–0.03% N6 (10:1 ratio of N6 to bisacrylamide) yielded gels with no focal deposits which yet supported rapid hESc adhesion and spreading and long term growth and survival. We have yet to establish that these conditions are reproducibly more "optimal" and so have not modified the original formulations given in Table 2. However, for researchers that have difficulty with gels prepared under these conditions, we recommend the following optimization: while maintaining by weight ratios of 10:1 N6 to bisacrylamide, vary total concentrations of N6 from 0.1 to 0.01% and final concentrations of photoinitiator from 0.1 to 0.2%. Particular attention should be paid to obtaining accuracy and reproducibility in weighing out and preparing N6 stocks as very little (0.1–1 mg) is typically called for and experimental errors on analytical balances for these weights are potentially significant. Larger excesses of N6 stocks may need to be made in order to ensure accuracy. To access the lower concentrations of N6 (0.01%) will necessitate serially diluting higher concentration stocks in 50% ethanol to obtain a workable concentration. In optimizing, researchers should monitor the production of focal deposits by phase contrast microscopy and correlate the results with the kinetics and extent of hESc binding, spreading, and long term growth and survival.

41. N6 is soluble in 50% ethanol at a final concentration of 0.3% but dissolution may be slow dependent on crystal size. Dissolution is aided considerably by using a sonicating water bath, and if one is available it is recommended that it be used. Succinimidyl esters are sensitive to hydrolysis. It is recommended that dissolution of N6 be delayed until required and that once dissolved one should be prepared to work quickly and efficiently to addition of the protein ligand to be coupled.

42. Adequately working the N6 solution by this back and forth motion across the gel appears to be important for achieving efficient and uniform surface functionalization. A thin surface layer of residual fluid on the surface of the gel might provide a "barrier" to access of the N6 to the gel surface and this can be overcome by adequate mixing using the above method. The flat edge of the 1 ml pipet tip can rest on the gel surface under its own weight while the investigator's hand, with a loose grip, moves it back and forth. Gels, even soft gels, will not be damaged, unless one applies downward force to the tip.
43. To remove entrapped air work the edge of a razor blade between the two coverglasses and lever the two apart, then allow them to settle together gently. The bubble(s) will often be extruded along with the liquid front. Failure to remove air bubbles will result in failure to functionalize the corresponding underlying gel surface.
44. If the number of gels to be processed is relatively small, all gels can be prepared for activation and then UV activated batchwise. For larger numbers, one of us prefers to process gels from application of N6 to UV activation to water wash (Subheading 3.3.2, steps 5–7) serially and then batchwise from this point.
45. There appears to be some variability in the efficacy of the long wavelength (365 nm) bulbs. Our laboratory first used longer wavelength bulbs in earlier experiments designed to photoencapsulate live cells in photoactivatable hydrogels, because they are less damaging to cells. However shorter wavelengths are likely to be more effective in activating Irgacure 2959 and bulb to bulb variation in manufacture may account for variable outputs at these shorter wavelengths when using the longer wavelength sources. Although we have less long term experience using it, the medium wavelength bulb may be the better choice for PA gel functionalization. One should, however, also pay close attention the effects this choice may have on the copolymerization reaction as described under Note 40.
46. All subsequent steps until completion of coupling of Matrigel are performed on ice. The cold temperature and mildly acidic pH of most water sources slows hydrolysis of succinimidyl esters. At no step between functionalization of the gel surface with N6 and completion of the coupling of Matrigel should the gels be exposed to amine containing buffers such as Tris as these will react more readily with the succinimidyl ester than water. Cold temperatures also minimize polymerization of Matrigel.
47. Other vessel formats for the reaction of activated gels with protein ligands are possible including static incubation of the protein ligand as a large drop pooled on the gel (minimizing the volume in cases where the protein ligand is rare and/or expensive).

Placing gels in "just-right" sized wells or dishes allows protein functionalization to culture of cells to be performed in the same vessel with a minimum volume of ECM ligand and without requiring further transfers.

48. 100 mM HEPES/100 mM NaCl pH 8.0 was optimized for coupling of Matrigel. Higher ionic strength and low temperature minimize polymerization of Matrigel allowing monomers to react as monomers with the gels surface. The same buffer seems to work well for Fibronectin and Collagen I. 50 mM HEPES pH 8.0 has been used successfully with Fibronectin and PBS can conveniently be used with Collgen I. We have quantitatively analyzed the conjugation of Matrigel, Collagen I, and Fibronectin to PA gels prepared by this method. All exhibit saturation coupling with one-half maximal coupling at 15–20 μg/ml and near saturation coupling at 100–150 μg/ml. More ligand appears to be able to be coupled to softer gels than stiff gels at the same concentration at the lower end of this range. However, using mammary epithelial cells as a model cell, conjugations using 20 μg/ml saturates some cellular responses (spread cell area/traction force per unit area; unpublished observations). Therefore, above about 20 μg/ml, although the gel has more "capacity" for ligand, this probably does not further affect the cellular response and use of high concentrations equalizes the disparity in surface coupling between softer and stiffer gels. For this reason, for consistency with earlier work, and because hESc are particularly sensitive in the unadhered state we have maintained the higher (possibly excessive) concentrations of Matrigel described in this method.
49. Glycine in this buffer provides a high concentration of a primary amine that will react with any unreacted succinimidyl ester.
50. Extensive washing is required to remove all traces of cytotoxic reagents associated with PA gel preparation. If cells initially attach robustly to gels but subsequently grow slowly or not at all and develop other morphological signs of cytotoxicity during prolonged culture, failure to adequately remove these reagents should be suspected and more extensive washing employed. Gels can be stored for at least a couple of weeks at 0 to +4°C at this point.
51. This provides an additional wash and soaks/impregnates the gel with medium. Antibiotics that we routinely use include 100 units/ml Penicillin G, 100 μg/ml Streptomycin sulfate, 50 μg/ml Gentamycin, and 250 μg/ml Fungizone. An overnight incubation in medium containing these antibiotics may sterilize the gels or if any microorganisms grow in spite of the presence of these antibiotics gels can be discarded before plating cells. A 2 or 3 day incubation can provide even stronger assurances against microbial contamination if the antibiotics

are only partially cytostatic. A similar incubation with hESc medium containing antibiotics prior to plating cells may also be helpful for their plating and growth. In this case we leave out bFGF and Y-27632 to reduce cost.

52. Both hESc colony fragments (mechanically or enzymatically generated) and single cells generated by trypsinization can be plated on PA gels functionalized with Matrigel. Plating colony fragments is faster than intitiating cultures from single cells as colony fragments are more resistant to cell death while unattached to matrix and have a shorter latency in reaching a desirable colony density (see Note 17). As a result such cultures generally require shorter periods of time of Y-27632 treatment (up to 3 days for the softest PA gels). However it is generally difficult to reproducibly control the seeding density. It is assumed that the reader is familiar with culturing hESc as colony fragments. As a result we include only the protocol for culture of tryptically generated single cells.

53. Tissue culture plates are coated using sufficient volume of 0.1% (w/v) gelatin to adequately cover the plate bottom (~0.1–0.2 ml/cm^2) and incubated either overnight at +4°C or for a minimum of 1 h at 37°C. Plates can be stored indefinitely at +4°C as long as they remain fully hydrated. This is aided by wrapping dishes in Parafilm but evaporation and condensation on the plate lid even at +4°C will eventually dehydrate the plate bottom, so we generally make only the number of plates needed within 1 week. Prior to use gelatin is aspirated and plates washed once with an equivalent volume of PBS.

54. Fibroblasts that have been plated longer than 2 days should not be used as recipients for newly passaged hESc. In our hands these "older" fibroblast cultures cause a high proportion of hESc colonies to differentiate around the colony margins. The responsible factor is probably directly or indirectly an insoluble fibronectin rich ECM that is progressively elaborated by these "older" cultures as such fibroblast cultures can be "reset" by trypsinization and replating in a new gelatin coated dish. Similarly significantly higher densities of fibroblast feeders also appear to promote the differentiation of hESc possibly by a similar mechanism. Prior to plating hESc on new fibroblast stock plates fibroblast feeders are washed with 1 ml of hESc medium without bFGF to remove residual serum.

55. We favor mechanical passaging as opposed to bulk enzymatic passaging as it leads to more uniformly undifferentiated stock cultures, is easier to generate feeder free Matrigel stock cultures, and when using the picking tool described below can be executed relatively quickly with high cell yields. To mechanically passage cells we do not favor drawn out glass cutting Pasteur pipets as these tend to dig into the plastic dish surface,

break easily, and result in greater cell damage and are unsuitable for picking large numbers of colonies in a reasonable time. Rather 10 μl plastic pipet tips are cut with a razor blade to generate a bevel at the end and mounted by friction on a 1 ml plastic pipet tip for a hand holder. The bevel is oriented upright and the sharp point is used to scrape colonies off the plastic surface while monitoring the process under a dissecting microscope. With some practice 50–100 colonies can be "bulk" passaged by this means in the same time that it takes to passage the same culture by enzymatic means.

56. We favor a daily regimen in which ½ volume spent media is removed and replace with ½ volume of fresh media as it appears at least anectodotally to better support dedifferentiated cultures presumptively by maintaining a more consistent level of undefined fibroblast secreted factors. This is possible in the early times after passaging (typically for the first 4–5 days) as the cultures have not yet reached densities at which each day would significantly reduce nutrients while producing inhibitory concentrations of waste products. However, one should daily monitor densities replacing all of the spent media if the media becomes significantly acidic overnight.
57. This secondary stock is required only to ensure complete removal of fibroblast feeders. One could maintain a feeder free Matrigel stock culture continuously rather than simultaneously maintaining feeder dependent stocks. We prefer to maintain a feeder free Matrigel stock for no more than one passage from the feeder dependent stock culture for greater consistency in maintaining undifferentiated stocks.
58. Matrigel coated dishes or activated coverglasses are prepared by diluting 1:100 (final protein concentration of ~80–100 μg/ml) in ice cold DMEM/F12 (or any protein free basal medium) using sufficient volume to cover adequately the plate bottom (~0.1–0.2 ml/cm^2) and incubated overnight at +4°C. Plates can be stored indefinitely at +4°C as long as they remain fully hydrated. This is aided by wrapping dishes in Parafilm but evaporation and condensation on the plate lid even at +4°C will eventually dehydrate the plate bottom so we generally make only the number of plates needed within 1 week. Prior to use solution is aspirated and plates washed once with an equivalent volume of PBS.
59. Up to 90 ml of fibroblast conditioned medium may be prepared from a culture in a single 60 mm tissue culture dish. One freezing vial of 1.2×10^6 passage 5 γ-irradiated CF1 fibroblasts is thawed and plated on a 60 mm gelatin coated tissue culture dish in 4 ml of fibroblast medium. The following day after cells have been plated, the medium is aspirated and the monolayer washed once quickly with 1 ml of DMEM/F12 (or other non

protein containing basal medium) to wash away traces of FBS in fibroblast medium and replaced with 9 ml of hESc media containing 4 ng/ml of bFGF and incubated in a tissue culture incubator at 37°C with 5% CO_2. This medium is collected 24 h later and replaced with the same repeating for a total of 10 days. Medium is collected in a single 100 ml medium bottle and stored at +4°C between collections. Conditioned medium may be stored at 4°C if it will be used within 1–3 weeks or if not used may be stored in smaller aliquots for long periods of time at −80°C. Conditioned medium when used for hESc culture is supplemented to a final concentration of 6 ng/ml with fresh bFGF.

60. Prepare to aspirate quickly as cells will begin to detach very quickly and if one waits too long there is a danger that a lot of cells will be aspirated away.

61. Y-27632 does not have to be preincubated with hESc prior to trypsinization as in (35). Y-27632 rapidly crosses the cell membrane and it is only when cells are released from all attachments that ROCK dependent cytoplasmic blebbing is activated ((36) and see Note 17). We add it during trypsinization rather than post-trypsinization as relatively long periods of trypsinization seem to be required to generate an adequate single cell suspension.

62. Fibroblast conditioned media is a convenient source of 10% FBS used to inactivate trypsin. Serum also facilitates formation of a well-formed cell pellet by "breaking up" a stubborn fibrous matrix that appears to be released from these cultures and preventing nonspecific cell binding to the tube walls. Subsequent wash steps remove traces of FBS that might result in differentiation of hESc.

63. The use of a 1.5 ml centrifuge tube and a swinging bucket rotor produces well-formed, well-defined cell pellets at the tube bottom that can be aspirated to very small residual volumes (10–20 μl) when using the aspirator described.

64. Surface area refers to the total surface area available for cell binding. This is the well or dish volume in which the gel has been placed not merely the surface area of the gel.

65. Adhesion, spreading, survival, and proliferation are significantly slower/lower on PA gels of $E \leq 1,050$ Pa introducing a lag in the culture growth relative to much stiffer gels. Plating at twice the density on $E = 1,050$ Pa and three to four times the density on $E = 140$ or 400 Pa gels can compensate somewhat for this lag.

66. Cells should become immobilized (to a mild mechanical disturbance such as tapping) homogeneously over the gel surface within this time frame and may even show evidence of protrusive activity on the stiffer substrates. After overnight incubation

most cells should be attached uniformly and spread to progressively larger extents over the range $E = 140–60,000$ Pa. These are the best indications that the gel surfaces have been adequately prepared. Failure of cells to bind quickly and/or uneven overnight plating and spreading suggest that an insufficient ligand density has been attained.

67. hESc, particularly on softer PA gels and shortly (1–2 days) after plating, are less well adhered than Matrigel coated tissue culture plastic. Maintaining one-half of the medium on the cells shields them from fluid shear forces that develop during feeding. Addition of new medium should however be as gentle as possible.

68. Y-27632, a small molecule inhibitor of Rho Kinase (ROCK) permits formation of colonies of hESC initiated from single cells by inhibiting hESc cell death (35). Strong, Y-27632 inhibitable, activation of actomyosin dependent cytoplasmic blebbing on removal of hESc from some (cell–matrix) or all (cell–cell and cell–matrix) adhesive interactions on the free surfaces of hESc suggests that ROCK is strongly activated in such cells and under negative feedback regulation from adhesive interactions ((36) and unpublished observations). Long term growth and survival of hESc colony fragments diminishes progressively with the softness of the PA gel and is most acute for $E < 1,000$ Pa despite evidence of early formation of nascent matrix adhesions (unpublished observations). Progression of nascent to mature adhesions, and long term growth and survival on soft PA gels can be rescued by Y-27632 suggesting that ROCK inhibition rescues hESc from default activation of cell death (an anoikis) indirectly by promoting stable cell adhesion (37). Y-27632 is thus needed during initial plating and for a variable time afterwards to permit colony outgrowth. Once stable adhesions have been established (and presumptively ROCK is returned to adhesion dependent feedback control), indicated by the formation of small multicellular colonies, it can be removed with only modest affects on growth and viability. However, Y-27632 also inhibits intercellular actomyosin contractility in established hESc colonies, suggested by colony compaction following removal (unpublished observations) or "loosening" following addition (38). The extent of compaction is dependent on the softness of the underlying substrate (barely detectable on very stiff substrates such as glass and tissue culture plastic and very pronounced on PA gels of $E \leq 1,000$ Pa) suggesting that permissive limits on this are set by cell–matrix adhesions. Although, hESc will continue to grow in the long term presence of Y-27632 across the range of PA gel stiffness, we are at present assuming that compaction heralds establishment of a "final" (substrate compliance

permitted) epithelial organizational state and that this represents the "natural" endpoint for analysis or as a starting point for subsequent treatment. Y-27632 is thus present only to facilitate establishment of viable colonies across the range of substrate compliances and must be removed at the earliest permissible point to allow completion of epithelial organization. Compaction of colonies after Y-27632 withdrawal typically occurs after 1–2 days depending on how rapidly and completely the inhibitor was withdrawn. The schedule of withdrawal indicated in step 10, Subheading 3.4.2 generally works for establishing viable hESc colonies from single hESc on Matrigel coated plastic or PA gels of $E > 1{,}000$ Pa and by 6–7 days will reduce Y-27632 concentrations to relatively ineffective levels (theoretically 1–2 μM). On these gels small colonies become established by 3–4 days so it is usually possible to completely remove Y-27632 at this point. Cells remain prone to fluid shearing from the surface so we recommend that removal be accomplished in a stepwise fashion as for feeding but compressed over a short space of time (typically a few hours to a day). We use pre-warmed medium for this purpose, and because, for the same reasons mixing by mechanical agitation is not recommended, we allow 15–30 min in a tissue culture incubator for the new addition of medium to mix by diffusion with the old. The latency for division is considerably longer for PA gels of $E < 1{,}000$ Pa and the cells take several days to spread. This generally necessitates longer periods of Y-27632 treatment before withdrawal and longer total time in culture. A suggested schedule is on day 2 and 3 after plating to feed with media containing 10 μM Y-27632, on day 4 and 5 with 5 μM Y-27632 and subsequently without inhibitor. Total growth is usually 8–10 days for these cultures. PA gels of $E \sim 1{,}000$ Pa are an intermediate case. Visual monitoring of the progress of colony growth is the best guide; we do not attempt to withdraw Y-27632 until colonies have reached a reasonable size and experimentally we try to compare cultures that have been deprived of Y-27632 for equivalent times paying less attention to the total time in culture. See Fig. 3 for representative images of hESc during various stages of hESc growth on soft and stiff Matrigel coupled PA gels.

69. Using colony fragments (or higher seeding densities) as opposed to single cells shortens considerably all of these Y-27632 treatment times presumably because cell–cell contact suppresses ROCK activation (36) and shortens the latency of spreading and resumption of growth on softer PA gels (unpublished observations).

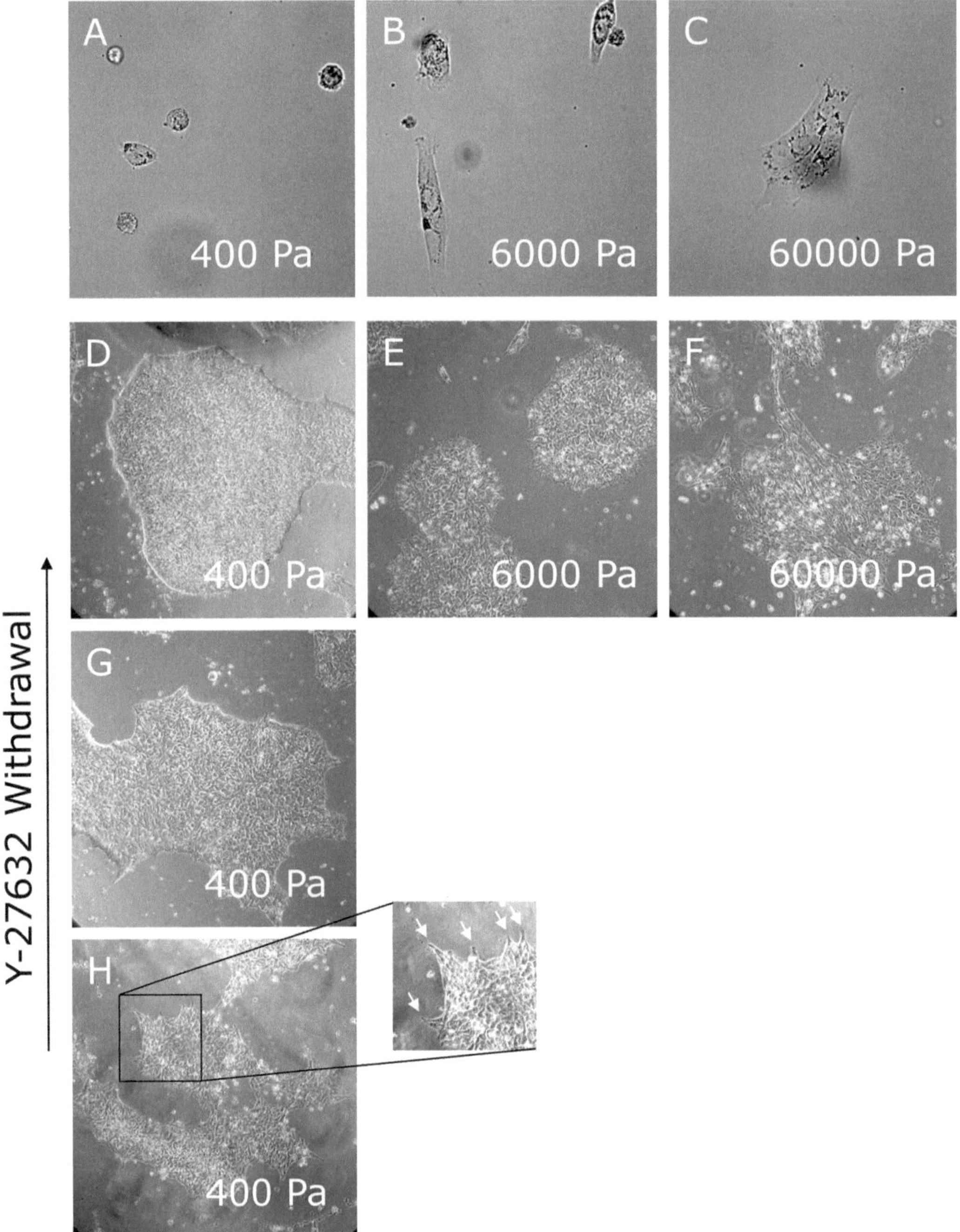

Fig. 3. Representative phase contrast images of hESc on soft ($E = 400$ Pa), intermediately stiff ($E = 6{,}000$ Pa), and very stiff ($E = 60{,}000$ Pa) gels 2 days after plating in the presence of 20 μM Y-27632 (**a**, **b**, and **c**) and after 8 days of growth following the protocol for graded withdrawal of Y-27632 (**d**, **e**, and **f**). hESc spread slowly on soft gels and to a progressively greater extent as gel stiffness increases for short times after plating. At late times following Y-27632 withdrawal colonies compact to a greater extent on softer gels under the influence of cell–cell adhesion and develop smooth edged boundaries. On stiff gels at comparable times, edge cells display many spiky actin rich protrusions and remain spread (Compare **d** and **f**). In the continuous presence of Y-27632, colonies of hESc on soft gels remain coherent and develop similar spiky protrusions (**h**; Day 7 of treatment; see *white arrows* in *boxed* projection). Over the course of several days following withdrawal of Y-27632 (**g**; Day 1 following withdrawal and **d**; 3 days following withdrawal), these protrusions are withdrawn as the colonies compact.

Acknowledgments

This work was supported by internal funding from the Dept of Surgery, UCSF, a CIRM Pilot Grant #RS1-00449-1, and NIH Grants R01CA138818-01A1 and 1U54CA163155-01 to VMW as well as a CIRM Bridges to Stem Cell Research Grant #TB1-01190 awarded to Humbolt State University to fund ARC.

References

1. Engler AJ, Sen S, Sweeney HL et al (2006) Matrix elasticity directs stem cell lineage specification. Cell 126:677–89
2. Engler AJ, Humbert PO, Wehrle-Haller B et al (2009) Multiscale modeling of form and function. Science 324:208–12
3. Cohen DM, Chen CS (2008) Mechanical control of stem cell differentiation. In: The Stem Cell Research Community (ed) StemBook. doi/10.3824/stembook.1.26.1, http://www.stembook.org
4. Pilot F, Lecuit T (2005) Compartmentalized morphogenesis in epithelia: from cell to tissue shape. Dev Dyn 232:685–94
5. Solnica-Krezel L (2005) Conserved patterns of cell movements during vertebrate gastrulation. Curr Biol 15:R213–28
6. Fernandez-Gonzalez R, Simoes Sde M, Roper JC et al (2009) Myosin II dynamics are regulated by tension in intercalating cells. Dev Cell 17: 736–43
7. Shook DR, Keller R (2008) Epithelial type, ingression, blastopore architecture and the evolution of chordate mesoderm morphogenesis. J Exp Zool B Mol Dev Evol 310:85–110
8. Brouzes E, Farge E (2004) Interplay of mechanical deformation and patterned gene expression in developing embryos. Curr Opin Genet Dev 14:367–74
9. Brouzes E, Supatto W, Farge E (2004) Is mechano-sensitive expression of twist involved in mesoderm formation? Biol Cell 96:471–7
10. Pouille PA, Ahmadi P, Brunet AC et al (2009) Mechanical signals trigger Myosin II redistribution and mesoderm invagination in Drosophila embryos. Sci Signal 2:16
11. Tesar PJ, Chenoweth JG, Brook FA et al (2007) New cell lines from mouse epiblast share defining features with human embryonic stem cells. Nature 448:196–9
12. Krtolica A, Genbacev O, Escobedo C et al (2007) Disruption of apical-basal polarity of human embryonic stem cells enhances hematoendothelial differentiation. Stem Cells 25: 2215–23
13. Xu C, Inokuma MS, Denham J et al (2001) Feeder-free growth of undifferentiated human embryonic stem cells. Nat Biotechnol 19:971–4
14. Li L, Arman E, Ekblom P et al (2004) Distinct GATA6- and laminin-dependent mechanisms regulate endodermal and ectodermal embryonic stem cell fates. Development 131:5277–86
15. Paszek MJ, Weaver VM (2004) The tension mounts: mechanics meets morphogenesis and malignancy. J Mammary Gland Biol Neoplasia 9:325–42
16. Paszek MJ, Zahir N, Johnson KR et al (2005) Tensional homeostasis and the malignant phenotype. Cancer Cell 8:241–54
17. Solon J, Levental I, Sengupta K et al (2007) Fibroblast adaptation and stiffness matching to soft elastic substrates. Biophys J 93:4453–61
18. Yeung T, Georges PC, Flanagan LA et al (2005) Effects of substrate stiffness on cell morphology, cytoskeletal structure, and adhesion. Cell Motil Cytoskeleton 60:24–34
19. Geiger B, Bershadsky A, Pankov R et al (2001) Transmembrane crosstalk between the extracellular matrix–cytoskeleton crosstalk. Nat Rev Mol Cell Biol 2:793–805
20. Nelson CM, Pirone DM, Tan JL et al (2004) Vascular endothelial-cadherin regulates cytoskeletal tension, cell spreading, and focal adhesions by stimulating RhoA. Mol Biol Cell 15:2943–53
21. Liu Z, Tan JL, Cohen DM et al (2010) Mechanical tugging force regulates the size of cell–cell junctions. Proc Natl Acad Sci USA 107:9944–9
22. Peerani R, Rao BM, Bauwens C et al (2007) Niche-mediated control of human embryonic stem cell self-renewal and differentiation. EMBO J 26:4744–55
23. Farge E (2003) Mechanical induction of Twist in the Drosophila foregut/stomodeal primordium. Curr Biol 13:1365–77

24. Sawada Y, Tamada M, Dubin-Thaler BJ et al (2006) Force sensing by mechanical extension of the Src family kinase substrate p130Cas. Cell 127:1015–26
25. Tschumperlin DJ (2004) EGFR autocrine signaling in a compliant interstitial space: mechanotransduction from the outside in. Cell Cycle 3:996–7
26. Kojic N, Chung E, Kho AT et al (2010) An EGFR autocrine loop encodes a slow-reacting but dominant mode of mechanotransduction in a polarized epithelium. Faseb J 24:1604–15
27. McBeath R, Pirone DM, Nelson CM et al (2004) Cell shape, cytoskeletal tension, and RhoA regulate stem cell lineage commitment. Dev Cell 6:483–95
28. Desprat N, Supatto W, Pouille PA et al (2008) Tissue deformation modulates twist expression to determine anterior midgut differentiation in Drosophila embryos. Dev Cell 15:470–7
29. Somogyi K, Rorth P (2004) Evidence for tension-based regulation of Drosophila MAL and SRF during invasive cell migration. Dev Cell 7: 85–93
30. Levental KR, Yu H, Kass L et al (2009) Matrix crosslinking forces tumor progression by enhancing integrin signaling. Cell 139:891–906
31. Pless DD, Lee YC, Roseman S et al (1983) Specific cell adhesion to immobilized glycoproteins demonstrated using new reagents for protein and glycoprotein immobilization. J Biol Chem 258:2340–9
32. Damljanovic V, Lagerholm BC, Jacobson K (2005) Bulk and micropatterned conjugation of extracellular matrix proteins to characterized polyacrylamide substrates for cell mechanotransduction assays. Biotechniques 39:847–51
33. Saha K, Keung AJ, Irwin EF et al (2008) Substrate modulus directs neural stem cell behavior. Biophys J 95:4426–38
34. Engler AJ, Rehfeldt F, Sen S et al (2007) Microtissue elasticity: measurements by atomic force microscopy and its influence on cell differentiation. Methods Cell Biol 83:521–45
35. Watanabe K, Ueno M, Kamiya D et al (2007) A ROCK inhibitor permits survival of dissociated human embryonic stem cells. Nat Biotechnol 25:681–6
36. Johnson KE (1976) Circus movements and blebbing locomotion in dissociated embryonic cells of an amphibian, Xenopus laevis. J Cell Sci 22:575–83
37. Shi J, Wei L (2007) Rho kinase in the regulation of cell death and survival. Arch Immunol Ther Exp (Warsz) 55:61–75
38. Harb N, Archer TK, Sato N (2008) The Rho-Rock-Myosin signaling axis determines cell-cell integrity of self-renewing pluripotent stem cells. PLoS One 3:e3001

Chapter 25

Isolation of Circulating Angiogenic Cells

Erin E. Vaughan and Timothy O'Brien

Abstract

Endothelial progenitor cells (EPCs) were first identified by Ashara et al. in 1997 (Asahara et al. Science 275:964–967, 1997) and were thought to contribute to angiogenesis and vasculogenesis. Since their discovery, circulating levels of EPCs were found to serve as biomarkers as low levels correlate with increased cardiovascular events and death from cardiovascular causes (Werner et al. N Engl J Med 353:999–1007, 2005; Fadini et al. J Am Coll Cardiol 45:1449–1457, 2005; Hill et al. N Engl J Med 348:593–600, 2003; Schmidt-Lucke et al. Circulation 111:2981–2987, 2005). Additionally, EPC dysfunction has been associated with diabetes mellitus and other disease states. However, recently there has been a great deal of controversy in the field over the exact definition and function of an EPC. To help classify EPCs, they have been divided into two distinct groups (1) circulating angiogenic cells (also referred to as early EPCs) and (2) endothelial colony forming cells (also referred to as late outgrowth EPCs). Circulating angiogenic cells are believed to represent a cell population enriched in monocytes and exert their angiogenic effects via paracrine and signaling mechanisms whereas endothelial colony forming cells are true EPCs and may enhance angiogenesis and vasculogenesis by incorporating into the newly forming vessels. Here the isolation and identification of circulating angiogenic cells are described.

Key words: Endothelial Progenitor Cells, Angiogenesis, Monocytes

1. Introduction

The discovery of endothelial progenitor cells (EPCs) introduced a new paradigm for endothelial repair and postnatal vasculogenesis (1). Since that time a significant amount of controversy has surrounded the exact definition and function of EPCs (reviewed in (2)). It is now well accepted that EPCs are actually a heterogeneous group of cells that exhibit a variety of functions depending on the method of isolation and culture. Various protocols and methods exist for both the identification and isolation of EPCs. Here, the isolation and identification of early EPCs, also referred to as circulating angiogenic cells, are addressed. Circulating angiogenic cells are spindle shaped and exhibit a low proliferative capacity. Studies

Kimberly A. Mace and Kristin M. Braun (eds.), *Progenitor Cells: Methods and Protocols*, Methods in Molecular Biology, vol. 916, DOI 10.1007/978-1-61779-980-8_25,

suggest that unlike late outgrowth EPCs, early EPCs do not directly form vessels in vivo but they do contribute to angiogenesis in a paracrine manner. Indeed, these cells, or alternatively conditioned media, have been shown to increase the angiogenic response in ischemic injury.

To obtain early EPCs from peripheral blood, mononuclear cells are isolated by density gradient centrifugation and these cells are plated on fibronectin coated plates for 4 days. After the 4 days non-adherent cells are removed and the remaining cells are considered circulating angiogenic cells (early EPCs). The cells can be stained with lectin and acetylated low density lipoprotein to confirm they are early EPCs.

2. Materials

2.1. Collection of Blood

1. 10 × 10 ml venous blood collection tubes coated with K_3 EDTA 15% solution (BD Biosciences).
2. 5 × 50 ml conical tubes.
3. 100 ml of Hanks Balanced Salt Solution.

2.2. Isolation of Mononuclear Cells

1. 50 ml conical tubes.
2. Pipettes.
3. Transfer pipettes.
4. Endothelial growth medium 2 Bullet Kit (Lonza) (see Note 1).
5. Ficoll-paque PLUS (Amersham).
6. Red Cell Lysis Buffer (Sigma).
7. Phospho-buffered Saline (PBS).

2.3. Seeding of Mononuclear Cells

1. Hemocytometer.
2. Trypan blue.
3. Fibronectin coated plates (BD Biosciences).
4. Endothelial Growth Media 2 (Lonza).

2.4. Identification of EPCs

1. Four-well chamber slides (Nunc) (Optional, see Note 6).
2. Human fibronectin (Sigma) (Optional, see Note 6).
3. 3% Paraformaldehyde (PFA).
4. PBS.
5. PBS + 3% Bovine serum albumin (BSA).
6. 1,1′-dioctadecyl-3,3,3′,3′-tetramethylindocarbocyanine-labeled Ac-LDL (DiI-Ac-LDL) (Invitrogen).
7. FITC-labeled *Ulex europaeus* agglutinin-I lectin (Sigma).

3. Methods

3.1. Collection of Blood

1. Collect 100 ml of venous blood in EDTA coated tubes and gently mix tubes at room temperature (see Note 2). Blood should be processed as soon as possible, if there are delays the tubes should be kept at room temperature with continuous rocking.
2. Aliquot 20 ml of blood into each 50 ml conical tube and dilute each aliquot (1:1) by adding 20 ml of Hanks Balanced Salt Solution.

3.2. Isolation of Mononuclear Cells

Warm EBM-2 to 36°C.

1. Pipette 15 ml of ficoll-paque into each of the 5 × 50 ml conical tubes.
2. Carefully overlay 35 ml of the diluted blood onto the 15 ml of ficoll ensuring that the blood and ficoll do not mix. Tilting the 50 ml conical tube and slowly adding the diluted blood aids in maintaining the barrier between the ficoll and the blood.
3. At room temperature, centrifuge the tubes at 600 × *g* for 30 min with no brake (deceleration set to 0) (see Note 3).
4. Upon careful removal from the centrifuge, four distinct layers should be observed:
   (a) Red blood cells will form a dark pellet at the bottom of the tube
   (b) Above the red blood cells a clear, ficoll layer will be found
   (c) A cloudy buffy coat containing mononuclear cells forms at the interface between the ficoll layer below and the serum above
   (d) The top yellowish layer is serum
5. Use a transfer pipette to carefully remove the cloudy mononuclear layer being careful to avoid collecting the serum or ficoll layers and transfer into new 50 ml conical tubes.
6. Centrifuge the buffy coat layers at 500 × *g* for 10 min at room temperature (the centrifuge brake can be turned on again for this and all subsequent steps).
7. The mononuclear cells are located in a lose pellet at the bottom of the tubes. Carefully remove the supernatant and discard.
8. Add 5 ml of red blood cell lysis buffer to each pellet and pipette up and down for a minute to ensure any contaminating red blood cells are lysed.

9. Centrifuge at 250 × *g* for 10 min.
10. Aspirate and discard the supernatant.
11. Wash cells in 5 ml PBS followed by centrifuging at 250 × *g* for 10 min.
12. Aspirate and discard the supernatant.
13. Repeat steps 12 and 13.
14. Resuspend the pellet in 10 ml of EBM-2 (cell pellets from the same donor can be combined).

3.3. Seeding of Mononuclear Cells

1. To determine the viability of the cells take a 50 μl aliquot of the cell suspension and mix with 50 μl of trypan blue. Use a hemocytometer to determine the number of live and dead cells and calculate the total number of cells obtained.
2. Resuspend the cells in EGM-2 at 1.25×10^7 cells/ml.
3. Pipette 4 ml of resuspended cells into each well of a six-well coated fibronectin plate.
4. Incubate cells at 5% CO_2 at 37°C.

3.4. Culture of EPCs

1. The medium is left on until day 4, this allows the EPCs time to adhere.
2. On day 4, aspirate and discard the medium.
3. Wash cells gently with 5 ml of PBS.
4. Repeat step 3.
5. Add 1 ml of EBM2
6. Medium should be changed every day to every other day after day 4 (see Note 5).

3.5. Staining for Acetylated LDL

1. Incubate the EPCs with 5 μg/ml DiI-acLDL for 4 h at 5% CO_2 at 37°C (see Notes 4 and 6).
2. Wash the cells 2× in PBS.
3. Fix the cells in 3% PFA for 15 min.
4. Wash cells for 10 min with gentle agitation in PBS containing 3% BSA.
5. Repeat step 4 twice (for a total of three washes).
6. Incubate cells in 10 μg/ml lectin-FITC in 3% BSA for 1 h at room temperature.
7. Wash cells 3× in PBS.
8. Cells which are dual positive are considered EPCs (see Note 5) (Fig. 1).

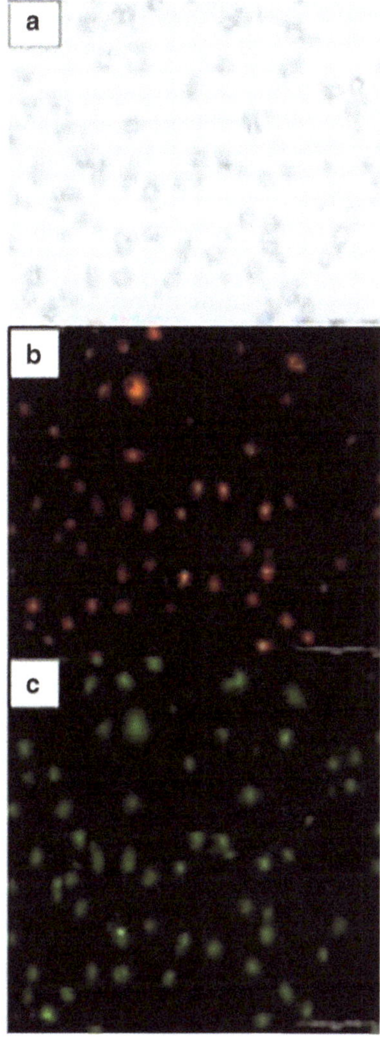

Fig. 1. (**a**) Early endothelial progenitor cells (EPCs) under light microscopy; (**b**) early EPCs stained with 1,1′-dioctadecyl-3,3,3′,3′-tetramethylindocarbocyanine acetylated low-density lipoprotein (dil-acLDL) (*red*); (**c**) early EPCs stained with fluoroscein isothiocyanate (FITC)-lectin (*green*); (from (3) with permission).

4. Notes

1. Add all supplements included in the bullet kit to the media before use.
2. All procedures should be performed in conformance with all institutional ethical requirements regarding human subjects.
3. Slowly starting and stopping the centrifuge is imperative to maintain separation between the layers.

4. Due to the current controversy in the field characterization of EPCs by additional methods such as FACS is often necessary.
5. Cells between days 4 and 7 are considered early EPCs (or circulating angiogenic cells). After day 7 early EPCs begin to die and the population of late (or outgrowth) EPCs becomes the prevalent cell type.
6. For easier viewing and conservation of reagents cells can be plated on four-well chamber slides coated with 10 μg/ml human fibronectin.

References

1. Asahara T, Murohara T, Sullivan A, Silver M, van der Zee R, Li T, Witzenbichler B, Schatteman G, Isner JM (1997) Isolation of putative progenitor endothelial cells for angiogenesis. Science 275:964–967
2. Pearson JD (2010) Endothelial progenitor cells–an evolving story. Microvasc Res 79: 162–168
3. Liew A, McDermott JH, Barry F, O'Brien T (2008) Endothelial progenitor cells for the treatment of diabetic vasculopathy: panacea or Pandora's box? Diabetes Obes Metab 10: 353–366

Chapter 26

Methods for Characterization/Manipulation of Human Corneal Stem Cells and their Applications in Regenerative Medicine

Francesca Corradini, Beatrice Venturi, Graziella Pellegrini, and Michele De Luca

Abstract

Cell therapy is an emerging therapeutic strategy aimed at replacing or repairing severely damaged tissues with cultured cells. Specifically, ocular burns cause depletion of limbal stem cells, which leads to corneal opacification and visual loss. Corneal stem cells are segregated in the basal layer of the limbus, which is the transitional zone of the epithelium located between the cornea and the bulbar conjunctiva. Autologous cultured limbal epithelial cells can restore damaged corneas. We sought to establish a culture system that allows preservation of limbal stem cells and preparation of manageable epithelial sheets. We outline some quality criteria, which assure the clinical performance of keratinocyte culture: evaluation of the number of holoclones within a cultured epithelial graft, proportion of aborting colonies, and percentage of cells expressing high levels of $\Delta Np63\alpha$.

Key words: Human stem cells, Keratinocytes, Holoclone, Limbus, Cornea, Squamous epithelia, Cell therapy, Visual acuity, Autologous graft, Regenerative medicine

1. Introduction

Regenerative medicine aims at the permanent restoration of damaged tissues. Keratinocyte stem cells are currently used to regenerate many types of squamous epithelia (1). The human ocular surface is covered with conjunctival and corneal squamous epithelia, which are formed by two phenotypically different cell types. Nevertheless, in several mammals, there is evidence that oligopotent stem cells are distributed throughout the mammalian ocular surface (2).

The conjunctival epithelium is well vascularized and contains unicellular mucin-secreting goblet glands, which contribute to the

Kimberly A. Mace and Kristin M. Braun (eds.), *Progenitor Cells: Methods and Protocols*, Methods in Molecular Biology, vol. 916, DOI 10.1007/978-1-61779-980-8_26,

maintenance of a tear film on the ocular surface. In humans, the cornea is covered with a flat epithelium, devoid of goblet cells, lying on the corneal stroma by the Bowman's layer. The clearness of the cornea, which depends on stroma avascularity and epithelial integrity, is essential to visual acuity. The narrow zone between the cornea and the bulbar conjunctiva is referred to as the limbus, the place where corneal (but not conjunctival) stem cells are located (3, 4).

Complete loss of the corneal–limbal epithelium leads to re-epithelialization by conjunctival cells: this is followed by neovascularization, chronic inflammation, recurrent epithelia defects, and stroma scarring, causing a pronounced decrease in visual acuity and severe discomfort. Unilateral defects of the corneal–limbal epithelium can be restored by the transplantation of autologous grafts of corneal epithelium generated by serial cultivation of limbal cells taken from the uninjured eye (5, 6).

Cultured limbal stem cells were found to include stem cells detectable as holoclones. The holoclone-forming cell is the smallest colony-founding keratinocyte, it has the highest proliferative capacity and it contains the stem cell of human epithelia described at least in epidermis, cornea–limbus, conjunctiva, and oral mucosa. During its clonal evolution, the holoclone produces paraclone-forming cells, which have a very limited proliferative capability, generate aborted colonies, and have the expected properties of transient amplifying (TA) cells. Meroclones have an intermediate proliferative and clonogenic potential and are a reservoir of paraclones (7).

Evaluation of the number of holoclones within a cultured epithelia graft is considered the best available quality control assuring the clinical performance of keratinocyte culture but would be very cumbersome as a standard test. The number of aborted colonies, which is inversely related to the number of holoclones, is easier to score. The proportion of aborting colonies rises slightly during the two stages of cultivation, but the mean value should not reach 10% of the total (8). Differently from both previous quality criteria, which are approved for all human epithelia described above, the identification of holoclones by immunodetection of ΔNp63α is an exclusive and simple method for determining the presence of an adequate number of stem cells restricted only to cultured limbal grafts (9, 10). Grafts containing <3% of cells expressing a high level of ΔNp63α have a high risk of failure and are currently not used to prepare limbal grafts for patient's treatment (11). However, other factors such as the severity of injury and the presence or absence of complications could also influence the clinical outcome of the corneal regeneration.

2. Materials

Some reagents could be changed according to the new rules of the European Community. New regulations on advanced therapies were proposed in 2003 in Europe: these therapies were classified as medicinal products and their manufacturing process should conform to the rules of "standard drugs." Competent authorities (Regulatory Authorities of each member State and European Medicinal Agency) evaluate manufacturing and safety of each reagent used within the manufacturing process as well as the final product. Some reagents commonly used for cell culture are reevaluated for risks related to virus and BSE transmission with a risk/benefit ratio, specific for each product at the time of marketing authorization which is needed for spreading the therapy in the European Territory.

2.1. 3T3-MCB Cultivation

1. Trypsin for 3T3 (Invitrogen, 0.25% trypsin-EDTA) diluted 1:2 in 0.02% EDTA (see Note 1).
2. 3T3 medium: Dulbecco's Modified Eagle's Medium (DMEM) supplemented with 10% calf serum (Euroclone), 4 mM L-Glutamine, 50 IU/ml penicillin/streptomycin (see Note 2).

2.2. Feeder Layer Preparation

1. Trypsin for 3T3.
2. 3T3 medium.
3. Kno medium: Dulbecco's Modified Eagle's Medium (DMEM) supplemented with 30% Hams's F12 medium, 10% fetal bovine serum (Gibco), 4 mM L-Glutamine, 50 IU/ml penicillin/streptomycin, 0.18 M adenine (Calbiochem), 5 μg/ml insulin (Sigma), 0.4 μg/ml hydrocortisone (Calbiochem), 2 nM triiodotironine (Sigma), 10^{-8} M colera toxin (Sigma) (see Note 2).
4. RAD-GIL (Gilardoni S.p.A.) preheated by the following irradiation cycles: 1 cycle at 110 kV/12 mA/2 min; 1 cycle at 150 kV/12 mA/3 min; and 1 cycle at 200 kV/12 mA/5 min.

2.3. Limbal Biopsy

1. Dulbecco's Phosphate-Buffered Saline (DPBS).
2. Trypsin for keratinocytes (Gibco, 0.5% trypsin-EDTA 10×) diluted 1:10 in DPBS (see Note 1).
3. Kno medium.

2.4. Primary Culture

1. Kno medium.
2. Kc medium: Epidermal growth factor (EGF, Austral Biologicals) is added to the Kno medium at 10 ng/ml 3 days after plating (see Note 2).
3. 24 multi-well plates (Corning).

2.5. Colony-Forming Efficiency Plating

1. Kno medium.
2. Kc medium.
3. 100 mm dish (Corning).

2.6. Freezing Human Primary Keratinocytes

1. Freezing medium: Dulbecco's Modified Eagle's Medium (DMEM) supplemented with 30% Hams's F12 medium, 10% fetal bovine serum (Gibco), 4 mM L-Glutamine, 50 IU/ml penicillin/streptomycin, 0.18 M adenine (Calbiochem), 5 μg/ml insulin (Sigma), 0.4 μg/ml hydrocortisone (Calbiochem), 2 nM triiodotironine (Sigma), 10^{-8} M colera toxin (Sigma), 10% glycerol (see Note 2).
2. Freezing Container "Mr. Frosty" (Nalgene).

2.7. Fibrin Gel Preparation

1. Trypsin for keratinocytes.
2. Kno medium.
3. Kc medium.
4. Tissucol (Baxter): It is composed by two frozen stock solutions: fibrinogen (1 ml) and thrombin (1 ml).
5. NaCl 10%.
6. NaCl 0.9%.
7. $CaCl_2$ 100 mM.
8. 144-cm^2 plate not treated for cell culture (Corning).
9. Sterile steel ring.
10. Grafting wash medium: Dulbecco's Modified Eagle's Medium (DMEM) supplemented with 4 mM L-Glutamine (see Note 2).
11. Sterile steel box.
12. Dispase solution: Weigh 250 mg of Dispase II (Roche) and melt in 100 ml of grafting wash medium. Heat until complete dissolution and filter with 0.45 μm filter unit (see Note 3).

2.8. Lifespan

1. Trypsin for keratinocytes.
2. Kno medium.
3. Kc medium.
4. 24-multi-well plates (Corning).

2.9. Clonal Analysis

1. Trypsin for keratinocytes.
2. Trypsin for 3T3.
3. Kno medium.
4. Kc medium.
5. 3T3 medium.
6. 96-multi-well plates (Corning).

2.10. CFE Coloration

1. 37% Formaldehyde (Sigma) diluted 1:10 in distilled water.
2. 2% Rhodamine B (Sigma) dissolved with 3.7% Formaldehyde fresh solution.
3. Water.

2.11. CFE Analysis

1. Stereomicroscope Stemi DV4 (Zeiss).

2.12. Immuno-cytochemistry

1. Microscope glass slides (Menzel-Glaser).
2. Micro cover glass (Menzel-Glaser).
3. Storage microscope slide boxes (Kartell).
4. Microscope slide staining containers (Kartell).
5. Shandon Cytospin4 (Thermo Scientific).
6. Cytofunnel (Thermo Scientific).
7. DMEM.
8. 1× Phosphate-Buffered Saline (PBS), pH 7.0.
9. Antibody dilution buffer (1× PBS, 2% BSA): Dissolve 1 g of BSA (Bovine Serum Albumin) into 50 ml of 1× PBS by vortexing. Store at 4°C and use within 24 h.
10. α-ΔNp63α-specific antiserum (Primm).
11. α-cytokeratin 19 (clone RCK 108, Dako).
12. α-cytokeratin 3 (clone AE5, Dako).
13. Donkey anti-mouse Alexa Fluor 488 (Invitrogen).
14. Paraformaldehyde 3%: Heat 80 ml 1X PBS to 60–65°C while stirring with 3 g paraformaldehyde and 2 g sucrose. Add 2 drops of 1 N NaOH and continue to stir to dissolve. Cool the solution and correct pH to 7.5 (see Note 4).
15. PBS 0.01% sodium azide: Dissolve 0.01 g of sodium azide (Sigma) into 100 ml of 1× PBS. Mix and store at 4°C.
16. PBS 0.5% Triton X-100: Dilute 1:200 Triton X-100 (Biorad) in 1× PBS.
17. Vectashield mounting medium with DAPI (Vector, USA).
18. Laser scanning confocal microscope; Zeiss LSM 510 confocal microscope.

2.13. Immuno-cytochemistry Analysis

1. Laser scanning confocal microscope: Zeiss LSM 510 confocal microscope.
2. Advanced analysis software: LSM 510 software.

3. Methods

Limbal stem cells grown on the fibrin matrix and in the presence of lethally irradiated 3T3-MCB cells and fetal calf serum: neither holoclone preservation, nor clonogenic ability and proliferative potential, nor ΔNp63α immunodetection have been tested in other culture conditions. This culture method has been used worldwide since the 1980s to treat patients with massive full-thickness burns (12); during the past 30 years, no adverse effects have been reported and this culture method has been approved for use by the Food and Drug Administration (FDA) in the USA and by regulatory authorities of Japan, Italy, and South Korea.

3.1. 3T3-MCB Cultivation

1. Perform 3T3-MCB cultivation from confluent culture ($5–6\times10^6$ cells/T175).
2. Remove all 3T3 medium.
3. Add 12.5 ml trypsin conveniently diluted and preheated.
4. Incubate for 5 min at 37°C.
5. Control cell detachment under the microscope and, if necessary, complete it mechanically with some gentle lateral pushes.
6. Obtain trypsin neutralization adding at least 12.5 ml 3T3 medium directly to the flask.
7. Collect cell suspension in a 50 ml tube.
8. Centrifuge cells at $200\times g$ for 5 min.
9. Resuspend the pellet in 3T3 medium.
10. Plate 3T3-MCB cell suspension diluted in 3T3 medium, avoiding 1:10 over dilutions (see Notes 5 and 6).
11. Feed the culture every 3–4 days with 3T3 medium.
12. Grow cells at 36.5°C, 6.5% CO_2, 96% humidified atmosphere.

3.2. Feeder Layer Preparation

1. Perform 3T3-MCB FL preparation from subconfluent culture ($3.5–4.5\times10^6$ cells/T175).
2. Remove all 3T3 medium.
3. Add 12.5 ml trypsin conveniently diluted and preheated.
4. Incubate for 5 min at 37°C.
5. Control cell detachment under the microscope and, if necessary, complete it mechanically with some gentle lateral pushes.
6. Obtain trypsin neutralization adding at least 12.5 ml 3T3 medium directly to the flask.
7. Collect cell suspension in a 50 ml tube.
8. Centrifuge cells at $200\times g$ for 5 min.

9. Resuspend the pellet in 20 ml Kno medium.
10. Irradiate cell suspension using RAD-GIL at 200 kV/ 12 mA/49.2 min.
11. Plate lethally irradiated 3T3-MCB cells (2.4×10^4 cells/cm^2) on plastic in Kno medium (see Note 6).
12. Keep cells at 36.5°C, 6.5% CO_2, 96% humidified atmosphere.
13. Use lethally irradiated 3T3-MCB cell preparation in the time slot 2 h–24 h after plating.

3.3. Limbal Biopsy

1. Limbal/corneal keratinocytes are obtained from ocular biopsies (2 mm^2) taken from organ donors who provide informed consent.
2. Samples are collected in Kno medium and are kept at 4°C until procedure (within 24 h).
3. Take off the ocular biopsy and dip it into a DPBS tube for few seconds.
4. Take off the ocular biopsy.
5. Treat the sample with trypsin at 37°C for ~30 min, shaking it gently.
6. Collect cell suspension in a 50 ml tube, named "I° trypsinization" (see Note 7).
7. Obtain trypsin neutralization by adding the same volume of Kno medium directly to the tube.
8. Centrifuge cells at $200 \times g$ for 5 min.
9. Resuspend the pellet in 2 ml Kno medium (see Note 7).
10. Count cells under the microscope and sign the number on the tube "I° trypsinization."
11. Meanwhile at step 8, repeat new round of trypsinization (steps 5–10) with fresh trypsin solution, until cell yield is empty.
12. Collect all cell suspensions in a single tube, following the increasing concentrations signed on the tubes (see Note 7).
13. Centrifuge the tube at $200 \times g$ for 5 min.
14. Resuspend the pellet in Kno medium (see Note 7).
15. The best yield obtained should be 17.3×10^3 cells/mm^2.

3.4. Primary Culture

1. Perform primary culture from limbal biopsy cell suspension.
2. Plate cell suspension (1.5×10^4 cells/cm^2) on lethally irradiated 3T3-MCB cell preparation in Kno medium (see Note 6).
3. Substitute Kno medium with freshly prepared Kc medium 3 days after plating.
4. Feed the culture every 48 h with Kc medium.
5. Keep cells at 36.5°C, 6.5% CO_2, 96% humidified atmosphere.

6. Meanwhile at step 2, determine efficiency of colony formation (CFE) by plating 10% limbal biopsy cell suspension in 100 mm plate in standard culture conditions; 12 days later fix and stain the colonies with Rhodamine B for the determination of the colony-forming efficiency (CFE) and the number of cell generations.

3.5. Colony-Forming Efficiency Plating

1. Plate cell suspension on lethally irradiated 3T3-MCB cell preparation in Kno medium (see Note 6).
2. Substitute Kno medium with freshly prepared Kc medium 3 days after plating.
3. Feed the culture every 48 h with Kc medium.
4. Keep cells at 36.5°C, 6.5% CO_2, 96% humidified atmosphere.

3.6. Freezing Human Primary Keratinocytes

1. Detach cells as described above.
2. Count cells under the microscope.
3. Centrifuge cell suspension (2×10^5–5×10^5 cells/vial) at $200 \times g$ for 5 min.
4. Aspirate the supernatant and carefully resuspend the pellet with 500 μl of freezing medium/each vial (see Note 7).
5. Put the vial in freezing container at −80°C (see Note 8).
6. 24 h later transfer the vials to liquid nitrogen.

3.7. Fibrin gel Preparation (See Note 9)

1. Thaw thrombin and fibrinogen stock solutions at 37°C in a thermostat bath until completely defrosted (10–15 min).
2. Dilute thrombin stock solution to a final concentration of 3 IU/ml of thrombin in 1.1% NaCl and 1 mM $CaCl_2$.
3. Mix fibrinogen stock solution with 1.16 ml saline solution (final concentration: NaCl 1.1% and $CaCl_2$ 1 mM).
4. Place the steel ring on the 144-cm^2 plate.
5. Dispense 300 μl of diluted thrombin solution and 300 μl diluted fibrinogen solution inside the ring.
6. Uniformly mix both the solutions, taking care not to form air bubbles.
7. Place the plate at room temperature for 10–15 min until a complete substrate polymerization is obtained.
8. Store the plate overnight at 4°C.
9. Culture cells on fibrin gel following standard culture condition: seed 2.5×10^5 lethally irradiated 3T3-MCB feeder layer cells and 8×10^4 human keratinocytes (see Note 6).
10. When fibrin-culture autograft is confluent, remove the steel ring and cut the perimeter of the fibrin ring.

11. Wash twice the fibrin-culture autograft with 30 ml of grafting wash medium.
12. Transfer the fibrin-culture autograft with two nippers to a sterile steel box in the presence of grafting wash medium and transfer to transplantation unit.
13. Incubate the remaining perimeter of fibrin-culture autograft with Dispase at 37°C for 15 min.
14. Remove the flaps from the fibrin gel.
15. Centrifuge the flaps at 200×*g* for 5 min.
16. Aspirate the supernatant and incubate with 10 ml of trypsin at 37°C for 15 min.
17. Neutralize the trypsin with 10 ml of Kno medium.
18. Centrifuge at 200×*g* for 5 min.
19. Aspirate the supernatant and resuspend the pellet with 3 ml of Kno medium (see Note 7).
20. Count cells under the microscope.
21. Plate a third culture.
22. Keep cells at 36.5°C, 6.5% CO_2, 96% humidified atmosphere.
23. Meanwhile at steps 9 and 21, determine efficiency of colony formation (CFE) by plating 200–1,000 cells in 100 mm plate; 12 days later fix and stain the colonies with Rhodamine B for the determination of the colony-forming efficiency and the number of cell generations.

3.8. Lifespan

1. Perform lifespan (serial cell propagations) from subconfluent keratinocytes primary culture.
2. Remove all Kc medium.
3. Add 1.5 ml trypsin conveniently diluted and preheated for few seconds.
4. Remove the trypsin.
5. Add 1.5 ml trypsin fresh solution conveniently diluted and preheated.
6. Incubate for 15 min at 37°C.
7. Control cell detachment under the microscope.
8. Resuspend gently and collect cell suspension in a 15 ml tube (see Note 7).
9. Obtain trypsin neutralization adding the same volume of Kno medium directly to the tube.
10. If necessary, repeat a new step of trypsinization, adding 1.5 ml trypsin fresh solution to the plate and incubating for another 5 min at 37°C.

11. Collect cell suspension in the same tube, adding the equal volume of Kno medium (see Note 7).
12. Centrifuge the total cell suspension at 200 × *g* for 5 min.
13. Resuspend the pellet very gently in Kno medium (see Note 7).
14. Count cells under the microscope.
15. Plate cell suspension (1.5×10^4 cells/cm^2) on lethally irradiated 3T3-MCB cell preparation in Kno medium (see Note 6).
16. Substitute Kno medium with freshly prepared Kc medium 3 days after plating.
17. Feed the culture every 48 h with Kc medium.
18. Keep cells at 36.5°C, 6.5% CO_2, 96% humidified atmosphere.
19. For serial propagation, cells are passaged as above, always at the stage of subconfluence, until they reach senescence (approximately, 80–120 cell doublings for ocular epithelia).
20. Meanwhile at step 15 of each cell passage, determine efficiency of colony formation (CFE) by plating 200–1,000 cells in suspension in a 100 mm plate in standard culture conditions; 12 days later fix and stain the colonies with Rhodamine B for the determination of the colony-forming efficiency and the number of cell generations.

3.9. Clonal Analysis

1. Perform clonal analysis from subconfluent keratinocytes primary culture.
2. Obtain cell suspension with at least 200 cells in 40 ml of Kno medium.
3. Inoculate single cells with 100 μl of Kno medium onto four "96-multi-well plates" containing 1×10^4 lethally irradiated feeder layer 3T3-MCB cells/well.
4. Substitute Kno medium with freshly prepared Kc medium 3 days after plating.
5. Feed the culture every other day with Kc medium.
6. After 7 days of cultivation, carefully score the plates under the microscope for the presence of holoclone/meroclone/paraclone-type single colony/well (see Note 10).
7. Treat each clone for 15′ at 37°C with 200 μl of trypsin.
8. Neutralize with 200 μl of Kno medium and collect cell suspension in a 15 ml tube (see Note 7).
9. Wash the well with 200 μl Kno medium and collect with the previous medium in the same tube.
10. Centrifuge cells at 200 × *g* for 5 min.
11. Aspirate the supernatant and resuspend in 800 μl of Kno medium (see Note 7).

12. Transfer 3/4 of cell culture suspension to a T25 flask and culture it in standard culture conditions (see Note 6).
13. Keep cells at 36.5°C, 6.5% CO_2, 96% humidified atmosphere.
14. 6 days later freeze the cells.
15. Meanwhile at step 12, culture the remaining 1/4 of cell suspension on 100 mm dish in standard culture conditions. 12 days later fix and stain with Rhodamine B for the classification of clonal type.

3.10. CFE Coloration

1. Perform CFE coloration with Rhodamine B on the 12th day after the plating.
2. Discard the Kc medium.
3. Add enough Rhodamine B solution to cover the plate.
4. Discard the Rhodamine B solution in a proper container the next day.
5. Rinse the plate with water several times, until the discarded liquid becomes clear.
6. Keep the plate on paper until dry.
7. Score the colonies under the microscope.

3.11. CFE Analysis

1. Determination of the colony-forming efficiency: Cells from each biopsy and from each cell passage of serially cultivated mass and clonal cultures are plated onto 3T3-MCB feeder layers and cultivated as above. Colonies are fixed 12 days later, stained with Rhodamine B, and scored under a dissecting microscope. Values are expressed as the ratio of the number of colonies to the number of inoculated cells. All colonies are scored whether progressively growing or aborted.
2. Determination of the number of cell generations: The number of cell generations is calculated using the following formula: $x = 3.322 \log N/No$, where N equals the total number of cells obtained at each passage and No equals the number of clonogenic cells. Clonogenic cells are calculated from the colony-forming efficiency data (see above), which are determined separately in parallel dishes at the time of cell passage. Considering all the passages of limbal biopsy cell cultivation, the best yield obtained as number of cell generations should be 80–120 cell doublings.
3. Classification of clonal type: The distinction among clonal types is based on the frequency of aborted colonies produced when the clone is transferred to the indicator 100 mm dish. After fixation with Rhodamine B, the colonies are scored under a low-power microscope (Fig. 1). Holoclone is the clonal type where 0–5% colonies are aborted colonies. Paraclone is the clonal

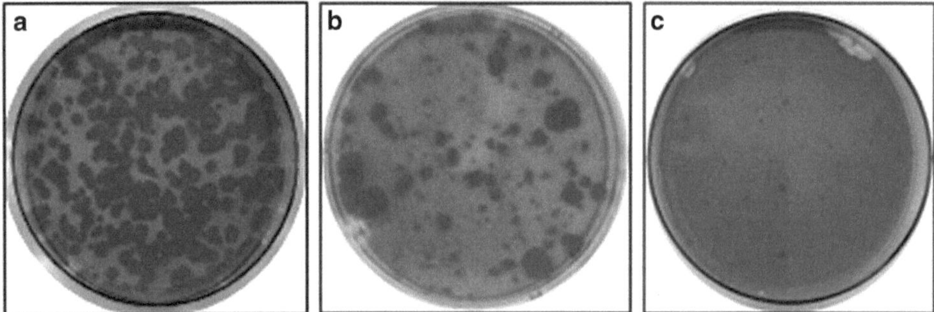

Fig. 1. *Corneal clonal types.* Daughter colonies obtained from a quarter of a colony generated by a single holoclone (**a**), meroclone (**b**), and paraclone (**c**). Limbal cultures and clonal analysis were performed as previously described (5).

type where no colonies are formed or all colonies are aborted colonies. Meroclone is the clonal type where more than 5% but less than 95% of the colonies are aborted colonies.

3.12. Immuno-cytochemistry

1. Obtain a suspension of corneal epithelial cells (ranging from 500 to 15,000 cells for each microscope glass slides) from primary culture.
2. Spin cell suspension at $200 \times g$ for 5 min.
3. Wash three times with 5 ml DMEM.
4. Resuspend the final pellet with 200 μl of DMEM for each microscope glass slide.
5. Dispense 200 μl of cell suspension into each cytofunnel.
6. Cytospin onto ThermoShandon at 1,000 rpm for 5 min.
7. Air dry slides in a chemical hood for 5 min.
8. Fix the cell spot with 200 μl of 3% paraformaldehyde for 10 min.
9. Wash slides three times shaking for 5 min in slide staining containers with 50 ml of 1× PBS (see Note 11).
10. Permeabilize with 1× PBS, 0.5% Triton X-100 for 10 min at 4°C.
11. Wash slides three times shaking for 5 min in slide staining containers with 50 ml of 1× PBS.
12. Incubate slides with 50 ml of 1× PBS, 2% BSA for 30 min at 37°C.
13. Prepare a humid chamber using wet paper on the bottom of a storage microscope slides box.
14. Dry slides around the spot and put it horizontally oriented in the humid chamber.

15. Incubate slides with 100 μl of primary antibody: α-cytocheratin19 is diluted 1:1,000 in 2% BSA and is incubated 1 h at room temperature; α-cytocheratin3 is diluted 1:1,000 in 2% BSA and is incubated 1 h at room temperature; ΔNp63α is diluted 1:500 in 2% BSA and is incubated 30 min at 37°C.
16. Wash slides three times, shaking for 5 min with 50 ml 1× PBS, 0.2% BSA.
17. Incubate slides with 1× PBS, 2% BSA for 15 min at 37°C.
18. Dry slides around the spot and put them horizontally oriented in the humid chamber.
19. Incubate each slide with 100 μl of donkey anti-mouse Alexa Fluor 488 diluted 1:1,000 in 1× PBS, 2% BSA for 30 min at 37°C.
20. Wash three times shaking for 5 min with 50 ml 1× PBS, 0.2% BSA.
21. Incubate slides with 1× PBS, 2% BSA for 15 min at 37°C.
22. Wash three times shaking for 5 min with 50 ml 1× PBS.
23. Dry slides around the spot.
24. Drop Vectashield mounting medium on the slide.
25. Cover with micro cover glass.
26. Analyze slides using laser scanning confocal microscopy.

3.13. Immunocytochemistry Analysis

1. Quantification of ΔNp63α: Since ΔNp63α has been recognized as a limbal stem cell marker, the accurate and fast determination of its abundance could be a powerful tool to evaluate the quality (defined by the stem cell content) of the epithelial sheet before it is grafted back onto the patient. The more reliable way to quantify ΔNp63α is by Quantitative Fluorescent Immunohistochemistry (Q-FIHC), which is a combination of laser scanning confocal microscopy techniques (Zeiss LSM 510 confocal microscope) and advanced analysis software (LSM 510 software) (13). Compared to conventional techniques, Q-FIHC allows the automated quantification of protein signals from a low number of cells. This is relevant when clinical samples from patient's biopsies are used. Quantitative analysis is performed through a tool of the LSM 510 software: the "Histo" function, which allows the user to obtain a table with cell area (πr^2) information and mean ΔNp63α Fluorescence Intensity values. On the basis of long lasting results from our group, only cultures that contain more than 3% ΔNp63α-bright cells with a small diameter (6–10 μm) are used to prepare grafts: indeed, we observed that a minimum of approximately 3,000 stem cells, detected as ΔNp63α-bright holoclone-forming cells, was required to achieve clinical success.

2. Expression of cytokeratin3 and cytokeratin19: Another important quality criterion to evaluate limbal stem cell culture is negativity to both cytokeratin3 and cytokeratin19 staining. Indeed, cytokeratin3 is expressed in corneal epithelium while cytokeratin19 is exclusively expressed in conjunctival stem cells. Therefore, when cells are analyzed with LSM 510 software, the percentage of ΔNp63α-bright cells must be the same as the percentage of K3/K19 negative cells.

4. Notes

1. Trypsin for cell detachment should be freshly prepared and conveniently preheated to 37°C.
2. Media must be filtered with 0.45 μm filter unit (Nalgene) and should be used preheated to 37°C within 48 h of preparation.
3. Dispase solution must be used the same day of preparation.
4. 3% PFA solution can be stored at −20°C for up to 6 months but it is best used freshly prepared.
5. Perform 3T3-MCB cultivation in the dilution range from 1:2 to 1:10 until the12th passage.
6. Seed cells in order to get an equal plating.
7. Every time we collect/resuspend a cell suspension with medium or trypsin, we pre-neutralize the pipette with the respective solution.
8. Freezing procedure of human primary keratinocytes is very sensitive to time: The transfer of cells to −80°C must be done within 5 min after adding the medium.
9. Fibrin gel preparation procedure is patented (European patent number: 1451302, USA patent number: 6,610,538). It is for research use only, not for commercial use.
10. It is important to note that classification of clones under the microscope provides a preliminary clue about clonal types; reliable holoclone identification is only possible by colony-forming efficiency assay. By scoring using a microscope, a holoclone colony is 0.5–1 mm^2 in area; its perimeter is nearly circular and this is particularly evident as the colony grows larger; it contains mainly small cells which may be concentrated near the perimeter. A meroclone colony grows progressively to microscopic size but does not reach the same size as the typical colonies produced by a holoclone; it has a wrinkled perimeter, suggesting some kind of heterogeneity within the colony, and such a colony will soon become terminal. A paraclone colony

is very tiny in area and the perimeter of the colony is drawn out into marked irregularities; the cells are large and flattened and this is clearest at the perimeter, which consists predominantly of a single layer of such cells.

11. Slides are best stored into 1× PBS, 0.01% sodium azide at 4°C for up to 6 months if they are not used immediately.

Acknowledgments

We thank Professor Howard Green, the father of Regenerative Medicine, for supplying 3T3-J2 cell line which were used in these years for the manipulation of corneal stem cells; Tung-Tien Sun who kindly gave us for years as a gift the antibody α-cytokeratin 3 for the selection of patients' cells. We thank all the collaborators of previous laboratories who supported us in the epithelia characterization, in particular Sergio Bondanza and Patrizia Paterna (IDI, Rome); Dr. Osvaldo Golisano, Dr. Enzo Di Iorio, and Dr.ssa Vanessa Barbaro (FBOV, Venice); the ophthalmologists Dott. Paolo Rama, Prof. Carlo Traverso, Dott. Alessandro Lambiase, and Dott. Stefano Bonini who helped us in the definition of clinical protocols.

We thank MIUR, Italian Ministry of Health, and European Community for the research funding.

References

1. De Luca M, Pellegrini G, Green H (2006) Regeneration of squamous epithelia from stem cells of cultured grafts. Regen Med 1:45–57
2. Majo F, Rochat A, Nicolas M, Jaoudé GA, Barrandon Y (2008) Oligopotent stem cells are distributed throughout the mammalian ocular surface. Nature 456:250–4
3. Schermer A, Galvin S, Sun TT (1986) Differentiation-related expression of a major 64 K corneal keratin in vivo and in culture suggests limbal location of corneal epithelial stem cells. J Cell Biol 103:49–62
4. Pellegrini G, Golisano O, Paterna P, Lambiase A, Bonini S, Rama P, De Luca M (1999) Location and clonal analysis of stem cells and their differentiated progeny in the human ocular surface. J Cell Biol 145:769–82
5. Pellegrini G, Traverso CE, Franzi AT, Zingirian M, Cancedda R, De Luca M (1997) Long-term restoration of damaged corneal surfaces with autologous cultivated corneal epithelium. Lancet 349:990–3
6. Rama P, Bonini S, Lambiase A, Golisano O, Paterna P, De Luca M, Pellegrini G (2001) Autologous fibrin-cultured limbal stem cells permanently restore the corneal surface of patients with total limbal stem cell deficiency. Transplantation 72:1478–85
7. Barrandon Y, Green H (1987) Three clonal types of keratinocyte with different capacities for multiplication. Proc Natl Acad Sci USA 84:2302–6
8. Pellegrini G, Rama P, Mavilio F, De Luca M (2009) Epithelial stem cells in corneal regeneration and epidermal gene therapy. J Pathol 217:217–28
9. Pellegrini G, Dellambra E, Golisano O, Martinelli E, Fantozzi I, Bondanza S, Ponzin D, McKeon F, De Luca M (2001) p63 identifies keratinocyte stem cells. Proc Natl Acad Sci USA 98:3156–61
10. Di Iorio E, Barbaro V, Ruzza A, Ponzin D, Pellegrini G, De Luca M (2005) Isoforms of DeltaNp63 and the migration of ocular limbal

cells in human corneal regeneration. Proc Natl Acad Sci USA 102:9523–8

11. Rama P, Matuska S, Paganoni G, Spinelli A, De Luca M, Pellegrini G (2010) Limbal stem-cell therapy and long-term corneal regeneration. N Engl J Med 363:147–55
12. Pellegrini G, Ranno R, Stracuzzi G, Bondanza S, Guerra L, Zambruno G, Micali G, De Luca M (1999) The control of epidermal stem cells (holoclones) in the treatment of massive full-thickness burns with autologous keratinocytes cultured on fibrin. Transplantation 68:868–79
13. Di Iorio E, Barbaro V, Ferrari S, Ortolani C, De Luca M, Pellegrini G (2006) Q-FIHC: quantification of fluorescence immunohistochemistry to analyse p63 isoforms and cell cycle phases in human limbal stem cells. Microsc Res Tech 69:983–91

Chapter 27

ALDH as a Marker for Enriching Tumorigenic Human Colonic Stem Cells

Anitha Shenoy, Elizabeth Butterworth, and Emina H. Huang

Abstract

Aldehyde dehydrogenase (ALDH) can be used as a marker to isolate, propagate, and track normal and cancerous human colon stem cells. To determine their tumorigenic potential, tissues obtained from proximal (normal counterpart) and distal (cancerous) colon of colon cancer patients are implanted into NOD-SCID mice. In parallel, $ALDH^{high}$ and $ALDH^{low}$ cells are isolated via Florescence Associated Cell Sorting (FACS) after the dissociation of distal and proximal colon tissues into a single-cell suspension. Flow cytometry for $ALDH^{high}$ and $ALDH^{low}$ cells is possible with the ALDEFLUOR assay. Following cell sorting, ALDH-enriched cells are tested for their tumorigenic potential *in vivo* as xenografts. Owing to cancer stem cell properties, $ALDH^{high}$ cells could be propagated *in vivo* by serial passaging of the human tissue as xenografts and *in vitro* as suspension cultures called sphere cultures. In this unit, all the above-mentioned methods to isolate and propagate colon cancer stem cells using ALDH as a stem cell marker are described in detail.

Key words: Colon cancer, Aldehyde dehydrogenase, Fluorescent activated cell sorting, Xenograft, Sphere cultures

1. Introduction

In the last decade ALDH1 has emerged as a potential universal marker for stem and progenitor cells in epithelial cancers. Our laboratory has shown that ALDH1 is a marker for both normal and cancerous colon stem cells (1). Immunostaining data by Huang et al. (1) show that rare epithelial cells at the base of the normal crypt express ALDH and that the ALDH expression expands in the epithelia of the cancerous colon. In cancer ALDH-expressing cells are no longer limited to the base of the crypt as the crypt like structure is usually disrupted, but can be found throughout the epithelium. Tumors were generated by injecting as few as twenty-five primary colon cancer cells expressing high levels of ALDH into NOD SCID mice. FACS analysis of subsequent serial passages

Kimberly A. Mace and Kristin M. Braun (eds.), *Progenitor Cells: Methods and Protocols*, Methods in Molecular Biology, vol. 916, DOI 10.1007/978-1-61779-980-8_27, © Springer Science+Business Media, LLC 2012

showed that the tumors were enriched with ALDH positive cells, which was associated with increased tumorigenicity and ability to propagate as spheres in culture. Thus ALDH may be used as a marker to identify and isolate both normal and tumorigenic colon stem cells. Further, we have demonstrated that ALDH1 serves as the marker to identify and isolate the tumor initiating cells in colitis (2). Tumor-initiating cells isolated from cancer and colitis can be propagated and expanded *in vitro* in suspension culture, called sphere culture, an essential *in vitro* tool needed for studying the mechanisms involved during the tumor initiation and metastasis of stem and progenitor cells. In this chapter, we share our protocols used (1) to isolate and implant, (2) to explant and dissociate xenografts and primary tissues, (3) to isolate ALDH+cells by FACS, and (4) to generate and propagate sphere cultures.

2. Materials

2.1. Subcutaneous Xenograft Implantation

1. Laminar flow hood.
2. Sterile surgical instruments (see Note 1) and sutures or Dermabond.
3. Nonobese diabetic severe combined immunodeficient mice (Nonobese diabetic IL2 γ receptor null mice may be best, see Note 2).
4. M199 medium.
5. Phosphate buffered saline (PBS).
6. Chlorohexadine and cotton swabs.

2.2. Xenograft Explantation and Dissociation

1. Collagenase type IV (Worthington Biochemical Corporation, Lakewood, NJ) (see Note 3).
2. M199 Medium.

2.3. Dissociation of Primary Tissues

1. Antibiotic–Antimycotic.
2. Disposable 10 mL serological pipettes (VWR).
3. Collagenase, from *Clostridium* (Sigma-Aldrich) (see Note 4).
4. Hank's balanced salt solution (HBSS).
5. Fetal bovine serum (FBS).
6. 40 μm and 70 μm sterile cell strainers.

2.4. Sorting for Aldehyde Dehydrogenase Expressing Cells (See Note 5)

1. 5 mL polystyrene 12×75 mm culture tubes with caps.
2. 2% fetal bovine serum in HBSS.
3. ALDEFLUOR kit (Stem Cell Technologies, Durham, NC).
4. PE Mouse IgG (BD Biosciences).

5. 1 mg/mL DAPI (4,6-diamidino-2-phenylindole) (Sigma Aldrich) reconstituted according to the manufacturer's instructions.

2.5. Plating and Culturing Sorted Cells as Colonospheres

1. 24-well low-attachment culture plates (Corning).
2. Defined medium (see Note 6), composition: DMEM F-12 50/50 Mix with L-glutamine and 15 mM HEPES, Glucose (6 mg/mL), Sodium Bicabonate (1 mg/mL), Glutamine (Glutamax) (2 mM), Albumin, from bovine serum (BSA) (4 mg/mL), Insulin Transferrin Selenium (ITS) (25 mg/mL), Progesterone (20 nM), Putrescine (9.6 μg/mL), Antimicrobial (see Note 7).

2.6. Immuno-histochemistry of ALDH on Colon Spheres

1. 4% Paraformaldehyde at pH 7.4.
2. 1% agarose in PBS.
3. 70% Ethanol.
4. Dako Target Retrieval Solution (Dako).
5. ALDH1 Antibody (BD Biosciences).
6. Secondary fluorescent antibody.
7. Mounting medium containing DAPI.

3. Methods

3.1. Subcutaneous Xenograft Implantation

1. Tissue that is to be implanted should be kept in sterile PBS or M199 medium on ice while the mouse is prepared for surgery.
2. Anesthesize the mouse according to your animal protocol. Place the mouse in the laminar flow hood and sterilize the flank (alcohol or chlorohexadine can be used) (see Note 8).
3. Implant the tissue into the flank of the mouse as follows. Use toothed forceps to gently pull skin over flank into a "teepee" shape. While holding the skin with the forceps cut a small incision (2–4 mm). Use the tip of the closed scissors to undermine a flap between the skin and the underlying muscle of the mouse. While still holding the toothed forceps with the skin away from the mouse body, use serrated micro-dissecting forceps to pick up the tissue implant and place it in the pocket. Be certain the tissue is firmly in place in the pocket! If the tissue is too near the incision it may explant slip out during suturing.
4. To close the incision load the sterile suture into the needle holder. A continuous running suture or several interrupted sutures are recommended. Use the toothed or serrated forceps to grip the top of the incision and pull it taut away from the mouse body. Begin sewing at the top of the incision. Pull the

suture through until only about 2–3 cm of the suture end remains. Make 3–4 square knots to secure the suture in place. The skin should be occluded with no gaps. Alternatively, stapling or surgical adhesives, such as Dermabond, may be used.

3.2. Xenograft Explantation and Dissociation

3.2.1. Recovering the Xenograft and Sacrificing the Mouse

Anesthetize the mouse according to your animal protocol. Ensure the mouse is fully anesthetized per your protocol before beginning surgery. Place the mouse inside the laminar flow hood. Cleanse the tumor-containing flank(s) and abdomen with an antiseptic solution. Recover the xenograft in the following manner.

1. Use forceps to hold the skin of the mouse and create a 2–3 cm incision down the midline of the mouse taking care not to puncture the wall of the abdomen. Working in the direction of the xenograft, carefully separate the skin from the underlying muscle until the tumor is revealed.
2. Once the xenograft tumor has been reached, separate the tumor from the underlying muscle.
3. Then separate the tumor from the skin of the mouse. A small piece of the tumor can be saved in 4% PFA for analysis including immunohistochemistry (see IHC section below) or routine hemotoxylin and eosin (H&E) staining.
4. If the tumor is to be serially passaged, cut a piece of the tumor that is approximately 3 × 3 mm and place it in PBS or M199 on ice until ready for implantation. The remainder of the xenograft should be placed in a 50 mL conical tube containing 5 mL M199.
5. Refer to your animal protocol for proper euthanasia procedures once the tumor has been recovered (see Note 9).

3.2.2. Mincing and Treatment with Collagenase Type IV

1. Use mincing scissors to mince the xenograft tissue from step 3 from Subheading 3.2.1 (see Note 10). Tissue should be minced for 15–20 min.
2. Once the tissue is adequately minced, add M199 medium until the total volume is 15 mL.
3. Add 1 mL (2,000 U/mL) Collagenase type IV (Worthington) to the 50 mL conical tube containing the minced tumor.
4. Incubate the tube(s) horizontally in a 37°C shaking water bath for 45 min.
5. After incubation, washes should be carried out under sterile conditions.
6. Add 20–30 mL HBSS to each conical tube.
7. Centrifuge at 250 × *g* for 5 min. Pipette or decant off the supernatant (see Note 11).
8. Resuspend the digested tissue with 15 mL of 20% FBS in HBSS to each conical tube.

9. Centrifuge and pipette or decant off supernatant as above.
10. Resuspend in 15 mL 2% FBS in HBSS.
11. Centrifuge, pipette or decant, and resuspend in 10 mL 2% FBS in HBSS again.
12. Place tubes on ice.
13. Filter one sample at a time while keeping the others on ice.

3.2.3. Filtering the Tissue

1. Two 50 mL conicals, at least one each 70 µm and 40 µm filters, and one 10 mL syringe should be used per sample.
2. Place the 70 µm filter in one 50 mL conical and pipette the digested tissue from Subheading 3.2.2. step 3, into the filter.
3. Use the plunger from a 10 mL syringe to grind the tissue against the filter (see Note 12).
4. Discard the filter and the syringe in an appropriate biohazard waste container.
5. Place the 40 µm filter in the second 50 mL conical tube and pipette the 70 µm filtrate through the 40 µm filter.
6. Discard the filter and recap the tube.
7. After filtering, centrifuge the flow through at 250 × *g* for 5 min. Resuspend the pellet in a known volume of 2% FBS in HBSS and count the live cells obtained (see Note 13).
8. Once cells are counted they can be frozen for future FACS analysis, sorted the same day or stored overnight in an ice bath at 4°C and sorted the next day.
9. To freeze the cells, centrifuge the cells at 250 × *g* for 5 min. Add 1 mL of freezing medium (90% FBS with 10% DMSO), resuspend the pellet, and place in a cryovial for routine freezing, if sorting the same day, aliquot one to two million cells into a fresh conical tube and bring the volume of 2% FBS/HBSS up to 2–3 mL. Follow the procedure for sorting in Subheading 3.4. Any remaining cells can be frozen for later use. If sorting the next day, bring the volume of 2% FBS/HBSS up to 3–5 mL. Store cells on ice at 4°C. Follow sorting procedure in Subheading 3.4 (see Note 14).

3.3. Dissociation of Primary Tissue

1. Transport primary tissues in 50 mL conical tubes of M199 medium.
2. Before dissociation wash all primary samples in a 50 mL conical tube. Rinse the samples with sterile PBS and discard the supernatant. (During the washing steps, sections of tissue can be taken for IHC, frozen sectioning, protein isolation, RNA isolation, etc. For IHC procedure, see Subheading 3.6).

3. Incubate the samples in enough antimicrobial solution to cover the sample at room temperature three times for 15 min each time, discarding the supernatant after each wash.
4. Incubate the samples in 0.014% bleach solution for 5 min and discard the supernatant.
5. Rinse the samples in PBS, discarding the supernatant, and place the samples in 5 mL of M199 medium. A portion of the tissue can be dissected and set aside for xenograft implantation (see Subheading 3.1).
6. Mince tissue according to step 1 in Subheading 3.2.2. Use a 10 mL serological pipette to pipette the tissue up and down (see Note 14).
7. Bring volume of M199 medium to 15 mL.
8. Add 1 mL collagenase type IV (1 mg/mL) and shake tubes containing tissue at 37°C for 45 min. Every 15 min pipette the tissue up and down using a serological pipette (see Note 15).
9. After the incubation follow wash and filtering steps 6–13 from Subheading 3.2.2 and Subheading 3.2.3 respectively.

3.4. Sorting for Aldehyde Dehydrogenase Expressing Cells (See Note 16)

1. Thaw cells for sorting at 37°C. Immediately after cells thaw transfer them to a 15 mL conical tube and add 2–3 mL sterile PBS.
2. Centrifuge at 250 × *g* for 5 min and wash two more times with 2% FBS in HBSS.
3. Resuspend cells in 2–3 mL of 2% FBS in HBSS and count the cells. If staining cells from culture or from a fresh harvest, proceed to step 4.
4. Blocking: Cells should be blocked with human immunoglobulins for 20 min on ice. After incubation wash 2× with 2% FBS/HBSS and resuspend the cells in the same volume of 2% FBS/HBSS.
5. Controls: We recommend DAPI, autofluorescence, a mouse control, a human control, and mouse IgG as controls. We recommend using 20,000 viable cells per control. The tubes are labeled as shown in Table 1.
6. Staining Procedure: ALDH staining is carried out using the ALDEFLUOR Assay according to the manufacturer's instructions. For separation of mouse and human cells during FACS, we suggest Epithelial Surface Antigen (Closely related to "Epcam"; Miltenyi Biotech) at a concentration of 2.5:100 (antibody purchased as liquid) and H2K (Southern Biotech) at a concentration of 0.5 μg/mL (for FACS settings (see Note 18)).
7. Flow cytometer and FACS settings: We use the FACS Aria Cell Sorter (BD Biosciences, San Jose, CA) and the FACS Diva Version 6.1.2 software. DAPI is used to discriminate live from

Table 1
Labelling scheme for tubes when sorting for Aldehyde Dehydrogenase expressing cells

Label	Tubes
A	DAPI only
B	Autofluorescence (no treatment, except washes)
C	Mouse control
D	Human control
E	DEAB (negative control for ALDEFLUOR assay)
F	ALDEFLUOR only
G	Sample
H	IgG

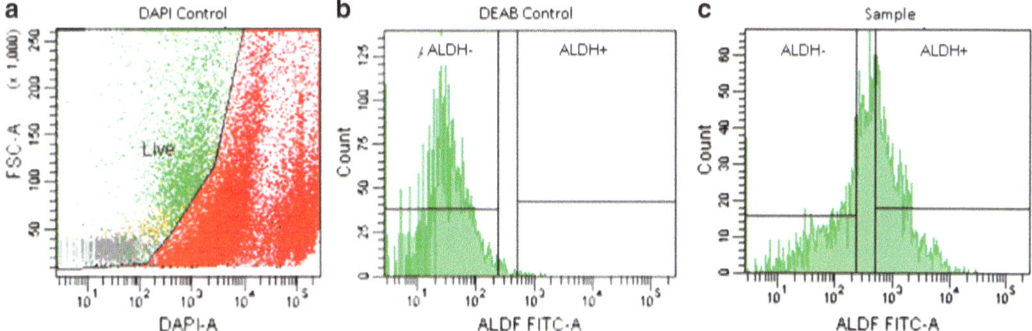

Fig. 1. *FACS ALDEFLUOR histogram* depicting (**a**) suggested DAPI control separating live and dead populations of cells, (**b**) DEAB (negative) control utilized in the ALDEFLUOR assay, and (**c**) example analysis of a sample for ALDH+ and ALDH− fractions.

dead cells. Fluorescence is excited at 355 nm and emission collected at 450 nm ±25 nm. A gate on the live cells is drawn on a Forward Light Scatter versus DAPI fluorescence plot, based on unstained cells (see Note 17, Fig. 1). Markers for sorting ALDEFLUOR+ cells are set based on the DEAB control sample, where approximately 99% of the sample is found to be negative. These markers are applied to the sample (tube G, Table 1). Sorting is carried out using a 100 μm nozzle and maximum flow rate is limited to 5,000 cells/s. All tubes (sample and sort collection) are maintained at 4°C during the sort.

8. After sorting, cells should be washed two times with sterile PBS to remove all serum from the cells. With each wash the cells should be centrifuged at 250 × *g* for 5 min. Cells can then be cultured or injected according to the instructions in Subheading 3.5.

3.5. Sphere Culture and Propagation

1. Culturing colonospheres (see Note 19): After sorting and washing (Subheading 3.4), 25,000 to 50,000 ESA+/ALDH+cells are plated in 24-well low attachment plate with 1 mL of defined medium (DM) (see Subheading 2.5 for the constituents) supplemented with Epidermal growth factor (EGF) (20 ng/mL) and Fibroblast growth factor-2 (FGF2) (10 ng/mL).
2. The cells are maintained in a 37°C incubator with 5% CO_2 and supplemented with the EGF (20 ng/mL) and FGF2 (10 ng/mL) every other day.
3. In 1–2 weeks, spherical clusters of cells are formed in suspension (Fig. 2). These spheres are manually plucked under the microscope in the hood using a 200 μl pipette and are transferred to a fresh well of a 24-well plate containing 1 mL of DM supplemented with EFG and FGF-2. If the number of spheres is less than 5, then the plucking is continued until there are a sufficient number of spheres that can be trypsinized (see Note 20) and passaged. At each passage, excess cells obtained from dissociated spheres can be routinely frozen using freezing medium (90% FBS+10% DMSO), then transferred to liquid nitrogen for long-term storage.
4. Injection of tumor cells into the flank of NSG mice: After sorting and washing (Subheading 3.4), cells should be resuspended in 50 μl of sterile PBS. 50 μl of matrigel is added and pipetted up and down to thoroughly mix the cells, PBS, and matrigel. The mixture is pulled into a syringe and can be injected into the flank of the mouse. These injections can be used to show tumorigenicity.

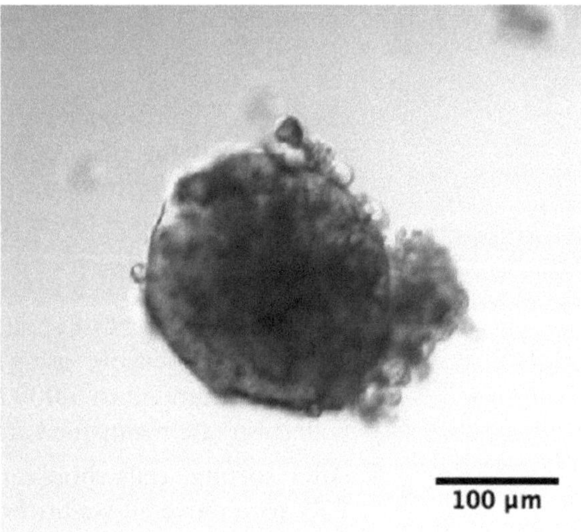

Fig. 2. *Colonosphere*. The sphere obtained from ALDH-enriched colon cancer cells in defined medium.

3.6. IHC for ALDH on Spheres and Primary Tissues

3.6.1. Fixation of Colonospheres

1. Pipette spheres into a conical tube.
2. Rinse culture plate with 1× PBS and add the rinse to the conical tube containing the spheres.
3. Centrifuge spheres at 250 × *g* for 5 min. Discard the supernatant and add 2–5 mL of 4% PFA to the conical tube.
4. Incubate on ice for 10 min.
5. Centrifuge spheres at 250 × *g* for 5 min.
6. Remove all but a small amount of the PFA (leave approximately 50 μl). Discard the waste PFA according to your institution's policies.
7. Pipette remaining PFA and spheres into a round-bottom glass tube and place on ice.
8. Heat 1% agarose until completely fluid. Allow the agarose to cool so that it is not very hot, but still remains liquid.
9. Use a 1 mL syringe (without a needle) to pull approximately 1 cc of 1% agarose into the syringe. Quickly dispense agarose into tube containing the spheres and pull the agarose/sphere mixture back into the syringe. Immediately place the syringe on ice and do not disturb it until the agarose has completely solidified (approximately 10–15 min).
10. Cut the tip of the syringe and dispense the plug *gently* into a 15 mL conical tube containing 5 mL of 4% PFA. Leave overnight at 4°C.
11. Discard PFA and add 70% ethanol to the plug. Incubate overnight at 4°C.
12. Paraffin-embed the plug for sectioning.
13. Section the paraffin block in 5 μm sections for staining.

3.6.2. Retrieval

1. Preheat Coplin staining jar containing 50 mL of Dako target retrieval solution to 95°C in a water bath. The water level should reach to approximately 1 cm below the lid of the container. Place the container in a rack or beaker to keep it stable. Allow the temperature of the Coplin jar to reach 95°C before carrying out the retrieval.
2. Quickly place slides in Coplin staining jar, loosely close the lid, and incubate 20 min at 95°C and 20 min on the bench.
3. Remove slides from retrieval solution and dip them in deionized water.
4. Rinse the slides with 1× TBS-T and place them in 1× TBS-T for 5 min.

3.6.3. Blocking and Primary Antibody Treatment

1. Block slides in TBS-T containing 2% horse serum for 30 min.
2. Apply the primary ALDH1 antibody at 0.5–1 μg/mL. Incubate overnight at 4°C.
3. Wash the slides three times for 5 min at room temperature in 1× TBS-T. Then apply the secondary antibody.

3.6.4. Secondary Antibody Treatment

Be sure to keep all reagents and slides in the dark as much as possible.

1. Apply the secondary antibody using 1× TBS-T with 2% horse serum as a diluent for 1 h at room temperature (we use donkey anti-mouse Alexafluor 488 at a concentration of 4 μg/mL).
2. Wash slides three times for 3 min at room temperature in 1× TBS-T. Use gauze to wipe away excess buffer being careful not to disturb the cells on the slide. Allow slides to dry completely in the dark before mounting.
3. Mount slides using a small amount (one drop is usually more than sufficient for an agarose plug) of mounting medium with DAPI. Slides can then be visualized.

4. Notes

1. Unless noted otherwise "surgical instruments" refers to toothed forceps, serrated microdissecting forceps, curved microdissecting scissors, needle holder, and operating (mincing) scissors.
2. Quintana et al. (3) showed that NOD/SCID mice underestimate the tumorigenicity of cancer cell lines. They show that use of "NSG" mice (NOD/SCID interleukin-2 receptor gamma chain null mice) in place of NOD/SCID mice increases the sensitivity of xenograft assays. The use of NSG mice allows a marked improvement for the assay to be able to detect tumorigenic cells. See also reference (4).
3. Collagenase should be stored at 4°C until reconstituted. It should be reconstituted with sterile PBS at a concentration of 2,000 U/mL. Once reconstituted it should be stored in 1 mL aliquots at −20°C.
4. Collagenase should be stored at −20°C and only the necessary amount reconstituted each time it is used. It should be reconstituted at 1 mg/mL with PBS and then filtered through a 0.22 μm filter to sterilize it before use.
5. All antibodies reconstituted and stored according to the manufacturer's instructions.
6. All liquid reagents and BSA are kept at 4°C, except progesterone, which should be stored at −20°C. All components should be added to the basal medium and mixed thoroughly before being passed through a 0.22 μm filter.
7. We use Antibiotic–Antimycotic solution from Gibco. Penicillin–Streptomycin or others may be used instead to prevent contamination of cell cultures.

8. Always refer to your institution's policies regarding mouse use and care including surgery policies such as the type of anesthesia your institution prefers. Your institution may also require the use of a heating pad be used through the duration of the surgery to prevent hypothermia of the mouse. Sterilization of the flank is essential as NSG mice are highly immune deficient and, therefore, susceptible to infection. The flank may be shaven to facilitate sterilization and implantation of the xenograft. Antibiotic treatment post-surgery is suggested to aid in the prevention of infection.
9. There are multiple ways to euthanize a mouse after the recovery of a xenograft. The method we use is to administer an overdose of anesthesia followed by confirmation using cervical dislocation. Any researcher attempting this or any other euthanasia procedure should be thoroughly trained in the technique(s) used.
10. An automated dissociator may be used for large batches of samples, but should not be used routinely. Automatic dissociation can decrease cell viability. Also, colon samples contain mucin, can clog the automated dissociator making cell sorting much more time consuming due to the increased viscosity of the cell suspension. Primary human samples should always be hand-minced as automated dissociators will become clogged with tissue and cannot adequately mince the sample.
11. Pipetting off the supernatant is recommended if the pellet does not stick well to the conical tube after centrifugation. Often normal samples, or tumors with a large quantity of mucin, will not pellet completely when centrifuged. In general, there is less cell loss involved with *all* samples when pipetting rather than decanting the supernatant.
12. Filtering the tissue aids in the release of epithelial cells from the underlying tissue and should be done thoroughly. After grinding the tissue on the filter it can be rinsed with the flow-through to collect as many epithelial cells as possible.
13. To count the cells, dilute them 1:10 with Trypan blue and use a hemocytometer to quantify them. Only count larger, round cells. Many tiny cells will be visible, but these are red blood cells and other cells that should not be included in the enumeration.
14. If sorting longer than 24–36 h after dissociation, freeze cells and thaw them right before sorting according to the procedures in Subheading 3.2.3, step 9 and Subheading 3.4.
15. It is necessary to pipette the tissue up and down in order to ensure the tissue is properly minced and to aid the release of epithelial cells during the incubation with collagenase.

16. The ALDEFLUOR kit detects the enzymatic activity of the family of aldehyde dehydrogenases, not exclusively ALDH1. All incubation steps should be performed on ice unless otherwise indicated to prevent cell death during the incubations.
17. If using ESA and $H2^K$ the following conditions may be used: ESA fluorescence (PE) and $H2^K$ fluorescence PE-CY5 are excited with a 488 nm laser. PE emission is collected at 575 ± 12.5 nm and PE-Cy5 emission is collected at 710 ± 25 nm. A gate for the ESA+ cells was applied using a ESA versus $H2^K$ dot plot.
18. Sorting notes: ESA fluorescence (PE) and $H2^K$ fluorescence PE-CY5 were excited with a 488 nm laser. PE emission was collected at 575 ± 12.5 nm and PE-Cy5 emission was collected at 710 ± 25 nm. A gate for the ESA+ cells was applied using a ESA versus H2K dot plot. ESA+ cells were plotted on a histogram showing FITC fluorescence, which was excited at 488 nm and emission collected at 530 ± 15 nm.
19. Sphere culture is initiated only after three or four passages of the human colon cancer tissue as xenografts in NOD-SCID mice as mentioned in Subheading 3.1. Passaging *in vivo* eliminates progenitors, as they are unable to propagate beyond a few passages. The passaging enriches the stem cell population, which can be readily propagated as spheres.
20. To trypsinize spheres, first centifuge them at 250 × *g* for 5 min at 4°C. Pipette or decant off the supernatant and resuspend the pellet in 1 mL of 0.25% trypsin with EDTA. The spheres are incubated in the trypsin with EDTA for 3–5 min and pipetted up and down every minute or so to aid dissociation of the spheres. The trypsinization is stopped using medium or PBS containing 10% FBS. The sphere cells are washed twice with plain PBS to remove any remaining serum from the cells. After the final wash, cells are resuspended in DM containing growth factors and are transferred into low attachment plate for further propagation.

Acknowledgments

The authors would like to thank Jason E. Cline for preparation of this chapter and Neal Benson for his expertise in flourescence-activated cell sorting. This work was supported by NCI R01 142808, the University of Florida and Shands Cancer Center, and the University of Florida Seed Fund (EHH).

References

1. Huang E, Hynes M, Zhang T, Ginestier C, Dontu G, Appelman H, Fields JZ, Wicha MS, Boman BM (2009) ALDH1 is a marker for normal and malignant human colonic stem cells and tracks stem cell overpopulation during colon tumorigenesis. Cancer Res 69:3382–3389
2. Carpentino JE, Hynes MJ, Appelman HD, Zheng T, Steindler DA, Scott EW, Huang EH (2009) Aldehyde dehydrogenase-expressing colon stem cells contribute to tumorigenesis in the transition from colitis to cancer. Cancer Res 69:8208–8215
3. Quintana E, Shackleton M, Sabel MS, Fullen DR, Johnson TM, Morrison SJ (2009) Efficient tumor formation by single melanoma cells. Nature 456:593–598
4. Ishizawa K, Rasheed ZA, Karisch R, Wang Q, Kowalski J, Susky E, Pereira K, Karamboulas C, Moghal N, Rajeshkumar NV, Hidalgo M, Tsao M, Ailles L, Waddell TK, Maitra A, Neel BG, Matsui W (2010) Tumor-initiating cells are rare in many human tumors. Cell Stem Cell 7:279–282

Chapter 28

Protocols for Investigating microRNA Functions in Human Neural Progenitor Cells

Sandra Almeida, Celine Delaloy, Lei Liu, and Fen-Biao Gao

Abstract

Human embryonic stem cells and induced pluripotent stem cells offer great hope for studies of pathogenic mechanisms of disease and cell-based therapies. One powerful approach to manipulate the behaviors of human stem cells and their progenies is through microRNAs (miRNAs), a class of small noncoding RNAs that regulate gene expression at the posttranscriptional level. Each miRNA may target up to hundreds of mRNAs; some are specifically expressed in progenitor cells and affect multiple cellular processes. Here we present experimental protocols for investigating the endogenous functions of specific miRNAs in the proliferation, survival, and migration of human neural progenitor cells derived from embryonic stem cells. These methods may be applicable to protein factors and neural progenitor cells derived from patient-specific induced pluripotent stem cells.

Key words: microRNA, Human stem cells, Neural progenitor cells, Proliferation, Migration, Apoptosis, Transplantation

1. Introduction

Human embryonic stem cells (hESCs) can develop along specific lineages into progenitor cells that in turn differentiate into diverse cell types, such as neurons, skin cells, and muscle cells (1, 2). Thus, hESCs and their progenies hold great promise for understanding human development and for regenerative medicine. This prospect is greatly enhanced by a technological breakthrough—the ability to use only a few transcription factors to directly reprogram human somatic cells into induced pluripotent stem cells (iPSCs) (3). Now, patient-specific iPSCs with global gene expression profiles similar to those of hESCs can be generated with relative ease (4, 5).

Both hESCs and hiPSCs can be differentiated into human neural progenitor cells (hNPCs) and subsequently into various

Kimberly A. Mace and Kristin M. Braun (eds.), *Progenitor Cells: Methods and Protocols*, Methods in Molecular Biology, vol. 916, DOI 10.1007/978-1-61779-980-8_28, © Springer Science+Business Media, LLC 2012

types of post-mitotic neurons using different protocols (6–8). Many attempts have been made to transplant hNPCs into animal models of human diseases, such as stroke (8), retinal degeneration (9), and Parkinson's disease (10). This therapeutic approach remains highly challenging, in part because we do not fully understand the characteristics and cellular behaviors of hNPCs. Much remains to be learned about the molecular mechanisms that control the proliferation, survival, migration, and differentiation of hNPCs in vitro and in vivo.

Among the important regulators of hNPCs are microRNAs (miRNAs), a class of noncoding RNAs of ~21–23 nucleotides that regulate mRNA translation and stability through base-pairing, mostly with sequences in the 3′ untranslated region (11, 12). miRNAs are excellent targets for molecular manipulations in hNPCs because of their small size and unique modes of action. For instance, although each miRNA is predicted to target hundreds of mRNAs (13), the effects of many miRNAs are mediated by one or a few key targets in a specific developmental process (14). Here, we describe in detail experimental approaches to knock down the expression of endogenous miRNAs in hNPCs and examine their physiological functions in the proliferation, survival, migration, and differentiation of hNPCs in vitro and ex vivo.

2. Materials

2.1. hNPC Culture and miR-9 Knockdown

1. Human ESCs (H9 line, WiCell Research Institute).
2. Mouse embryonic fibroblasts (MEFs) (PMEF-CFL), previously inactivated with mitomycin C (Roche Applied Science).
3. Human ESC medium: Dulbecco's modified Eagles's medium (DMEM)/F12 (Gibco) supplemented with 20% KnockOut Serum Replacement (Gibco), 1× nonessential amino acids (Gibco), 4 ng/mL human basic fibroblast growth factor 2 (FGF2; Gibco), and 0.1 mM 2-mercaptoethanol.
4. Neural induction medium (NIM) consisting of 33% F12 nutrient mixture (Gibco), 66% DMEM (Gibco), 1× N2 supplement (Gibco), 1× nonessential amino acids solution, 10 ng/mL FGF2, and 2 μg/mL heparin (Sigma).
5. Dispase (Gibco) is prepared at 1 mg/mL in DMEM/F12 and incubated at 37°C for 15 min to dissolve completely. It should be filter-sterilized and used fresh.
6. Poly-L-ornithine solution (Sigma).
7. Laminin (Sigma) is dissolved at 1 mg/mL in distilled water.
8. 24-well plates coated with Ultra-Low Attachment Surface (Corning).

9. miRCURY LNA microRNA inhibitor hsa-anti-miR-9, 5 nmol (Exiqon) and miRCURY LNA microRNA inhibitor negative control A, 5 nmol (Exiqon).
10. Lipofectamine 2000 (Invitrogen).
11. Opti-MEM I reduced-serum medium (Invitrogen).
12. miRNeasy kit (Qiagen).
13. Turbo DNA-free kit (Ambion).
14. TaqMan microRNA reverse transcription kit (Applied Biosystems).
15. Reagents for TaqMan microRNA assays: hsa-miR-9 and RPL21/small nucleolar RNA (snoRNA) (Applied Biosystems). Each TaqMan assay kit provides primers for RT (5×) and qPCR (20×).
16. TaqMan Universal PCR Master Mix, No AmpErase UNG (Applied Biosystems).

2.2. Cell Proliferation and Viability Assays

1. In Situ Cell Death Detection Kit, TMR Red (Roche Applied Science).
2. Matrigel (BD Bioscience).
3. Fixation solution: 4% paraformaldehyde in phosphate-buffered saline (PBS), pH 7.4, freshly prepared.
4. Permeabilization solution: 0.1% Triton X100 in 0.1% sodium citrate, freshly prepared.
5. DNase I recombinant (Roche Applied Science) solution: 300 U/mL in 50 mM Tris–HCl, pH 7.5, 1 mg/mL bovine serum albumin.
6. Glass coverslip.
7. Vectashield mounting medium (Vector Laboratories).
8. CytoTox 96 nonradioactive cytotoxicity assay (Promega).
9. 5-Bromo-2′-deoxyuridine (BrdU) is dissolved at 1 mg/mL in water. Combine 10 mg of BrdU with 10 mL of water and filter sterilize with a 0.22-μm filter. Dispense in 1-mL aliquots and freeze at −20°C.
10. PBS/1% Triton X-100.
11. 1 M and 2 M HCl.
12. 0.1 M borate buffer: Prepare 200 mL of 0.2 M sodium borate and 200 mL of 0.2 M boric acid. Add boric acid to sodium borate solution until pH 7.4 is reached.
13. Blocking solution: Combine 0.1 g bovine serum albumin (BSA), 0.5 mL donkey serum (Sigma), and PBS/Triton X-100 to make 10 mL final volume.
14. BrdU antibody (Abcam).

15. Alexa Fluor 594 donkey anti-mouse (Molecular Probes).
16. Hoechst 33258 (Invitrogen) is dissolved at 1 mM in distilled water.
17. Cell proliferation reagent WST1 (Roche Applied Science).

2.3. Migration Assays

1. GFP-FUGW plasmid (Addgene).
2. PEG-it Virus precipitation solution (SBI System Biosciences).
3. E14.5 C57BL/6 pregnant mice (Charles River Laboratories).
4. Autoclaved dissecting tools.
5. Penicillin–streptomycin solution.
6. Rinse solution: Hanks' balanced salt solution (HBSS) (Gibco) containing 100 U/mL penicillin/streptomycin.
7. Embedding medium: 2.5% low-melt agarose (Sigma) in slicing medium.
8. Slicing medium: DMEM/F12 (Gibco) and glucose (final concentration, 6.5 mg/mL).
9. Vibratome for brain slices. Razor blade cleaned with ethanol before use.
10. Millicell cell culture insert (Millipore).
11. Nutrition medium: DMEM/F12, 10% Characterized Fetal Bovine Serum (HyClone) and 100 U/mL penicillin/streptomycin.
12. Standard Wall Borosilicate Tubing (Sutter Instrument).

3. Methods

3.1. In Vitro Differentiation of Neural Precursors from hESCs

1. Human ESCs are cultured on an MEF feeder layer (Fig. 1); the medium should be refreshed daily. To obtain efficient and reproducible differentiation of hESCs and to prevent premature differentiation or cell death, colonies must be split before they become too large or start to fuse with each other.
2. When it is time to split the colonies, remove the medium from three wells of a 6-well plate containing hESC colonies (see Note 1).
3. Add 0.5 mL of hESC medium and 0.5 mL of dispase (final concentration, 0.5 mg/mL) to each well.
4. Incubate in a CO_2 incubator for 5–10 min until the colonies start to curve up.
5. Remove the dispase and add 1 mL of hESC medium to each well. Gently scrape cells with a cell lifter and pool the suspension into a 15-mL conical tube. Be careful to not break up the colonies.

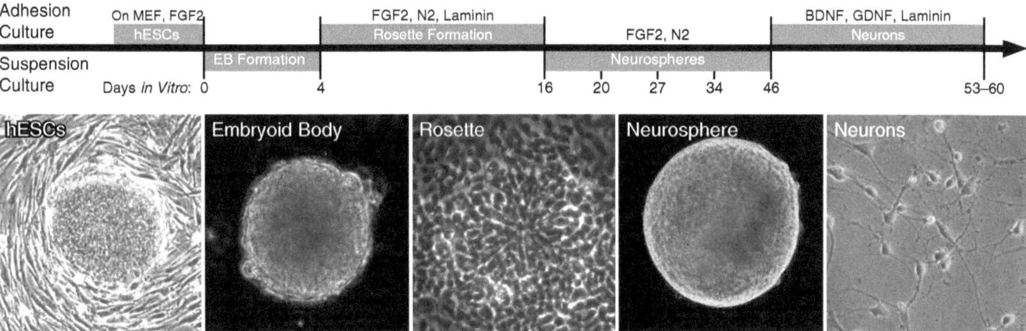

Fig. 1. Neural differentiation of hESCs. H9 hESCs were maintained on MEF feeder cells in the presence of FGF2. Day in vitro (DIV) 0 refers to the day hESCs were first dissociated from MEFs and cultured in suspension. Neural differentiation of hESCs started with the formation of EBs, followed by rosette formation at 13–16 DIV in the presence of FGF2 and N2 supplements. Rosette structures resemble neuroectodermal cells. Neural progenitors in rosettes were positive for SOX2 and PAX6. For neurosphere formation, rosette-forming cells were dissociated and expanded in suspension cultures, and the resulting neurospheres were cultured at 16–46 DIV. For terminal neuronal differentiation, neural progenitors were dissociated from neurospheres and exposed to BDNF and GDNF at 43–60 DIV. Scale bar, 100 μm. (Reproduced from ref. (8) with permission from Elsevier Science).

6. Centrifuge at 200 × *g* for 2 min at the room temperature.
7. Aspirate supernatant from hESC pellet. Resuspend pellet in hESC medium without FGF2 and centrifuge again.
8. Remove medium and resuspend cell pellet in 8 mL of hESC medium without FGF2.
9. Culture cells for 4 days in a T25 flask; change half of the medium daily (see Note 2). Human ESC cell aggregates will form embryoid bodies (EBs) in 24 h.
10. On day 5, collect the EBs into a 15-mL tube and centrifuge at 50 × *g* for 1 min at room temperature.
11. Wash EBs with 5 mL of NIM and centrifuge again.
12. Resuspend EBs in 24 mL of NIM and dispense 8 mL of the suspension per T25 flask, each freshly pre-coated with poly-L-ornithine/laminin (see Note 3).
13. EBs attach to the pre-coated plastic substrates in 2 days. After that, replace 5 mL of culture medium with fresh NIM every other day. Cells will begin to migrate out from the EBs, and in a few days elongated cells will appear and form rosettes.
14. On day 16–18 of differentiation culture (11–13 days after plating the EBs), isolate neuroepithelial cells in the rosettes from the surrounding cells. Wash the rosette culture once with PBS and add 4 mL of NIM and 1 mL of dispase (final concentration, 0.2 mg/mL).
15. Incubate in a CO_2 incubator for 15 min. Retracted rosette clumps will begin to come off the plate; if necessary, tap sides of the flask and incubate for additional 5–10 min.

16. Collect rosette clumps from the three T25 flasks into a 15-mL tube and triturate them with a 5 mL pipette but without breaking up the clumps. Centrifuge at 50 × *g* for 1 min at room temperature and aspirate medium. Add 10 mL of NIM and centrifuge again to wash away the dispase and single cells.
17. Resuspend the cells in 6 mL of NIM, transfer to a T25 flask, and incubate in a CO_2 incubator for 2–3 h to allow nonneural cells to attach to the flask.
18. Collect the floating cells into a 15-mL tube and centrifuge at 50 × *g* for 1 min. Resuspend the pellet in 3 mL of NIM.
19. Mechanically dissociate them into smaller clumps (around ten cells) with a 5 mL pipette (see Note 4). Divide cells in suspension into six wells of a 24-well plate coated with Ultra-Low Attachment Surface.

3.2. Transfection with Anti-miR-9 LNA Probe (See Note 5)

1. Prepare LNA/Lipofectamine complexes. For one well of a 24-well plate: mixture A, combine 2 μL of hsa-anti-miR-9 or scrambled control (see Note 6) (final concentration, 100 nM) and 50 μL of Opti-MEM I. For mixture B, in a separate tube, combine 2 μL Lipofectamine with 50 μL of Opti-MEM I. Mix well and incubate 5 min at room temperature.
2. Add mixture A to mixture B, mix well, and incubate for 20 min at room temperature.
3. Apply the complexes to the cells. Incubate at 37°C in a CO_2 incubator overnight.
4. The next day, add 0.5 mL of NIM. Move the plate back to the CO_2 incubator. Assay for miR-9 knockdown 24 h later and at later stages (see Note 7).

3.3. Analysis of miR-9 Expression

1. Extract total RNA with the miRNeasy kit according to the manufacturer's instruction.
2. Purify RNA with the Turbo DNA-free kit following manufacturer's instruction.
3. Determine RNA concentration by measuring absorbance at 260 nm in a spectrophotometer (see Note 8).
4. Reverse transcribe 50 ng of total RNA with the TaqMan microRNA reverse transcription kit and miR-9-and RPL21/snoRNA-specific RT primers (TaqMan microRNA assays) in the same RT reaction (see Note 9).
5. Incubate the reaction tube at 16°C for 30 min, 42°C for 30 min, and 85°C for 5 min in the thermal cycler. Store reaction tubes at –20°C if not proceeding to the next step.
6. For qPCR amplification, prepare a qPCR master mix for each primer. For each 20 μL reaction, add 1 μL of TaqMan MicroRNA Assay (20×; forward primer, reverse primer, and

probe), 10 μL of TaqMan Universal PCR Master Mix (2×) and 4 μL of nuclease-free water.

7. Dilute the RT reaction 1:12 (see Note 10).
8. Transfer 5 μL of diluted RT reaction into each well of a 96-well qPCR plate (each sample should be run in triplicate). A no-template control reaction should also be prepared to evaluate background signal.
9. Add 15 μL of the master mix (miR-9 or RPL21) to each well.
10. Seal the plate with the optical adhesive cover and centrifuge the plate for 2 min.
11. Load the plate into the instrument and run the following thermal cycling conditions: 95°C for 10 min followed by [95°C for 15 s and 60°C for 60 s] for a total of 40 cycles. Quantify at the threshold detection line (Ct value) and normalize the Ct of miR-9 to that of human snoRNA located within an RPL21 gene intron.

3.4. Evaluation of Cell Viability with the TUNEL Assay

1. The day after transfection (or up to 6 days after; see Subheading 3.2), transfer hNPCs (whole or dissociated neurospheres) to Matrigel (1:100)-coated wells (24-well plate) and allow cells to attach for 3–4 h (see Note 11).
2. Remove medium from cells and rinse with PBS.
3. Fix cells in 0.3 mL of 4% paraformaldehyde for 1 h at room temperature (see Note 12).
4. Rinse cells with PBS and incubate with 0.3 mL of permeabilization solution for 2 min on ice.
5. Rinse cells twice with PBS.
6. Prepare positive control: incubate two wells with DNase I recombinant for 10 min at room temperature to induce DNA strand breaks. Rinse with PBS.
7. Prepare TUNEL reaction mixture (500 μL) by adding total volume of enzyme solution (vial 1, In Situ Cell Death Detector Kit) to 450 μL of Label Solution (vial 2, In Situ Cell Death Detector Kit). Mix well with a pipette.
8. Add 50 μL of TUNEL reaction mixture to each well (including positive control). For the negative control, add 50 μL of label solution (vial 2) to each well (prepare two wells).
9. Cover plate with parafilm and incubate in a humidified atmosphere for 1 h at 37°C in the dark.
10. Rinse three times with PBS.
11. Add a drop of mounting medium and place a glass coverslip on top. Samples are ready to be analyzed under a fluorescence microscope (see Notes 13 and 14) See example in Fig. 2a.

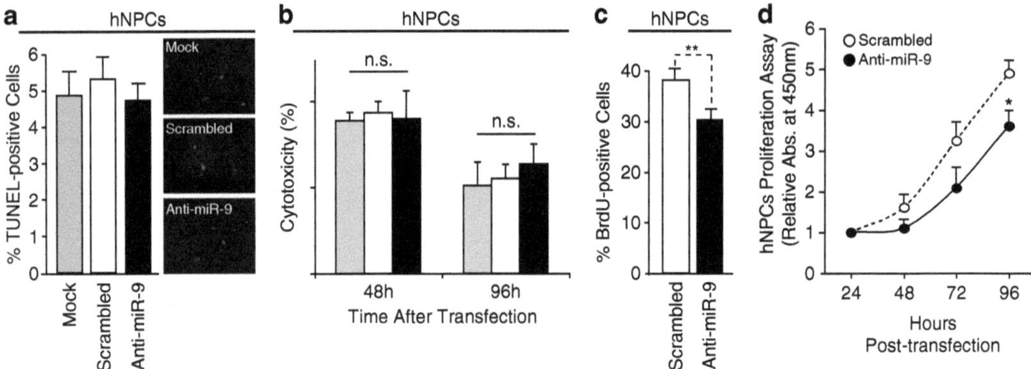

Fig. 2. Assays for studying the survival and proliferation of hESC-derived hNPCs. (**a**) Loss of miR-9 does not affect the survival of human neural progenitors as shown by TUNEL staining. Three representative images of TUNEL staining are presented on the right. (**b**) LDH assay to measure the cytotoxic effect of miR-9 knockdown on hNPCs. Cells were transfected with scrambled LNA probe or with anti-miR-9 LNA probe at 16 DIV, and cytotoxicity was measured 48 h and 96 h after transfection. *n.s.* not significant. Values are mean ± SEM. (**c**) BrdU incorporation was measured to indicate the rate of proliferation of hNPCs with or without miR-9 knockdown. The percentage of BrdU-positive cells at 3 days after transfection is shown. Values are mean ± SEM. $^{**}P<0.01$ ($n=3$). (**d**) WST-1 assay to measure the proliferation of hNPCs after transfection of the LNA-anti-miR-9 or scrambled probe at 16 DIV of neuronal differentiation. Values are mean ± SEM of 8–16 wells per time point and per conditions. $^{*}P<0.05$. (Reproduced from ref. (8) with permission from Elsevier Science.).

3.5. Evaluation of Cell Viability with the LDH Assay (See Note 15)

1. The day after transfection (or up to 6 days; see Subheading 3.2), remove medium and add 0.5 mL of fresh NIM to each well. Add medium to an empty well as a background control.
2. Twenty-four hours later, move the contents of each well (medium and neurospheres) to a 1.5-mL centrifuge tube and centrifuge at 5,000 × g (Microcentrifuge) for 2 min.
3. Collect each supernatant in a new tube (released LDH) and add 0.5 mL 1× lysis buffer (CytoTox 96 Cytotoxicity Assay kit) to each cell pellet (intracellular LDH). Mix well.
4. Incubate the tubes containing the cell pellet at –70°C for 20 min. Thaw at 37°C for 2–3 min and centrifuge the tubes at 10,000 × g (Microcentrifuge) for 4 min.
5. Transfer 50 μL of supernatant/cell lysate of each sample to a 96-well flat-bottom plate. For the background controls, transfer 50 μL of the medium/1× lysis buffer. All samples should be done in duplicate.
6. Add 12 mL of room temperature Assay Buffer (CytoTox 96 Cytotoxicity Assay kit) to a bottle of Substrate Mix (CytoTox 96 Cytotoxicity Assay kit). Invert and shake gently to dissolve the substrate. Protect the substrate solution from light and use immediately.
7. Add 50 μL of the substrate solution to each well of the plate. Protect plate from light and incubate at room temperature for 30 min.

8. Add 50 μL of Stop Solution (CytoTox 96 Cytotoxicity Assay kit) to each well. Read absorbance at 490 nm.
9. Subtract background values from the sample readings and determine cytotoxicity by calculating the percent of LDH released into the medium out of the total LDH activity (the sum of intracellular and released LDH activity) (Fig. 2b).

3.6. BrdU Staining

1. The day after transfection (or up to 6 days after; see Subheading 3.2), transfer hNPCs to Matrigel-coated wells (24-well plate) and allow cells to attach for 3–4 h (see Note 11).
2. Dilute BrdU stock solution in NIM to obtain a 1 mM BrdU solution.
3. Add 10 μL of the 1 mM BrdU solution per mL of culture medium to each well (final concentration, 10 μM) (e.g., if the well volume is 0.5 mL, add 5 μL to each well). Incubate for 4 h.
4. Remove medium from cells and rinse once with PBS.
5. Fix cells with 4% paraformaldehyde for 30 min at 4°C.
6. Wash cells three times for 5 min each with PBS/1% Triton X100.
7. Incubate cells in 1 M HCl for 10 min on ice. This step breaks open the DNA structure of the labeled cells.
8. Incubate in 2 M HCl for 10 min at room temperature followed by 20 min at 37°C.
9. Remove HCl and add 0.1 M borate buffer for 12 min at room temperature.
10. Wash cells three times for 5 min each with PBS/1% Triton X-100.
11. Incubate with blocking solution for 30 min at room temperature.
12. Dilute anti-BrdU antibody 1:400 in blocking solution and incubate at 4°C overnight.
13. Wash cells three times for 5 min each with PBS/1% Triton X-100.
14. Dilute Alexa Fluor 594 anti-mouse antibody 1:500 in blocking solution and incubate at room temperature for 2 h. Protect plate from light.
15. Wash cells for 5 min with PBS/1% Triton X-100 and twice for 5 min each with PBS.
16. Incubate cells with Hoechst (final concentration, 1 μM in PBS) at room temperature for 5–10 min.
17. Rinse cells with PBS.
18. Add a drop of mounting medium and place a glass coverslip on top. Samples are ready to be analyzed under a fluorescence microscope (Fig. 2c).

3.7. WST1 Proliferation Assay

1. Twenty-four hours after transfection (see Subheading 3.2), collect hNPCs and seed them in a 96-well plate (8–16 wells per condition, see Note 16) in a final volume of 100 μL/well. Add medium to an empty well to be used as a background control. Prepare a new plate for each time point (see Note 17).
2. Add Cell Proliferation Reagent WST1 (10 μL/well) and incubate the cells for 1 h in the CO_2 incubator. During the incubation, viable cells convert WST1 to a water-soluble formazan dye.
3. Shake the plate for 1 min on a shaker.
4. Read absorbance at 450 nm.
5. Subtract background value (medium without cell) from the sample readings. Normalize the average absorbance for each time point to the value on day 1 after transfection. The absorbance is directly correlated with the cell number (Fig. 2d).

3.8. In Vitro: Three-Dimensional Cell Migration Assay (See Note 18)

1. Dilute Matrigel (see Subheading 2.2, Item 2) 1:1 in NIM (see Note 19).
2. Put 70 μL of Matrigel mixture in each well of a 96-well plate. Incubate at 37°C overnight to polymerize.
3. Three days after transfection (see Subheading 3.2), collect hNPCs forming neurospheres from each condition into 15-mL tubes and centrifuge at 50 × g for 1 min at room temperature.
4. Resuspend neurospheres in a small volume of NIM, just enough to keep them in suspension and place a small volume (not more than 5 μL) on top of the polymerized Matrigel. Rock the plate back and forth. Check plate under the microscope to make sure the neurospheres are evenly distributed, with no more than three per well.
5. Apply 70 μL of Matrigel mix (see Note 19) on top of the hNPC layer. Incubate at 37°C.
6. Take pictures of the hNPCs at desired time points (e.g., 2.5 h, 24 h, 48 h, 96 h) (see Note 20). The migration phenotype can be quantified by measuring the distance from the hNPC to the nucleus of the most distant cell at various times after transfection or/and by counting cells that migrated out of the hNPCs, see example in Fig. 3.

3.9. Ex Vivo: Migration of Transplanted hNPCs in Embryonic Brain Slices (See Note 21)

1. Prepare GFP-FUGW lentivirus (see Note 22). Cotransfect the vector encoding GFP, the HIV-1 packing vector Δ8-9, and the VSVG envelope glycoprotein into 293TN producer cells. Concentrate the pseudoviral particles using PEG-it virus precipitation solution following manufacturer instructions to obtain ultrahigh titer preps without toxic reactions.
2. Transfect hNPCs as indicated in Subheading 3.2, Items 1–3. At the same time, transduce the cells with high titer virus to obtain at least 30% infectivity (see Note 23).

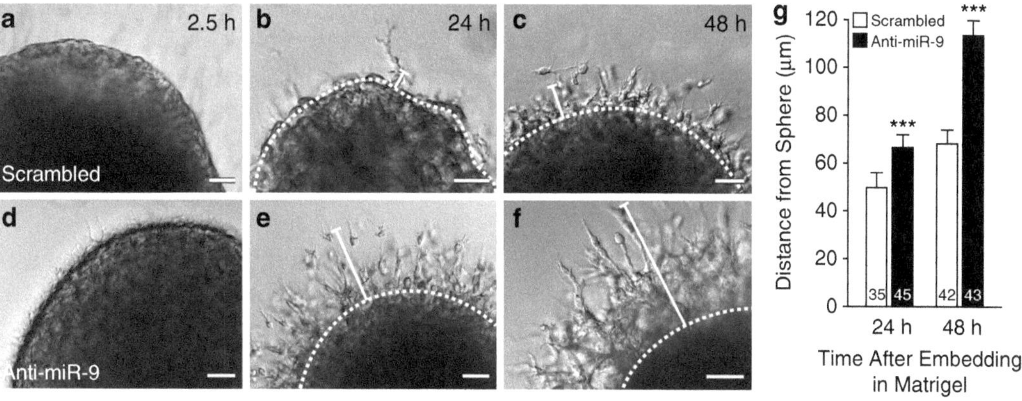

Fig. 3. 3D Matrigel migration assay to examine the migratory behaviors of hNPCs. (**a–c**) hNPCs transfected with scrambled LNA probe and cultured in a 3D matrix. (**d–f**) hNPCs transfected with anti-miR-9 LNA probe and cultured in a 3D matrix. hNPCs were dissociated from rosettes at 16 DIV, transfected, cultured for 3 days in suspension, and embedded in Matrigel. (**g**) Quantification of migratory behaviors of hNPCs in Matrigel. The distance from the edge of the neurosphere to the most distant nucleus of the outmigrating cells was quantified. Values are mean ± SEM. (Reproduced from ref. (8) with permission from Elsevier Science.).

3. Incubate for 24 h and replace the medium with fresh NIM.
4. Two days after transfection, anesthetize a pregnant female and sacrifice by cervical dislocation. Remove the embryos and place them in ice-cold PBS.
5. Carefully remove the brains from the embryos and transfer them to a Petri dish containing ice-cold rinse solution and rinse brains once.
6. Embed brains in embedded medium in a Petri dish and immediately put them on ice.
7. Cut out a block containing a single embedded brain. Orient the embedded brain and mount it on the Vibratome stage with cyanoacrylate (e.g., SuperGlue) so that coronal forebrain sections will be made.
8. Cut coronal forebrain slices 250 μm thick with the Vibrotome (at a high frequency and low speed) and keep them in ice-cold slicing medium.
9. Transfer the slices to culture dishes with ice-cold slicing medium. Select the three best forebrain slices from each brain under the microscope (use ~5 embryos per condition). Wash them twice with sterile slicing medium and once with the nutrition medium. With a cut 1 mL disposable transfer pipette, individually place slices onto the membrane of a tissue culture insert in a 6-well plate containing 1.2 mL of nutrition medium per well (see Note 24). Up to six slices may be placed on a single membrane.

10. Place plates in the CO_2 incubator for ~1 h. Meanwhile, prepare the neurospheres for transplantation by washing them twice in PBS and once in nutrition medium.
11. Prepare transfer pipettes by pulling glass pipettes (see Note 25). Cut pipettes by hand with a sharp scalpel blade under a dissecting microscope where the diameter is ~150 μm. A clean neat break is essential.
12. Under dissection microscope, carefully select neurospheres on the basis of size and load one at a time into a pipette. Make sure that the tip of the pipette does not touch anything to avoid contamination.
13. Transfer a membrane insert containing the forebrain sections to a 10-cm Petri dish without culture medium and place on the dissection microscope stage. Blow gently into the mouth capillary until a single neurosphere has been expelled into the medial ganglionic eminence.
14. Replace the insert into the 6-well plate and incubate in CO_2 incubator. Refresh the medium every other day for 4 days (see Note 26).
15. Visualize under the fluorescence microscope at desired time points. The migration of transplanted hNPCs in mouse brain slices can be quantified as the maximum distance covered. For each transplanted neurosphere, select the five cells that migrated farthest from the neurosphere and measure the distance to the edge of the neurosphere. Calculate the average.

4. Notes

1. For simplicity, all the volumes in this section are given for three wells of a 6-well plate. We recommend using at least a full 6-well plate. The procedure for isolating hNPCs was adapted from ref. (15).
2. Floating hESC aggregates are transferred to new flasks in the first 2 days to eliminate contaminating MEFs that attach to the flask.
3. T25 flasks are first coated with 2.5 mL of poly-L-ornithine at 37°C, overnight, washed three times with 5 mL of PBS, and coated with 20 μg/mL laminin (1 mL) for 4 h at 37°C. Flasks are washed three times with PBS just before use.
4. Breaking up the clumps is necessary for better transfection efficiency; however, disaggregated individual cells will die. Use a pipette to gently break up the clumps and watch the suspension closely to obtain best transfection condition.

5. It is important to transfect the multipotent hNPCs immediately after isolating the rosettes and generating the small-clump suspension, before the clumps form neurospheres, which are more difficult to transfect.

6. Proper controls are important for any knockdown experiment. It is always necessary to verify that the observed phenotype is not reproduced when using a negative control (scrambled) LNA knockdown oligonucleotide. A custom-designed mismatch control LNA knockdown oligonucleotide can be used to assess the specificity of the miRNA knockdown.

7. In the early stages of planning miRNA knockdown experiments, it is important to profile its expression and ensure that the miRNA is actually expressed in the cells (See ref. (8) for details on miR-9 expression in hNPCs). At 16 days of hESC differentiation, hNPCs are more homogenous since they were derived shortly after dissociation from the rosette structures. The ability to study miR-9 function in early-stage hNPCs is an advantage of using hESC-derived hNPCs in culture. First, one can determine the specific roles of miR-9 at particular developmental stages. Second, the lower expression level of miR-9 than at later stages does not diminish its functional significance at this particular developmental stage. The same miRNA may have different functions at different developmental stages of various NPCs. Ultimately it is important to determine the efficiency and the time point of the transient knockdown. Knockdown studies, including miRNA-target interactions and rescue experiments, are more suitable to reveal miRNA endogenous and physiologically relevant functions. Overexpression of miRNAs in certain cellular contexts can have dramatic nonphysiological effects.

8. Besides absorbance at 260 nm (A_{260}), the A_{260}/A_{280} ratio should be calculated to verify the purity of the sample. This value should be in the range 1.8–2.0. A value <1.8 indicates contamination that could interfere with the qPCR.

9. The endogenous control for real-time quantification of miRNA with TaqMan microRNA assays we chose shares similarities with the miRNA, such as RNA stability and size. This assay uses human snoRNA located within an intron of the RPL21 gene. The suitability of this snoRNA as endogenous control was validated across a wide variety of tissues and cell lines (Applied Biosystems Application Note).

10. The dilution factor must be adapted depending on the level of expression of the miRNA studied.

11. As an alternative to Matrigel-coated plates, poly-D-lysine and laminin-coated coverslips (BD Bioscience) can be used. This assay is easier, and better pictures can be obtained, if cells are attached to the plate/coverslip. To facilitate the counting

labeled cells, neurospheres can be mechanically dissociated with dispase before the assay.

12. Paraformaldehyde is toxic and carcinogenic and should be disposed properly.
13. If a coated glass coverslip was used in Subheading 3.4, step 1, add a drop of mounting medium to a glass slide and place the coated glass coverslip with cells facing down on top of the slide.
14. Use an excitation wavelength of 520–560 nm (maximum 540 nm; green) and detection at 570–620 nm (maximum 580 nm, red).
15. The amount of LDH released into the medium is proportional to the number of dead cells.
16. Several assays must be performed for each condition, as hNPCs cannot be dissociated into single cells for this proliferation assay. Variability is observed due to distribution of hNPC aggregates in suspension.
17. As a good start, perform the WST1 proliferation assay every 24 h for 4–5 days after transfection.
18. Cellular behavior in the three-dimensional (3D) migration assay more accurately reflects the in vivo situation.
19. Allow the Matrigel aliquot to thaw overnight at 4°C. Tips for pipetting Matrigel and the medium used for dilution must be at 4°C to avoid polymerization of the Matrigel.
20. Filamentous processes will start to come off of the hNPCs within the first 2.5 h when miR-9 activity is inhibited. Under control conditions, that will happen within the first 24 h. After 24 h, cells will migrate from the neurospheres into the matrix, in chains, or as individual cells.
21. Organotypic culture is more relevant, both anatomically and physiologically, than cell culture in dish to study neuronal migration. The maintenance of cell–cell interactions in brain slices allows analysis of cell migration under conditions that resemble those in vivo.
22. Transducing hNPCs with the gene encoding green fluorescent protein (GFP) makes it possible to analyze details of differentiation, morphology, and direction control mechanisms of cells migrating in embryonic brain slices.
23. Infectivity is determined after 72 h by fluorescence microscopy for GFP expression. Optimization must be performed depending on the number of hNPCs and the virus preparation. It is important to check the transduction efficiency by plating a portion of neurospheres from each condition in a Matrigel-coated Petri dish. Determine the percentage of GFP-positive hNPCs at the time you look for their migration in the forebrain slices.

24. Prepare the 6-well plate with membrane inserts before starting to remove the embryos. Add 1.2 mL of medium to each well and place it in the CO_2 incubator.
25. Hold each end of a glass capillary between your thumb and forefinger. Position the center of the capillary horizontally over the top of a Bunsen flame. Roll the capillary back and forth between your thumb and forefinger over the flame. When the glass gets red in the center, remove the glass from the flame and immediately pull your arms apart so that the micropipette lengthens.
26. Only a thin layer of medium is needed on the surface of the slice to keep it humid. Tissues must be well exposed to the air (CO_2 enriched) to be healthy.

Acknowledgments

S.A. and C.D. contributed equally to this chapter. The authors thank S. Ordway for editorial assistance and lab members for discussions. This work was supported by grants from the Tau Consortium (F.B.G) and the NIH (RO1 NS066586 to F.B.G.).

References

1. Jones JM, Thomson JA (2000) Human embryonic stem cell technology. Semin Reprod Med 18:219–223
2. Keller G (2005) Embryonic stem cell differentiation: emergence of a new era in biology and medicine. Genes Dev 19:1129–1155
3. Yamanaka S (2007) Strategies and new developments in the generation of patient-specific pluripotent stem cells. Cell Stem Cell 1:39–49
4. Okita K, Ichisaka T, Yamanaka S (2007) Generation of germline-competent induced pluripotent stem cells. Nature 448:313–317
5. Soldner F, Hockemeyer D, Beard C, Gao Q, Bell GW, Cook EG, Hargus G, Blak A, Cooper O, Mitalipova M, Isacson O, Jaenisch R (2009) Parkinson's disease patient-derived induced pluripotent stem cells free of viral reprogramming factors. Cell 136:964–977
6. Cho MS, Lee YE, Kim JY, Chung S, Cho YH, Kim DS, Kang SM, Lee H, Kim MH, Kim JH, Leem JW, Oh SK, Choi YM, Hwang DY, Chang JW, Kim DW (2008) Highly efficient and large-scale generation of functional dopamine neurons from human embryonic stem cells. Proc Natl Acad Sci USA 105:3392–3397
7. Dimos JT, Rodolfa KT, Niakan KK, Weisenthal LM, Mitsumoto H, Chung W, Croft GF, Saphier G, Leibel R, Goland R, Wichterle H, Henderson CE, Eggan K (2008) Induced pluripotent stem cells generated from patients with ALS can be differentiated into motor neurons. Science 321:1218–1221
8. Delaloy C, Liu L, Lee J-A, Su H, Shen F, Yang GY, Young WL, Ivey KN, Gao F-B (2010) MicroRNA-9 coordinates proliferation and migration of human embryonic stem cells-derived neural progenitors. Cell Stem Cell 6: 323–335
9. Englund-Johansson U, Mohlin C, Liljekvist-Soltic I, Ekström P, Johansson K (2010) Human neural progenitor cells promote photoreceptor survival in retinal explants. Exp Eye Res 90:292–299
10. Anderson L, Caldwell MA (2007) Human neural progenitor cell transplants into the subthalamic nucleus lead to functional recovery in a rat model of Parkinson's disease. Neurobiol Dis 27:133–140
11. Ambros V (2001) MicroRNAs: tiny regulators with great potential. Cell 107:823–826

12. Fabian MR, Sonenberg N, Filipowicz W (2010) Regulation of mRNA translation and stability by microRNAs. Annu Rev Biochem 79: 351–379
13. Bartel DP (2009) MicroRNAs: target recognition and regulatory functions. Cell 136: 215–233
14. Herranz H, Cohen SM (2010) MicroRNAs and gene regulatory networks: managing the impact of noise in biological systems. Genes Dev 24:1339–1344
15. Li X-J, Zheng SC (2006) In vitro differentiation of neural precursors from human embryonic stem cells. Methods Mol Biol 331:169–177

INDEX

A

Adipose-derived stem cells (ASC) 49, 51–52, 54, 55
  comparison to bone marrow-derived stem/progenitor cells.......219
  differentiation to Schwann cells.......47–56
  isolation and expansion.......50–51
Aggregation chimeras
  embryo aggregation.......264, 265, 267
  embryo fertilization.......264
  embryo harvesting.......264
  embryo implantation.......267
  identification of chimeric offspring.......267–268
  imaging and analysis.......265
  immunostaining.......265
  microdissection.......265, 268, 269
  tissue fixation.......265
Aldehyde dehydrogenase (ALDH).......373–384
  Aldefluor® kit.......374, 378, 379, 384
  sorting for expressing cells.......374–375, 378–379
Anesthesia.......53, 151–153, 221–224, 227, 237, 383
Angiogenesis.......226–227
  gene expression arrays.......220, 222, 223
  in vitro assays.......220, 223
Animal husbandry.......264
Annexin V.......247, 256
Anterior ventral blood island (aVBI).......141, 142, 144, 147–148, 150
Apoptosis.......60, 77, 82, 88, 158, 182, 244, 247, 256, 328
ASC. *See* Adipose-derived stem cells (ASC)
Asymmetric cell division.......100, 111–124
aVBI. *See* Anterior ventral blood island (aVBI)
Axolotl.......199
  neurospheres.......198, 200
  spinal cord regeneration.......197, 198, 200–201
  transgenic.......201

B

Bacterial culture.......115, 306–308, 314
Blastocyst.......275–285
  chimaerae.......278–283
  culture.......276–282
  imaging.......275–285
  injection.......267
  obtaining from pregnant mice.......280
Bone marrow-derived cells.......219
Bromodeoxyuridine (5-bromo-2'-deoxyuridine, BrdU).......168, 245, 247–249, 257, 389, 394, 395,
Bulge.......244, 253, 254, 256

C

Caenorhabditis elegans (C. elegans)
  assymetric cell division.......111–124
  dissection, of embryos.......114
  growth and maintenance.......114–115
  image analysis.......122–123
  imaging.......118–122
  live imaging.......111–124
  mounting, of embryos.......116–118
  RNAi.......114–116
Calvarium bone marrow.......231–233, 239
CCO. *See* Cytochrome *c* oxidase (CCO)
C2C12-Myogenic cell line.......4
$CD34^+$ cells.......219, 254, 256, 303, 305, 306, 310–315
$CD49f^+$ cells.......244
Cell cycle analysis.......246–247, 253–256
Cell therapy.......359
Chicken embryo
  dissection, of inner ear.......127–138
  electroporation.......128–130, 132–134
  inducible gene expression *in ovo*.......129, 131–132, 134–135
  organotypic culture.......135
Chimera.......263–272, 283. *See also* Aggregation chimeras
Clonal.......67, 289–290
  analysis.......185, 189–192, 360, 366–368
  mutations.......182, 291
Colon cancer.......373, 380, 384
Colony-forming cell assay.......313–314
  Confocal microscopy. *See also* Imaging
  laser-scanning confocal (LSC) microscopy.......120–121, 361

Kimberly A. Mace and Kristin M. Braun (eds.), *Progenitor Cells: Methods and Protocols*, Methods in Molecular Biology, vol. 916, DOI 10.1007/978-1-61779-980-8, © Springer Science+Business Media, LLC 2012

Colony-forming cell assay (*con't*)
spinning-disk confocal (SDC) microscopy........121–122
swept field confocal (SFC) microscopy......................122
Z scan..124
Corneal stem cells
clonal analysis..360, 368
colony forming efficiency (CFE)......................364–366
feeder layer preparation 359, 362–363
fibrin gel preparation.........................360, 364–365, 370
freezing, of human primary keratinocytes................360, 364, 370
immunocytochemistry.......................................368–369
lifespan (serial cell propagations).......................365–366
limbal biopsy..359, 363
primary culture 359
3T3-MCB cultivation 359, 362, 370
Coverglass
activation with glutaraldehyde................. 322, 326–327, 329, 337, 338
as a polyacrylamide gel support........................321–323, 327–329, 331–333, 339
treatment with Rain-x/Enviro-coat........... 323, 329, 330
Cryosectioning 34–35, 39
Cytochrome *c* oxidase (CCO)................. 290, 292, 294, 297

D

Danio rerio. *See* Zebrafish
Density gradient centrifugation......................................352
Drosophila
GAL4/UAS.....................................100,–102, 106, 107
in situ cell picking.................................... 100, 106, 107
neural stem cells..99–108
transcriptome analysis..99–108

E

EBs. *See* Embryoid bodies (EBs)
EDL. *See* Muscle
Embryoid bodies (EBs)........ 82, 84, 87–89, 91, 94, 275, 391
Embryonic stem cells (ESCs)
culture on mouse embryonic fibroblasts (mEFs)..89, 94
culture on polyacrylamide gels..........................317–348
differentiation into embryoid bodies (EBs)................82, 84, 87, 88, 91, 94
differentiation to endothelial progenitor cells (EPCs)..81–95
differentiation to human neural progenitor cells (hNPCs)...387
Endothelial progenitor cells (EPCs)
acetylated low density lipoprotein staining........352, 355
culture.. 83–84, 354
differentiation from embryonic stem cells 81–95
isolation from blood..351–353
isolation from embryoid bodies (EBs).................82, 84, 87, 88, 91, 94
isolation using FACS...81–95
lectin staining 352
marker expression and analysis 82
Epiblast (EPI) 275, 276, 281, 318–320, 328
Epidermal stem cells
apoptosis...247, 256
cell cycle analysis 246–247, 253–256
disaggregation of mouse epidermis........... 246, 252–253
epidermal wholemount.....................................243–259
immunofluorescence.......... 17, 18, 20–21, 246, 253–256
terminal differentiation...15–21
ESCs. *See* Embryonic stem cells (ESCs)

F

FACS. *See* Fluorescence activated cell sorting (FACS)
Fate mapping..207, 208
Feeder layer.................... 84–89, 94, 335, 359, 362–367, 390
Fluorescence activated cell sorting (FACS)................60, 61, 64, 65, 72–75, 77, 78, 82, 84, 89–92, 94, 107, 167–178, 221, 222, 225, 228, 234, 239, 240, 244, 246, 253–256, 259, 356, 373, 374, 377–379
Fluorescent probes..232, 233
Fluorophores, red-shifted...232
Fly .... 101, 102, 104, 105. *See also Drosophila*
Frog...................... 142, 144, 145, 203, 207. *See also Xenopus*

G

Gastrula.. 141, 204, 207, 213
Geltrex™...5, 10, 12
Genomic DNA
analysis...298
extraction of DNA from cells 294
laser capture microdissection 293
PCR.. 293, 306–298
Germ cells 157–161
GFP. *See* Green fluorescent protein (GFP)
γ-irradiator.. 234, 239, 335
Gold-coated slides..16, 19
Graft................................ 48, 203–214, 358, 360, 365, 369. *See also* Transplantation
Gr-1⁺CD11b⁺ cells
FACS..221, 222, 225, 228
gene expression analysis..222
injection into wounds 227
isolation from wounds 221, 222
Green fluorescent protein (GFP) 25, 90, 106, 107, 113, 121, 123, 148, 182, 184–187, 189, 191, 193, 194, 232, 236, 238, 265, 267, 268, 270–272, 276, 281, 283, 284, 396, 400
GST protein purification................................ 307, 309, 315

H

Hair cell 128, 129, 131, 132
Hair follicle (HF) 243, 244, 247, 253, 257
Hematopoietic stem and progenitor cells (HSPC)
  cell sorting 233, 235
  harvesting, of total bone marrow 235
  imaging, *in vivo* 233–234, 237–239
  irradiation 237, 239
  isolation 233
  labelling 232, 237
  lineage depletion 233–237
  transplantation 240
HF. *See* Hair follicle (HF)
H9 line, human embryonic stem cells 388
hNPC. *See* Human neural progenitor cells (hNPC)
Holoclone 358, 362, 366–370
Human neural progenitor cells (hNPC)
  culture 388–389
  differentiation from hESCs 390–392
  migration assays 390
  miR expression 392–393, 399
  miR transfection 392
  proliferation assays 396, 400
  viability assays 389

I

Imaginal disc 105
Imaging. *See also* Confocal microscopy
  C. elegans embryos 118, 120
  differential interference contrast (DIC)
    microscopy 112, 118–119
  epidermal wholemounts 250–252
  epifluorescence microscopy 111, 120, 150
  fluorescence live imaging 119–120
  hematopoietic stem cells 231–240
  image analysis 122–123
  ImageJ 113, 122, 151
  intravital microscopy 231, 232
  in vivo imaging 231–240
  lung aggregation chimeras 264
  mice 142
  photobleaching 118–121, 123
  phototoxicity 119, 120, 123
  stereomicroscope 114, 117, 130, 132, 133, 135
  time-lapse 112, 113, 118–120, 122, 124
  two-photon microscopy 232, 235, 236
  Xenopus embryos 144
  Xenopus tadpoles 153–154
  zebrafish 142
  z scan 124
Immunostaining 34–35, 39, 160–162,
    206–207, 210, 212, 265, 268, 373
Inner ear 127–138
  dissection 135, 136, 138
  organotypic culture 135
In ovo electroporation 127–138
In situ cell picking 107

K

K562 human leukemia cell line 305, 306, 310, 311

L

Lactate dehydrogenase (LDH) assay 394, 395, 400
Larva 101, 104, 105, 182, 183, 189
Laser capture microdissection (LMD) 293
LDH. *See* Lactate dehydrogenase (LDH) assay
Limbus 358
Lineage tracing 289, 290
Live imaging 100, 111–124, 129, 239, 275–285
LMD. *See* Laser capture microdissection (LMD)
Low adhesion culture plates, preparation of 61–62
Lung alveolar stem cells
  culture 23–30
  differentiation potential (*in vitro* assessment) 26, 28
  engraftment, kidney capsule 26, 28–29
  enrichment 25–28
  immunofluorescence 26
  isolation 23–30
  sorting 25–28

M

MACS™ separation system 235
Mammary stem cells
  isolation
    from human breast tissue 66–67
    from mouse breast tissue 69–70
  mammosphere culture 60, 63, 65, 67–68
  organoids 66–67, 74, 75
  PKH26 labeling 60, 61, 63–65, 67–68, 71–78
  sorting 64, 73, 74, 78
  sphere-forming efficiency (SFE) 65, 73, 76, 78
  suspension culture 61–65, 68–71, 76
Mammosphere culture 60, 63, 65, 67–68
Matrigel 12, 319–321, 323, 327,
    329–334, 341–344, 346, 347, 380, 389,
    393, 395–397, 399, 400
MEFs. *See* Mouse embryonic fibroblasts (MEFs)
Melanocytes and melanocyte stem cells
  ablating 183, 184, 187–189
  birthdating 184, 187, 189
  clonal analysis 185, 189, 191, 192
  counting 181, 184–187
  lineage analysis 181–194
  reverse labeling, using PTU 184, 187, 188
  Tol2 transposable element 183
  X-ray induced clones 185, 192–193
  in zebrafish 181–194

Mesenchymal stem cells
cell tracking ... 32, 34, 38
Co-culture ... 32, 34, 37–38, 44
3D dynamic culture ... 33
differentiation to Schwann cells ... 51–52, 55
3D microenvironment ... 31
3D spheroid culture ... 32
3D static culture ... 33–34
flow cytometry ... 35, 39–40
fluorescent dye labeling ... 250
immortalisation ... 32
immunostaining ... 34–35, 39
isolation and expansion ... 50
rotary culture system ... 33, 37
spinner flasks ... 33, 36–37, 44
viability assay ... 32, 38
Microcapillary needle pulling and beveling ... 101
Micro-contact printing ... 16, 17
Microenvironment ... 16, 31, 32, 232, 233, 235
Micromanipulator ... 103–105, 130, 134, 146, 185, 191, 282
Micro-patterned substrates ... 15–21
MicroRNA (miRNA)
miR expression ... 387–389, 392–393, 399
miR transfection ... 392
Micro-ruby ... 143, 146, 148, 205, 209, 210, 213
Microsatellite
analysis ... 291, 293, 298–300
Genescan preparation ... 299
PCR ... 291, 293, 298, 299
Microscopy. *See* Imaging
Migration ... 48, 148, 150, 151, 158, 388, 390, 396–398, 400
Mitochondrial DNA (mtDNA)
analysis ... 290–292, 294–297
extraction of DNA from cells ... 294
histochemistry ... 290, 292, 294–297
laser capture microdissection ... 293
PCR ... 290, 292, 295
sequencing ... 295–297
Mitomycin C ... 17, 83–85, 93, 94, 335, 388
Monocytes ... 220
Monomeric cherry protein (mCherry) ... 232
Mononuclear cells ... 233, 314, 352–354
Morphogenesis ... 128
Morpholino ... 142, 148, 150
Morulae
aggregation ... 282
decompaction ... 282
imaging ... 280, 282
obtaining from pregnant mice ... 280
Mouse
CAG::mRFP1 transgenic strain ... 276
C57BL/6 ... 83
C57BL/6-Tg(CAG-EGFP) 1Osb/J transgenic strain ... 264
CD1 strain ... 280
FVB/NCrl ... 264
housing ... 239
injection ... 29
mdx (model of Duchenne muscular dystrophy) ... 5
non-obese diabetic severe combined immunodeficient (NOD/SCID) ... 382
non-obese severe combined immunodeficient diabetic IL2 γ receptor null (NOG) ... 374
Pdgfrα$^{H2B-GFP}$ transgenic strain ... 276
tail ... 247–249
tissue dissociation ... 65, 70
wounding ... 223
Mouse embryonic fibroblasts (MEFs) ... 87, 94, 388, 391, 398
culture conditions ... 388
freezing ... 83
medium ... 53
mitomycin C inactivation ... 388
Muscle
digestion ... 9
dissection ... 5
extensor digitorum longus (EDL) muscle ... 6, 7
plating of fibres ... 10
purification of fibres ... 5
soleus muscle ... 8
Mutations
in genomic DNA ... 290, 293, 297–298
in mitochondrial DNA (mtDNA) ... 290–292, 294–297
Myotubes ... 4, 8, 9, 11

N

Neoblast ... 168
Neural stem/progenitor cells
in Axolotl ... 200
isolation from *Drosophila* larval brain ... 103, 105
in *Xenopus* ... 141–154
Neuroepithelial cells ... 100–102, 104, 106, 108, 203, 204, 391
Neurospheres ... 197–202, 391, 393, 394, 396–400
Neurula ... 141, 147, 150
NF-Ya ... 303–315
Niche
bone marrow ... 231–240
stem cell ... 231–240
NOG. *See* Mouse
N-succinimidyl acrylamidohexanoic acid (N6) ... 321–326

O

Oocyte progenitor cells, in zebrafish
confocal immunofluorescence ... 160
dissection ... 161
immunostaining ... 160–162

mounting, of ovaries 162–163
tissue fixation and permeabilization........ 159
Otic placode/Otic cup, electroporation of 127–138

P

PAR proteins 112, 122
pASCs. *See* Planarian adult stem cells (pASCs)
Phenylthiocarbamide (PTU) 154, 183, 184, 187–189
PKH26, lipopilic fluorescent dye 60
Planarian adult stem cells (pASCs)
cell dissociation 169
FACS 169
radiation 169
in regeneration 167–169, 172, 176, 177
staining 176
X1 cells 169, 174
X2 cells 169, 174
Planarians
culture 170, 171
hydroxyurea (HU) treatment 170
irradiation 170
PMEF-CFL, mouse embryonic fibroblasts 323, 388
Polyacrylamide substrates
for growth of hESCs 318
matrigel coupling 347
protein coupling 321
PolyHEMA (2-hydroxyethylmethyl methacrylate) 61–64, 68, 69, 71, 72, 74, 84, 88, 94
Polymerase chain reaction (PCR) 35, 36, 40, 42, 100, 102, 103, 106–108, 222, 226, 268, 290–300, 305, 307, 308, 312, 313, 389, 392, 393, 399
Pressure injector 185, 191
Primitive endoderm 275–285, 319, 328
Primitive myeloid cells
development 142, 144
imaging 150–152
Progenitor cells
angiogenic 219, 220, 227
CNS 214
endodermal 147, 208
endothelial 81–95
epidermal 21, 243–259
epithelial 82, 90
hematopoietic 231, 233–234, 238, 303–315
lung alveolar 26, 27, 30
mammary 60
myeloid 141–154
neural 387–401
oocyte 157–164
Proliferation 5, 9–12, 31, 48, 54, 84, 100, 101, 107, 112, 159, 176, 177, 198, 263, 305, 308, 311, 312, 314, 388–390, 394, 396, 400
Protein transduction 304, 305
PTU. *See* Phenylthiocarbamide (PTU)

R

Radial glial cells 197
Real time microscopy 111
Regeneration 15, 47, 49, 83, 122–138, 167–169, 172, 176, 177, 182, 183, 188, 197–202, 220, 253, 358
Regenerative medicine 81, 357–371, 387
Reverse transcription 100, 103, 106, 222, 226, 307, 313, 389, 392
RNA interference (RNAi) 43, 112, 114–116, 167, 170, 171, 176

S

Salamander. *See* Axolotl
Satellite cells 3–12
Sca1[+] cells 235, 238
Schwann cells, differentiation from adipose-derived cells 49
SFE. *See* Mammary stem cells
Siliconization of plasticware 52–53
Sphere cultures 374
Spinal cord
dissection and dissociation 200
lesion 48, 198, 200
neurosphere implantation 200
Stem cells
colonic 373–384
corneal 357–371
embryonic 48, 81–95, 275, 317–348
epidermal 15–21, 243, 244, 246–247, 253–256
epithelial 63
hematopoietic 48, 231–240
keratinocyte 357
limbal 358, 362, 369, 370
lung alveolar 23–30
mammary 59–78
melanocyte 181–194
mesenchymal 31–44, 47, 48, 50, 53–54
myogenic 3
neural 99–108, 197, 198, 202
planarian 167–178
radial glial 197
tumorigenic 373–384
Succinate dehydrogenase 290
Suspension culture 16, 61–65, 68–71, 76, 82, 87–88, 367, 374, 391

T

Tail 17, 117, 150, 151, 161, 197–202, 237, 239, 245, 247–250, 268

TAT fusion protein....305
Tissue dissociation....66, 69, 71
Tol2 transposon vectors....128, 131, 132
Transcription....4, 49, 82, 100, 103, 106, 142, 222, 226, 304, 307, 313, 318, 387, 389, 392
Transcriptome analysis....99–108
Transient amplifying cells....358
Transplantation
- anterior ventral blood island....147–148
- autologous graft....358
- bone marrow....303
- corneal stem cells....357
- embryonic brain slices....396–398
- hematopoietic stem and progenitor cells (HSPC)....233, 236
- human neural progenitor cells....388, 396–398
- kidney capsule....28
- lung alveolar stem cells....28
- neuroepithelial layer....100
- tail vein....237
- xenograft....374

Tricaine....143, 145, 151, 152, 159, 160, 184–187, 192, 205
Tumorigenic human colonic stem cells
- colonospheres....375
- dissociation of primary tissues....374
- immunohistochemistry of ALDH on colon spheres....375
- sorting for aldehyde dehydrogenase expressing cells....374–375, 378–379
- subcutaneous xenograft implantation....374–376
- xenograft explantation and dissociation....374, 376–377

Tumor-initiating cells....374
TUNEL (Terminal deoxynucleotidyl transferase dUTP nick end labeling)....257, 393, 394

V

Vasa....159, 160, 162–164
Visual acuity....358

W

Western blotting....306, 310
Wholemount
- confocal imaging....246, 250–252
- epidermis, from mouse tail....247
- imaging....263–272
- immunolabelling....245, 248–250
- lung, of mouse aggregation chimeras....265
- ovaries, of zebrafish epidermal wholemount....162

X

Xenograft....374–376, 378, 382–384
Xenopus
- eggs....152
- embryo labeling....144
- embryos....142, 144–146
- fertilization....145
- imaging....150–152
- injection....146
- maintenance....203
- mating....143, 145, 149, 152
- priming....145
- tadpoles....153
- transplants....146

X-ray....185, 192–194, 221

Z

Zebrafish
- *alb*$^{b4/b4}$, Albino mutant line....185
- melanocytes....181–194
- melanocyte stem cells....181–194
- *mlpha*^{j120}, mutant line deficient in melanophilin....182, 184, 186, 187
- oocyte progenitor cells....157–164
- oogonia....157, 159, 160, 164
- ovary....157–164
- Tg(fTyrp1>eGFP)^{j900}, transgenic line expressing eGFP....182

Zona pellucida....267, 280, 282

MIX
Papier aus verantwortungsvollen Quellen
Paper from responsible sources
FSC® C105338

If you have any concerns about our products,
you can contact us on
ProductSafety@springernature.com

In case Publisher is established outside the EU,
the EU authorized representative is:
Springer Nature Customer Service Center GmbH
Europaplatz 3, 69115 Heidelberg, Germany

Printed by Libri Plureos GmbH
in Hamburg, Germany